Toward a Genetics of Language

Toward a Genetics of Language

Edited by

Mabel L. Rice
University of Kansas

LEA LAWRENCE ERLBAUM ASSOCIATES, PUBLISHERS
1996   Mahwah, New Jersey

Lawrence Erlbaum Associates, Inc., Publishers
10 Industrial Avenue
Mahwah, NJ 07430

cover design by Gail Silverman

Library of Congress Cataloging-in-Publication Data

Toward a genetics of language / edited by Mabel L. Rice.
    p.   cm.
    Includes bibliographical references and index.
    ISBN 0-8058-1677-1 (hbk. : alk. paper). — ISBN 0-8058-1678-X
(pbk. : alk. paper)
    1. Language disorders—Genetic aspects. I. Rice, Mabel.
RC429.T68 1996
616.85′5042—dc20   96-3264
                                                           CIP

Printed in the United States of America
10 9 8 7 6 5 4 3 2 1

CONTENTS

ABOUT THE CONTRIBUTORS

Shanley E. M. Allen Max-Planck-Institut für Psycholinguistik

Linda Brzustowicz Center for Molecular and Behavioral Neuroscience, Rutgers University

Martha Crago School of Human Communication Disorders, McGill University

John C. DeFries Department of Psychology, University of Colorado

Jill de Villiers Department of Psychology, Smith College

Jeffrey W. Gilger Department of Speech-Language-Hearing, University of Kansas

Dianne L. Lefly Department of Psychology, University of Denver

Laurence B. Leonard Audiology and Speech Sciences, Purdue University

Jon F. Miller Department of Communicative Disorders, University of Wisconsin–Madison

Bruce F. Pennington Department of Psychology, University of Denver

Elena Plante Department of Speech and Hearing Sciences, University of Arizona

David Poeppel Biomagnetic Imaging Laboratory, University of California, San Francisco

Mabel L. Rice Child Language Doctoral Program, University of Kansas

Shelley D. Smith Center for Genetic Research in Communication Disorders, Boys Town National Research Hospital

Catherine E. Snow Graduate School of Education, Harvard University

J. Bruce Tomblin Wendell Johnson Speech and Hearing Center, University of Iowa

Kenneth Wexler Department of Brain and Cognitive Sciences, Massachusetts Institute of Technology

OF LANGUAGE, PHENOTYPES, AND GENETICS: BUILDING A CROSS-DISCIPLINARY PLATFORM FOR INQUIRY

Mabel L. Rice
University of Kansas

This is a good time to be considering the question of the genetics of language. Since the 1980s, important new advances have been made in the fields of genetics, behavioral genetics, linguistics, language acquisition, studies of language impairment, and brain imaging. Although these advances are each highly relevant to the determination of what a child is innately prepared to bring to language acquisition, the contributing fields of endeavor have traditionally been relatively self-contained, with little cross-over communication. The collection of papers in this volume, and the conference at which they were originally presented (*Toward a Genetics of Language*, Nov. 11–13, 1993, University of Kansas), were developed with the belief that there is considerable value to be gained in the creation of a shared platform for a dialogue across the disciplines. What the authors share is the conclusion that discrete evidence of a linkage between a particular gene or combination of genes and an individual's grammatical ability does not currently exist, although there is considerable evidence pointing toward that probability. They also share the expectation that a genetic contribution of some sort will ultimately be identified as an important factor in language acquisition; what is disputed is what this contribution might be, how to interpret its effects, and how to specify essential evidence.

This volume was prepared with an audience of scientific peers and colleagues in mind. It provides a summary and overview of current developments in genetics, behavioral genetics, linguistics, language acquisition, studies of language impairment, and brain imaging. In addition to scientists working on the topic of the genetics of language acquisition, it will be useful for scholars interested in, but not yet working on, the topic, who wish to learn about new developments in the various fields of inquiry involved. The volume will also be of value to advanced students who wish to become familiar with emerging developments bearing on the genetics of language. The chapters are well suited to advanced graduate level instruction, by virtue of the controversial issues involved and the challenging scientific issues addressed. The volume is not intended to resolve issues of intervention or treatment, although these issues are not ignored. Practitioners will find new concepts and theories, new findings, and new directions of inquiry that portend future directions for intervention. At the same time, it is clear that it is premature to turn these developments into blueprints for practice.

In this introduction, a background for the topic is provided and reasons why it is timely to consider the chapters collected here are mentioned. There is a need for awareness of new developments across several fields and a convergence of a collective body of knowledge that can serve as a basis for further gains in understanding. Key questions to be addressed are identified. In the spirit of moving toward possible answers, possible criteria for a workable phenotype for Specific Language Impairment (SLI) are laid out. A preview of the chapters is then provided.

Time of Genetic Breakthroughs

In this age of the human genome project and great advances in the technology of molecular genetics, it seems that every week brings a new announcement of an important breakthrough in the mapping of human genes to human diseases and biological processes. The breakthroughs extend to human behavior as well. In a recent volume published by the American Psychological Association (APA), McClearn (1993) asserted, "At one time, the demonstration of a genetic influence in a behavioral phenotype was noteworthy (and, per se, publication-worthy). Now, it is almost a foregone conclusion that some degree of heritability will be found; the issue is its magnitude" (p. 38). In the same volume, the list of psychological phenomena thought to have a significant degree of heritability includes the following: intelligence and its components, school achievement, organic dementias (such as Alzheimer's disease, Huntington's chorea, alcoholism), functional psychopathology (such as affective disorder), traits of personality (such as extroversion vs. introversion), attitudes (such as conservatism vs. traditionalism), delinquency-criminality, and vocational interests. Note that language is not among the phenomena listed. The scope of the phenomena included does, however, suggest that genetic influences on behavior are pervasive (although the evidence is far from conclusive; see Mann, 1994). It would seem that an inherited contribution to language acquisition would be a foregone conclusion, that the kinds of evidence to be generated are self-evident, and that breakthroughs are just around the corner.

Recent developments in the genetic underpinnings of dyslexia add to this sense of imminent breakthrough. Two of the contributors to this volume, Shelley Smith and Bruce Pennington, are members of the team who have just announced the discovery of a quantitative trait locus (QTL) for dyslexia (see Smith, Pennington, & DeFries and Lefly & Pennington, this volume). This is crucially important documentation that a particular cluster of genes can collectively and/or individually influence the probability that an individual will have the condition of dyslexia. Dyslexia is known to be highly associated with language impairments of a certain sort. Can a QTL for language then be not far behind a QTL for dyslexia?

An Impending Breakthrough for Language?

Now is a time of rumors about investigations of language genes. Preliminary statements made at the annual convention of the American Association for the Advancement of Science (AAAS) in 1993 were picked up by the popular and scientific press as premature announcements of a gene for language (one of the participants was Steven Pinker, who gave a lively description of the reports and a disclaimer for the assertion in his recent book). At the 1994 AAAS convention, a panel of scholars (including Plante, Rice, Smith, & Tomblin, this volume) reviewed the available evidence and noted that a linkage of discrete genes to language was plausible, but yet to be established. This situation could change quickly, however. At the rate of scientific discovery, it is possible that by the time of publication of this volume, credible claims of genetic linkage will have appeared.

The tone of this conference and collection of papers is somewhat cautious, though. At the conference it was noted that there are important breakthroughs in the related disciplines that should be common knowledge but have not yet spread across disciplinary boundaries. For example, in the recent Plomin and McClearn (1993) volume, the advances in evidence for the genetic contributions to language were not included, perhaps because these findings are emerging in the study of language impairments, which are reported in journals devoted to language impairment. That evidence was brought out at the conference and in the chapters here. The import is that language acquisition (normal and impaired) can be added to the list of psychological phenomena for which there is evidence of heritability. Likewise, the impending discovery of the QTL for dyslexia was not familiar to the scholars of language acquisition, although it is a development of great import for investigations of the genetics of language. That line of investigation was a valuable contribution to the discussions and to this collection of chapters.

It was obvious from the conference interactions, and also in the commentaries of this volume, that a multidisciplinary approach will be necessary for determination of the role of genetics in language acquisition. The necessary scientific expertise to solve the problem does not exist within any one discipline. Although cross-disciplinary communication does not immediately ensure solutions, it does push up some central themes for consideration. Some of the key questions that emerged in interactions among the participants follow.

Key Questions to Be Addressed

What Is Language? As the current debate about the genetics of intelligence exemplifies (cf. Gould, 1994; Herrnstein & Murray, 1994), the definition of the capacity that is thought to be passed from one generation to the next is of vital importance. The linguist's view of language can be very different from the psychologist's (cf. chapters by de Villiers, Snow, & Wexler, commentaries by de Villiers and Snow, this volume). Many linguists view language as a specialized module in the human mind that is configured in certain highly circumscribed grammatical relationships that are consistent with observed linguistic universals. Within this view, the measurement of language is the measurement of grammatical principles, inferred from performance on highly specified grammatical tasks. Developmental psychologists tend toward a different view, in which language is thought to be an unspecialized part of a general cognitive ability that is intrinsically bounded by performance demands, memory, on-line processing, conceptual development, and social insights. Within this view, the measurement of language is meaningfully organized in terms of performance dimensions, such as comprehension versus performance, and in terms of surface descriptions of linguistic components, such as isolated grammatical markers, counts of sentence constituents, or counts of the number of different vocabulary items. This aptitude can be operationalized as performance on a standardized test, with a single summary score of performance.

A synthesis or amalgamation of these opposing viewpoints does not seem to be immediately forthcoming; investigators interested in studying the genetics of language acquisition will have to make some choices as to the model to follow. A choice is unavoidable, because the model of language will determine the means of measurement and interpretation of the capacity that is thought to be inherited.

Exploration of the inherited aspects of language will, on the other hand, help clarify the nature of the human language capacity and its origins. The model of language, then, is central to the model of inheritance and vice versa; the scientific challenge is how to work on both ends of the puzzle simultaneously.

Where Is Variation? Essentially, the search for a genetic contribution to language requires the identification of variation in an individual's linguistic capability that corresponds to variation at the level of genes. What is striking about language acquisition is that from the earliest ages, grammatical variation across individual children is minimal. For example, most children by age 4 use grammatical markers, such as tense morphemes, in almost every context in which they are required. For this reason, normative grammatical development does not lend itself readily to the determination of genetic variation, at least for those grammatical structures that are fundamental and obligatory for the organization of the grammar. An alternative strategy is to focus on those properties of language that are not invariant, such as vocabulary size. This strategy has yielded heritability estimates (cf. Gilger's chapter, this volume). It is not clear, however, if this dimension is part of the same innate linguistic capacity as is grammar. The most interesting question is how to account for the apparently relative invariant aspects of language acquisition that are attested to across languages and across children in widely varying environmental circumstances. Rice and Wexler (ch. 8, this volume) argue that a promising strategy is to look for variation in grammatical competencies, such as tense-marking, where invariance is expected.

Unexpected grammatical variation is evident in a clinical condition in which individuals show markedly atypical delays in language acquisition, even though the putative prerequisite competencies are evidently sufficiently intact to support language acquisition. These individuals do not demonstrate clinically significant deficits in hearing acuity, cognition, or socioemotional development, but their language acquisition is nevertheless delayed in emergence and shows interesting and unexpected variation in grammatical markers. This condition has been in the literature for more than 60 years (e.g., Ewing, 1930). In the modern literature, the condition has been labeled *specific language impairment* (SLI) (see Watkins & Rice, 1994, for a recent review of the literature). Although the etiology of this condition has long been of interest, it remains unknown. Recent evidence reveals a high heritability for SLI which has, in turn, inspired investigators to more precisely explore the nature of the impairment and methods for identifying individuals who are affected. Several of those investigators are included in this volume (i.e., Crago, Leonard, Miller, Plante, & Tomblin and Rice & Wexler). The search is on to identify the ways in which individuals with SLI vary from the expected normative developmental trajectory for language and how those variations may co-occur with yet-to-be-identified genetic variation.

What Is the Behavioral Phenotype/Disease Definition?

The behavioral phenotype is the behavioral manifestation of the effect of the underlying genetic influence. This can be at the level of a capacity that is measured in quantitative terms (such as high vs. low reading aptitude) or, more traditionally, at the level of disease categories (dyslexic or schizophrenic). At the most simplistic level, this means that it should be possible to measure a given behavior or set of behaviors evident in an individual's linguistic repertoire that would correspond to the effect of a specific gene or genetic locus. In the case of a diagnostic category, this would require that individuals be identified as affected or unaffected, SLI or non-SLI. At the level of a quantitative phenotype, it would require specification of a linguistic dimension or trait on which individuals vary according to the amount of this dimension or trait they possess.

As noted by Brzustowicz in her chapter, and in almost every other chapter as well, the question of the phenotype is of central significance for genetic studies.

The problem is that it is not clear how we should measure the phenotype for language acquisition. Should we look for grammatical performance, closed aspects of language, vocabulary size, performance under conditions of processing demand, the use of politeness markers, or a composite language measure? Will the phenotype be the same across different ages and stages of language acquisition? Is it to be expected that the phenotype will always be directly observable?

Much of the success in the investigation of the genetics of language hinges on the definition of phenotype. This definition is, in turn, intrinsically linked with the choice of language model and the expected loci of variation. As many of the chapters here will attest, investigators are sorting out the criteria for determination of a plausible phenotype as they move toward an investigation of possible genetic linkages. This issue is further discussed in a following section.

What Is the Path From Genotype to Phenotype? What Are the Intervening Biological Mechanisms? Genes encode for proteins, and proteins are involved in building the central nervous system, including the cortex of the brain where language functioning is thought to reside. Modern genetics has advanced rapidly in part because investigators sidestepped the issue of intervening biological pathways between genes and overt manifestation of disease. Significant discoveries have occurred in the demonstration of linkages between molecular variations of DNA and observed clinical differences, such as those associated with Huntington's disease, muscular dystrophy, and cystic fibrosis. The actual intervening biological mechanisms that generate the symptomatology are still poorly understood, although in a growing number of cases, such as phenylketonuria (PKU) and Tay-Sachs disease, answers are forthcoming.

As several of the participants in this volume point out, a full understanding of innate contributions to language acquisition will require an account of how innateness relates to biological processes and mechanisms that, in turn, influence linguistic competence and performance. The available biological evidence is relatively meager and controversial, as is made clear in the chapter by Poeppel, but it is nevertheless possible that differences in cortical structure or functioning may prove to be as informative a phenotype as behavioral variations in grammatical competence, a possibility argued by Plante in her chapter.

What Is the Environmental Contribution? Even the most ardent advocates for an innate contribution to language acquisition do not deny the potential role of environmental factors for many aspects of language. The environmental influences that have received the most attention in the language acquisition literature are the ways in which adults talk to children and respond to children's language. What is interesting in the context of this volume is that "environment" can also be defined in biological terms. For example, the uterine environment is an important source of possible variation in individual competencies, although this variation is not genetically derived. Provision of treatment for PKU at birth is an environmental influence of crucial significance. On the other hand, what can seem to be environmental effects may not be unambiguously so. It is pointed out by Gilger (this volume) that traditional studies of environmental influences, such as mother-child interactions, present an inherent confound of environmental and genetic effects, insofar as mothers contribute to their children's linguistic ability by virtue of common genetic material and common language experiences. The point to be emphasized here is that consideration of the inherited aspects of language acquisition brings a better understanding of environmental influences as well.

Are There Implications for Language Intervention? It is sometimes asserted that the risk of discovering a genetic etiology to language impairment is that such a discovery would lead inevitably to the conclusion that language intervention is not warranted. That is not the conclusion advocated here. Instead, there are reasons to believe that genetic advances will bring positive information to language intervention. What a genetics account could supply would be a more complete and accurate view of the possible etiology of language impairments. Among the positive clinical effects would be a diminution of the undesirable consequences of ascribing causality to poor parenting skills when that is not the case. With a better appreciation of innate influences, we can de-emphasize child-rearing practices as a causal factor and emphasize individual differences in children's language aptitude. In most instances, this would be a positive outcome that would allow parents and clinicians to get on with the ways in which they can construct an environment that is optimal for an individual child's language acquisition.

There is an important distinction between the minimal conditions necessary for the activation of language acquisition in most children and the maximal conditions necessary for the support of language acquisition in children with language impairments. More than most children, children with limited language aptitude, such as those with SLI, need early identification so environmental enhancements can be implemented at the earliest ages. They can and do benefit from these early interventions (cf. Rice & Wilcox, 1995). What will enhance the design of intervention environments is identification of the aspects of language acquisition that are more or less constrained by innate influences (cf. de Villiers, this volume). Attention can focus on a judicious combination of efforts toward the aspects of language likely to show the most gain from intervention and persistent attention to the aspects that are likely to be most resistant. In this way, the intervention process will be able to move forward with maximal efficiency and minimal frustration on the part of the learner and the teacher.

This volume focuses on the issues involved in going forward with programmatic investigations of possible genetic influences on language acquisition and language impairment, in the belief that treatment implications can be positive and more fully elaborated in future investigations.

The Phenotype Issue: Possible Criteria for a Workable Phenotype for Specific Language Impairment

It is easier to identify key questions than to specify possible solutions. In the spirit of moving toward solutions, I offer here some criteria that have been invoked in the discussions throughout this volume as desirable for the specification of a phenotype for language impairment, accompanied by comments that highlight some of the challenges involved and previews of some of the discussions in the chapters that follow. Because SLI is both the diagnostic category featured in this volume and one that I have thought about for some time, that is the clinical category selected for illustration, although it must be acknowledged that this category is not the only clinical category of language impairment for which there is a plausible genetic etiology. In the context of current knowledge, these criteria can be thought of as desirable standards to achieve, even if somewhat beyond our present reach.

The Phenotype Is Consistent With Universal Features of Language. This criterion states that any possible phenotype of language impairments should be consistent with attested linguistic representation and performance across the set of human languages. Put most succinctly, language should be in the criteria for a phenotype of language impairments. As noted previously, there is no consensus on how to define language. At the same time, there is a growing understanding of many of the important details of linguistic representations and the ways these are manifest in diverse languages. A fully satisfactory genetics model of language acquisition and impairments must be able to predict the details of linguistic representations and linguistic performance. As many of the chapters in this volume illustrate, there is a growing commitment among investigators to describe language impairments in terms of precise linguistic distinctions that can be compared across languages. Such specificity allows for careful delineation of the nature of the impairment, and for clear differentiation of the role of related processes. For example, it is now clear that there is an important differentiation between late onset of language and particular areas of subsequent grammatical deficit. Children with SLI are more than just delayed in the emergence of their grammar. Instead, they continue to have relatively long-term incomplete specification of important dimensions of their grammar, especially in the domain of morphosyntax. This important fact can be obscured if the means of measurement are not suitably precise. Another observation is that, in some important ways, children's performance on general measures of intellectual performance does not fully predict the details of grammatical acquisition. Instead, grammatical competence follows a developmental course at least partially independent of general cognitive development. This fact can go undetected if the means of measuring language is not attentive to grammatical distinctions. What is needed is a phenotype that can be reconciled with the linguistic facts.

The Phenotype Must Yield to Reliable Measurements. Two approaches to measurement have been widely used. One is a *categorical phenotype*, in which individuals are identified as being affected (i.e., whether an individual has the condition of SLI). The other is a *continuous phenotype*, in which an individual is identified on the basis of a numerical score.

The primary way of arriving at a categorical phenotype is to follow existing clinical diagnostic categories. There are major risks to this approach. Just as it has proven difficult to identify individuals affected by alcoholism for studies of genetic etiology (Holden, 1994; Plomin, Owen, & McGuffin, 1994), it is likely that there will be considerable variation in clinical standards for identification of SLI. This is because the disorder is inherently complex, there is no general consensus about the best means of measurement; the conventional clinical assessment measures gloss over many of the important linguistic details, and there is considerable possible variation in surface symptomology.

Another complication is that diagnostic categories may well change over time. The child who, as a preschooler, was identified as SLI is likely to be reclassified as dyslexic or learning disabled if that child goes on to encounter difficulties with reading achievement in elementary school. So, the probability of an SLI diagnosis is greater with younger children, where the criteria are better understood, than with older children or young adults, where less is known about expected language performance.

A further complication in the real world of service provision is that clinical criteria are often influenced by educational or health service provision policies which are, in turn, driven by economic forces (for example, the criteria for service eligibility are often set to levels that correspond to the number of available service providers or to categories established for payment of services). The obvious risks are that SLI may be underrepresented in many clinical caseloads for reasons related to economic factors, or because the techniques necessary for identification may be too specialized or too time-consuming. The conclusion is that it is unlikely that a reliable categorical phenotype will emerge from the diagnostic records of service providers.

At the level of a continuous phenotype, the best source of guidance is to be found in the definitions of dyslexia. Dyslexia can be defined in terms of an age-based criterion or an IQ-based criterion. For an age-based definition, individuals are regarded as dyslexic if their reading levels are below those expected for their age; for IQ-based definitions, identification depends on a discrepancy between IQ level and reading level. According to Gilger (this volume), it does not seem to make much difference for genetic analyses as to which definition is used; either definition yields a genetic etiology. The question of age-discrepant versus IQ-discrepant definitions of reading disorder has parallels in the determination of a language impairment. Either IQ- or age-referencing are possible ways to define a continuous phenotype. In his chapter, Miller notes that IQ referencing may not be fully satisfactory in that IQ alone does not seem to fully predict profiles of language impairment. Fey, Long, and Cleave (1994) pointed out that IQ also does not predict intervention outcomes, at least within a broad range of IQ performance. Bishop (1994) argued that the traditional IQ level for exclusionary diagnosis should be relaxed for investigations of heritability. Recent work by Rescorla and her colleagues suggested that the condition of SLI may first manifest itself according to age-discrepant criteria, but as the children mature it may be evident in a discrepancy between performance IQ and language performance (Rescorla et al., 1995).

The point to be made here is that relatively little evidence is available as to how to define the continuous phenotype for language impairment and whether it will be possible to do this in a way that is reliable from one time of measurement to the next. The current work in dyslexia shows one way in which this can be done, but what remains to be seen is if that method is equally applicable to language impairment.

The Phenotype Differentiates Affected From Nonaffected Individuals.
Brzustowicz and Tomblin, among others in this volume, point out that a phenotype definition needs to have the properties of specificity and selectivity, that is, this criterion should identify the individuals who are affected and not identify the individuals who are not affected.

Some of the complexities can be revealed in a brief consideration of the limitations of composite measures derived from standardized tests. In the studies of dyslexia, standardized tests of reading achievement have proven to be useful for the formation of discrepancy scores. Any such discrepancy measure for language impairment will presumably be based on some form of standardized test measurement. Yet, there are important limitations to global estimates of language acquisition. In her chapter, de Villiers differentiates between open and closed aspects of language acquisition; in his chapter, Wexler highlights certain properties of grammar that are thought to be part of an innate grammatical acquisition mechanism. These distinctions are not captured in omnibus language tests. If some aspects of language acquisition prove to be more likely to be influenced by the environment than others, and the two categories are lumped together in standardized subtests, the resulting test scores will be unsuitable for determination of inherited versus noninherited contributions to language acquisition.

What is needed is a program of investigation in which specific phenotypic definitions of language impairment are compared to more global definitions, to determine the specificity and selectivity of such criteria and their potential value as markers of genetic influence. These studies must be relatively large-scale epidemiological investigations in order to carry out the statistical analyses necessary for estimating prevalence and determining specificity and selectivity. Such a program of investigation is in progress under the direction of Bruce Tomblin (1994). In this project, Tomblin explores psychometrically constructed phenotypes based on children's performance on standardized tests of language and intelligence. In addition, alternative definitions will be evaluated. One of those is tense-marking as a clinical marker of SLI, in a collaborative project with my lab. Other investigators are exploring processing phenotypes. The point to be highlighted here is that this program of investigation is representative of a growing appreciation of the need to re-examine the clinical markers of SLI and how these descriptors may be evaluated as possible phenotypes.

The Phenotype Shows Variation Where None Is Expected. A highly desirable linguistic phenotype would be one that could serve as a pathognomonic marker, that is, as a symptom typical of the condition of SLI. To date, this diagnostic bullet has proven elusive to identify. Much of what is known about individuals with SLI is in the form of evidence of relatively low performance, where the standard for "relative" can vary according to age or other criteria. The "relatively low" designation is referenced to a distribution of expected variation across individuals. Affected individuals are defined as those at the extremely low end of observed variation.

An alternative is put forth in the chapter by Rice and Wexler, who argue that a phenotype can be defined in terms of linguistic variation where variation is not to be expected. In this case, the normative situation is invariant; almost all individuals demonstrate mastery. This is true for obligatory grammatical structures, that is, grammatical structures that are required for the formulation of grammatical sentences. For example, when speakers generate sentences such as "the dog barks" and "she is barking," they must say "barks" and not "bark" in the first sentence and "is" as well as "barking" in the second. These grammatical forms are obligatory markers of tense. Almost all 5-year-old English-speaking children know this requirement. Presumably, the relative invariance of this knowledge is attributable to powerful mechanisms of grammatical acquisition. These mechanisms are likely to be part of an inherited contribution to language acquisition.

Yet children with SLI, who are receptively and expressively impaired, do not seem to know the obligatory nature of tense-marking by 5 years of age (Rice & Wexler, in preparation; Rice, Wexler, & Cleave, 1995). This gap in linguistic knowledge shows promise as a pathognomonic marker, in that children who do not know tense-marking by 5 years of age are missing a fundamental part of the adult grammar that is available to almost all of their age peers. This is an important discovery because it is the first identification of such a discrete part of the grammar to be characteristically affected in children with SLI and, therefore, a discrete ability to be targeted for measurement as a marker. A further advantage is that this marker has an explicit tie to the adult end-state grammar. In psychometric terms, there is strong face validity for this marker; children who are 5 years of age and who do not know the obligatory properties of tense-marking have a significant language impairment. Such an extreme variation in language acquisition is a prime candidate for phenotypic status.

The Phenotype Is Relatively Resistant to Environmental Effects. Some aspects of language acquisition are more amenable to environmental effects than others. Although this assertion is more or less agreed on by most scholars of language acquisition, the interchange between de Villiers and Snow in this volume shows that there is dispute over where to draw the line and what the line would mean. When viewed from the perspective of possible genetic studies, it seems that this is an issue that could be illuminated by careful consideration of environmental effects. There are two possible ways to go about this. One is the use of twin studies. Following the conventional logic that identical twins share an environment and the same genes, whereas fraternal twins share an environment but only some of the same genes, it would be possible to predict that members of identical and fraternal twin pairs would be similar on properties of language that are influenced by environmental factors. On the other hand, for properties of language thought to be resistant to environmental effects, identical twin pairs should be more similar than fraternal twin pairs. To the best of my knowledge, twin studies currently available have not differentiated predicted linguistic outcomes in this way.

Another possible method is that of intervention studies. The logic here is that, all else being equal, a target linguistic competence that is predicted to be relatively immune to environmental effects will show less of an intervention effect than would a target linguistic competence that is predicted to be amenable to intervention. The details of the "all else being equal" stipulation, it must be noted, will be bothersome, but the rationale has some appeal.

A third strategy would be to combine the two possible ways, in a study of intervention with fraternal and identical twin pairs. In this design, interventions targeting linguistic forms with little environmental effect should show similar outcomes for pairs of identical twins, but less similar outcomes for pairs of fraternal twins, if the individuals in these twin pairs in fact have different aptitudes for the targeted linguistic knowledge. As far as I know, such an investigation has not yet been attempted.

The Behavioral Phenotype Applies Over the Age Span. Linguistic performance changes from the early stages of language acquisition, through the preschool and childhood years, and on into adult status. Furthermore, the rate of changes varies from one aspect of language to another. Much more is known about the early stages through the preschool years, than about the subsequent periods of change. There is simply no satisfactory measure of language change over the lifespan. Thus, if the definition of a phenotype focuses on the kinds of language structures and functions that an individual demonstrates, it is necessary to change the definition with age. At the same time, relatively little is understood about underlying psycholinguistic processes that may exercise constant influence on language performance over the lifespan. In order to fill in these gaps, a life-span perspective will be needed.

The Phenotype Can Be Specified in Terms of Biological Mechanisms and Functioning. A desirable feature of specification of the phenotype in terms of biological mechanisms and/or functioning is that it may be possible to avoid the age-related complexities noted in the previous item. In this volume, Plante argues for a phenotype defined in terms of cortical structures. Although this work is at the early stages of investigation, the possible import is highly significant. It would be a major advance to know the cortical characteristics of individuals with SLI. The caution is that in these early investigations, SLI is defined in terms of diagnostic category, which, as noted earlier, is likely to show considerable variation in symptomology. What is needed is evidence of relatively specific linkages between linguistic performance/impairment and cortical location/functioning (see Poeppel, this volume, for further elaboration of this point).

A Preview of the Chapters

In the chapters that follow, the issues highlighted earlier show up in ways related to the diverse perspectives of the author(s). The areas of expertise represented by the contributors to the volume include medical genetics, behavioral genetics, linguistics, language acquisition, language impairments, brain science, and developmental psychology. In order to capture the liveliness of the cross-disciplinary dialogue that unfolded in the conference setting and the subsequent enhancements of each speaker's material, the participants were invited to provide commentary on the papers of other presenters. Those commentaries are included and follow their respective chapters. The commentaries are highly recommended reading in that they often add substantially to the information and interpretations presented in the chapters.

The chapters are ordered as follows: The initial section features chapters on genetics. In the first chapter, Brzustowicz provides an overview of key developments in medical genetics and lessons to be drawn for the investigation of complex behaviors. This is followed by a chapter by Smith, Pennington, and DeFries, in which they report on the linkage analyses of complex traits, with particular emphasis on studies of dyslexia. The next chapter is by Lefly and Pennington, in which they report on a longitudinal study of children at high family risk for dyslexia. The final chapter in this section is by Jeffrey Gilger, in which he surveys the ways in which behavioral genetic research can help clarify language acquisition and disorders.

The second section turns to linguistics and language acquisition. Wexler provides an overview of contemporary linguistics and models of children's language acquisition. The next chapter by de Villiers differentiates between what could be open versus closed in language acquisition, drawing upon her extensive work with *wh*-questions.

The third section takes up the investigation of children with language impairments. Tomblin's chapter reports on important new epidemiological and heritability evidence. Rice and Wexler present the case for a particular grammatical phenotype, that of an Extended Optional Infinitive (EOI) stage for SLI. Leonard provides a contrasting view of SLI, drawing on cross-linguistic studies to argue for a processing capacity limitation. Crago and Allen report on evidence from children learning Inuktitut, a language interesting for its complex surface morphology. Miller completes the chapters in this section with consideration of possible nongrammatical markers of SLI and the related syndromic conditions of Down Syndrome and fragile X syndrome.

The fourth section addresses the cortical structures and functioning that may or may not underlie variations in language performance. Plante argues for phenotypic variability in brain-behavior relationships in individuals with SLI. Poeppel provides an informative critique of the limitations of PET studies of phonology.

The final section includes a chapter by Snow, who argues for an interactionist account of language acquisition in which environmental input plays a decisive role.

In this brief sketch of the order of the chapters and topics, what is missing is the contribution of the commentaries, in which there is cross-referencing within and across the content groupings. Readers may well wish to work their way through the volume by following the track of the commentaries to the chapters authored by the commentators. If so, that would only further attest to the premise that the question of the genetics of language does not easily yield to one particular paradigm or perspective. This volume, it is hoped, will further lead toward a satisfactory model of the genetics of language which, in turn, will yield complementary information about the contribution of environmental factors.

REFERENCES

Bishop, D. (1994, June). *Is specific language impairment a valid diagnostic category? Genetic and psycholinguistic evidence.* Invited presentation at the 15th Annual Symposium on Research in Child Language Disorders, Madison, WI.

Ewing, A. W. G. (1930). *Aphasia in children.* London: Oxford University Press.

Fey, M. E., Long, S. H., & Cleave, P. L. (1994). Reconsideration of IQ criteria in the definition of specific language impairment. In R. V. Watkins & M. L. Rice (Eds.), *Specific language impairments in children* (pp. 161–178). Baltimore: Brookes Publishing Co.

Gould, S. J. (1994, November 28). Curveball. *The New Yorker,* pp. 139–148.

Herrnstein, R., & Murray, D. (1994). *The bell curve: Intelligence and class structure in American life.* New York: Free Press.

Holden, C. (1994). A cautionary genetic tale: The sobering story of D2. *Science, 264,* 1696–1697.

Mann, C. C. (1994). Behavioral genetics in transition. *Science, 264,* 1686–1689.

McClearn, G. E. (1993). Behavioral genetics: The last century and next. In R. Plomin & G. E. McClearn (Eds.), *Nature, nurture, and psychology* (pp. 27–54). Washington, DC: American Psychological Association.

Pinker, S. (1994). *The language instinct: How the mind creates language.* New York: Morrow.

Plomin, R., & McClearn, G. E. (Eds.). (1993). *Nature, nurture, and psychology.* Washington, DC: American Psychological Association.

Plomin, R., Owen, M. J., & McGuffin, P. (1994). The genetic basis of complex human behaviors. *Science, 264*, 1733–1739.

Rescorla, L. A., Ratner Bernstein, N., Pharr, A., Pavluk, D., Torgenson, D., & Skripak, J. (1995, March). *Phonetic profiles of typically developing and language-delayed toddlers.* Poster presentation at the biennial meeting of the Society for Research in Child Development, Indianapolis, IN.

Rice, M. L., & Wexler, K. (in preparation). *Tense as a clinical marker of specific language impairment in English-speaking children.*

Rice, M. L., Wexler, K., & Cleave, P. (1995). SLI as a period of Extended Optional Infinitive. *Journal of Speech and Hearing Research, 38*, 850–863.

Rice, M. L., & Wilcox, K. A. (Eds.). (1995). *Building a language-focused curriculum for the preschool classroom: A foundation for life-long communication.* Baltimore: Brookes Publishing Co.

Tomblin, B. (1994). *Morphosyntactic diagnosis of specific language impairments.* Grant from NIDCD.

Watkins, R. V., & Rice, M. L. (1994). *Specific language impairments in children.* Baltimore: Brookes Publishing Co.

ACKNOWLEDGMENTS

This book began as a conference, which was held in the fall of 1993 at the University of Kansas, under the sponsorship of the Merrill Advanced Studies Center (MASC). Located in the Schiefelbusch Institute for Life Span Studies, the MASC supports conferences to bring together scholars of diverse disciplines for the consideration of topics related to life span development. As the study of human development has become increasingly fractionated into specialized areas of inquiry, there is a need for interdisciplinary forums that set the occasion for specialists in one field to understand emerging developments and new concepts in other related fields. The consideration of a familiar problem from the perspective of another discipline can lead to new insights. This is especially so when there are simultaneous breakthroughs in areas of study that have been, traditionally, somewhat isolated from each other. Emerging paradigms in each of the areas can develop in ways that are highly relevant to the other areas of investigation. In the isolation of their own highly focused endeavors, scholars can be unaware of, or unable to fully benefit from, new advances that could help crack long-standing unresolved problems.

The intent of the MASC is to sponsor occasions in which scholars can come together for the purpose of communication about emerging concepts and insights. The intent is not so much to provide highly developed action plans for further research as to share what is known and not known about particular issues in a way that can be assimilated by scholars working in related fields. The process of learning about new developments can set the occasion for new understandings of familiar problems, and the identification of new methods and techniques for arriving at more satisfactory answers. This is especially so when there is an opportunity to comment on and react to the work in other fields, and to query, challenge, elaborate on, and mull over new information.

The expected outcome of the MASC conferences is the publication of a volume of collected papers. The preparation of formal papers for publication serves as an extension of the occasion, when authors reflect on their own work from the enhanced perspective gained from the consideration of other scholars, and when they reflect more fully on the positions espoused by colleagues working in different paradigms.

The topic of this volume, the genetics of language acquisition, was well suited to a cross-disciplinary conference. The participants were chosen from the fields of language impairment, language acquisition, behavioral genetics, linguistics, and brain imaging. Each of these areas of inquiry is undergoing significant change, with new developments moving at a brisk pace, and yet there is no clear interface between them on the important topic of the genetics of language acquisition. At the conference, each participant delivered a paper, which was followed by a lengthy period of time for full discussion. In addition, there were many opportunities for informal discussion among participants over the course of the conference. The upshot was that the views of the presenters and the subsequent papers were, in many instances, influenced by the feedback provided by other participants. The spirit of the cross-disciplinary dialogue is captured in the commentaries in this collection of papers. The assignment of discussants was made with diversity of paradigms in mind; as much as possible, discussants were paired with authors whose perspectives differed from their own. The commentaries, in many cases, highlight the points of disagreement and convergence of perspective across the paradigms represented among the participants.

The conference was made possible with the generous financial support of Virginia and Fred Merrill of Shawnee Mission, Kansas. We express our deep appreciation for their understanding of the value of such occasions. The conference enjoyed the support of the Schiefelbusch Institute for Life Span Studies at the University of Kansas, and the Mental Retardation Research Center. In particular, we wish to acknowledge the many ways in which the current Director, Stephen R. Schroeder, and the former Director, Richard L. Schiefelbusch, contributed to the success of the conference. The Advisory Board of the MASC also played an important role, with particular appreciation to Kathleen McCluskey Fawcett, Associate Provost for Academic Services, who contributed her time and talent. A number of people carried out the logistical tasks for the conference with competence and poise. Our gratitude goes to Patsy Woods, Mary Howe, Carolyn Roy, and Janice Chazdon, and to Patricia L. Cleave and C. Melanie Schuele, doctoral students at the time of the conference. Special thanks go to Patsy Woods for her assistance in the editorial chores of compiling the volume. And much gratitude goes to Su Dong Chen and Sean Redmond who prepared the subject and author indexes.

GENETICS

1

LOOKING FOR LANGUAGE GENES: LESSONS FROM COMPLEX DISORDER STUDIES

Linda Brzustowicz
Rutgers University

Recent advances have revolutionized the field of human genetics. Laboratory techniques developed over the past decade have provided methods for locating the actual genetic defects responsible for numerous inherited disorders. Because the statistical techniques first used to analyze this new laboratory data required assumptions about how diseases were inherited, most of the early findings described disorders that followed simple inheritance patterns, such as those described by Gregor Mendel. Complex genetic disorders do not clearly follow these simple Mendelian patterns and have proved more challenging to study. Because studies of complex disorders use many of the same research methods as studies of simple Mendelian disorders, it is useful to first review the recent advances in the study of simple disorders. This review covers the principles of linkage analysis and the challenges of complex disorders. The second section of the chapter develops strategies for disease definition.

BACKGROUND GENETICS

Mendelian Inheritance

Much of our understanding of patterns of inheritance stems from the work Gregor Mendel presented in 1865. Mendel conducted breeding experiments with garden peas, following the expression of physical characteristics through generations of offspring, determining experimentally that in peas yellow color is the dominant trait, and green is the recessive. Mendel observed that when he crossed two strains that differed in seed color, one that produces green peas and one that produces yellow peas, all of the hybrid offspring produced yellow peas. Interestingly, even though these hybrid offspring produced peas of the same yellow color as one of their parents, the offspring and yellow parent plant were genetically different from one another. Another interesting finding was that when the yellow-producing parent plants were crossed with themselves, all of the offspring produced yellow peas. Finally, when plants from the yellow peas of the green–yellow cross were crossed with themselves, a very different picture emerged. While 75 percent of these offspring still produced yellow peas, 25 percent produced green peas. Thus the ability to produce green peas, even though not expressed in the offspring of the original cross, was preserved and passed on to the next generation. This ability of two traits, such as yellow and green color, to pass through generations and retain their ability to produce the original physical characteristics is called Mendel's first law, or the law of independent segregation.

We now understand the biological mechanisms that underlie this process. Most higher forms of life—including garden peas and humans—have two copies of most of their genes and are referred to as diploid organisms. The two copies of a particular gene may be identical or they may be different. The different forms, or alleles, of a particular gene may produce differences in a particular physical feature. In the foregoing example of peas, we were examining the effects of a single gene that controls pea color and following the interaction of two different forms of the gene: one allele that produces green peas and one allele that produces yellow peas. Because each pea plant has two copies of this color gene, there are three possible genetic situations, or genotypes: two green alleles, two yellow alleles, or one green and one yellow allele. Because yellow is dominant to green, only two physical appearances, or phenotypes, exist: yellow or green. Pea plants that have two yellow or two green alleles are called homozygous and will produce peas of their particular color. The heterozygous plants, those with one yellow and one green allele, will produce peas of the dominant color, yellow. Note that plants with the same phenotype (yellow peas) may actually have different genotypes (homozygous for yellow or heterozygous for yellow and green). This explains why peas that look alike may not reproduce alike.

In sexual reproduction, diploid organisms inherit a haploid genome (containing one copy of each gene) from each parent. Haploid gametes (sperm and egg) are produced through the process of meiosis. During meiosis, the two copies of each gene are lined up through the precise arranging of the chromosomes. The two copies are then segregated into different gametes (Fig. 1.1). Returning to Mendel's pea experiment, the original parents were homozygous for the yellow allele and homozygous for the green allele. All the offspring of this cross inherit one yellow allele from one parent and one green allele from the other parent. So although they will produce yellow peas, these yellow offspring will all be yellow–green heterozygotes. What will happen when we cross two of these

MEIOSIS

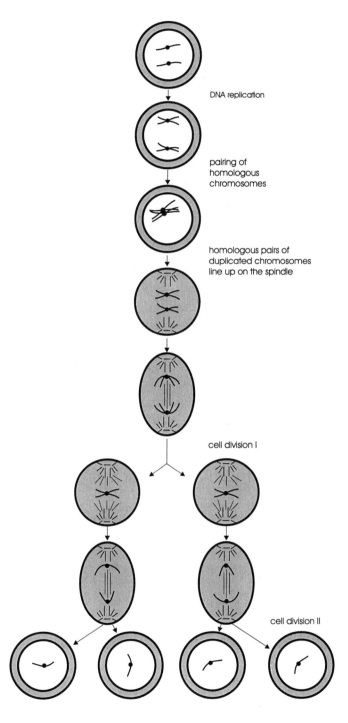

DNA replication

pairing of
homologous
chromosomes

homologous pairs of
duplicated chromosomes
line up on the spindle

cell division I

cell division II

FIG. 1.1. Meiosis.

heterozygotes? Each parent may produce gametes that contain either a yellow or a green allele. Hence, four types of combinations are possible. Listing the contribution from the first parent first and the second parent next, the four possible genotypes of the offspring are: yellow–yellow, yellow–green, green–yellow, and green–green. The corresponding phenotypes of these plants would be yellow peas, yellow peas, yellow peas, and green peas, or the 75 percent yellow and 25 percent green result originally described by Mendel (Fig. 1.2).

Turning from benign color variations to disease states, one can appreciate how very different patterns will exist if a disease-producing allele is dominant or recessive to the normal allele. Diseases that can be caused by a defect in only one copy of a gene will follow a dominant pattern of inheritance, while those requiring defects in both copies will show a recessive pattern. Humans have approximately 100,000 genes, arranged on 23 chromosome pairs. The first 22 are called autosomes, while the last pair are called the sex chromosomes, due to their role in sex determination. Women have two copies of an X chromosome, while men have one X and one Y chromosome. If a disease allele is located on a sex chromosome, the pattern of inheritance will be related to the sex of the individual. The commonly described patterns of inheritance are autosomal dominant, autosomal recessive, X-linked dominant, and X-linked recessive. Figure 1.3 depicts family trees, or pedigrees, which illustrate each of these patterns, or modes, of inheritance.

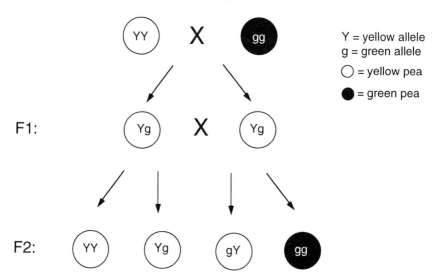

FIG. 1.2. Heterozygote cross. Expected results of a cross between homozygous yellow (YY) and homozygous green (gg) pea-producing plants, and of a cross between the F_1 hybrid offspring (Yg). The yellow allele (Y) is dominant to the green allele (g). All F_1 hybrid offspring are yellow in color, but are heterozygous for color genes. The F_2 offspring are 25% green and 75% yellow in color, although they represent three distinct genotypes. The re-emergence of an original parental phenotype (green color), not seen in the F_1 generation, illustrates Mendel's first law of unit inheritance.

Autosomal Dominant

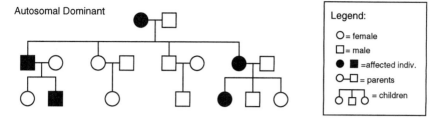

Legend:

O = female
□ = male
● ■ = affected indiv.
O—□ = parents
♂♂♂ = children

Autosomal Recessive

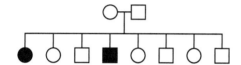

X-Linked Dominant

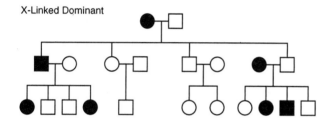

X-Linked Recessive

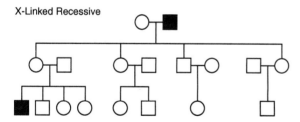

FIG. 1.3. Pedigrees demonstrating Mendelian patterns of inheritance.

Autosomal dominant disease alleles are often quite rare, so one parent is usually heterozygous for the disease-producing allele, while the other parent is homozygous for normal alleles. All of the children who inherit the disease-producing allele from the affected parent (50%, on average) will be affected. Approximately half of the children of these affected individuals will likewise be affected, and so on through the generations.

Alleles that produce autosomal recessive disorders tend to be more frequent in the population. These disorders typically do not show such a family history.

They tend to manifest in a single generation, when both parents are healthy but are carriers of (i.e., heterozygous for) a disease-producing allele. The 25% of their children that are homozygous for this allele will have the disease, the 50% who are heterozygous will be carriers (like their parents), and 25% will be homozygous for the normal allele. If the affected individuals have children with a homozygous normal individual, all of their children will be carriers, but phenotypically normal. If they have children with a carrier, half their children will be carriers and half will be affected. If two individuals affected by the same recessive disorder have children, all of their children will be affected.

X-linked patterns are slightly more complex, due to the fact that women normally have two copies of X-linked genes (one copy from each parent), while men normally have only one copy (inherited from their mothers). X-linked dominant traits have patterns similar to autosomal dominant traits, with the exception that affected men never transmit the disease to their sons but always transmit it to their daughters. X-linked recessive traits, however, look quite different, because although women still need two defective copies of the gene to have the disease, men only need to inherit one defective copy (since they only have one copy) to inherit the disease.

Methods of Genetic Study

While certain molecular genetic techniques are quite new, other methods of genetic study have been in use for many years. These questions answered by these different types of studies are summarized in Table 1.1. Linkage studies, which seek to locate the actual gene responsible for a particular disorder, are ordinarily undertaken once it has been firmly established that a disorder is genetic, and usually when something is known about the mode of inheritance. Before focusing on linkage studies, we briefly review the methods that can establish whether a disorder is genetic. A more detailed description of all these methods can be found in Khoury, Beaty, and Cohen (1993).

The first step in determining whether a disorder is genetic is usually to determine whether it is familial. Family studies compare the rates of illness in relatives of an affected individual (the proband) to rates of illness in the general population. An increased rate in the relatives of a proband suggests that the

TABLE 1.1
Questions Answered by Genetic Studies

Question	Study Type
Is the disorder familial?	Family studies
Is the disorder inherited?	Twin and adoption studies
How is the disorder inherited?	Segregation and pedigree analysis
Where is (are) the abnormal gene(s)?	Linkage analysis
What is the genetic defect?	Molecular analysis

disorder is familial. Although genetic disorders are familial, familial problems may not be genetic but due instead to environmental factors such as rearing practices. For example, although certain families may have an unusually high concentration of doctors or lawyers, it is unlikely that these situations are due to genetic factors. And while genetics may play a role in language development, the language one first learns to speak would also seem to be familial rather than genetic.

If a disease appears to be familial, the next step is to determine whether it is actually genetic. Two classic approaches to this question are adoption studies and twin studies. In adoption studies, children who might be at high genetic risk, on the basis of being born into a family with elevated rates for a disorder, are examined after being raised in a family with no elevated rate. Likewise, children at low genetic risk who are raised in families with a high rate of disease can be examined. If the disorder is genetic, the adopted children would be expected to show disease at a rate consistent with their family of origin. If the disorder is only familial, their rate of disease should instead reflect that of the family in which they were raised. In twin studies, rates of disease are compared between identical and fraternal twins. Since identical twins are identical genetically, they would be expected to be more similar to each other (i.e., either both have the disease or neither have the disease) than fraternal twins, who are genetically no more similar than any two siblings. If the disorder is familial but not genetic, the identical twins would be expected to be no more similar than fraternal twins or any other sibling pair.

Segregation analysis and pedigree analysis are methods used to investigate the specific mode of inheritance. In segregation analysis, the patterns of disease in many families are compared to what would be expected under different models of inheritance, with the goal of identifying the most likely pattern. Segregation analysis is limited by the assumption that the same mode of inheritance will be operating in all families. Pedigree analysis similarly seeks to identify a pattern of inheritance but uses only a single, large, multigeneration family. Although the risk of multiple modes of inheritance is greatly reduced, this approach is limited by the possibility that the identified pattern will be applicable only to the studied family.

Gene Linkage and Genetic Recombination

To understand how gene linkage studies work, we must return to the biological mechanisms underlying inheritance. We have already discussed how, during meiosis, the two copies of each gene are lined up and segregated into different gametes. Mendel's second law, independent assortment, states that the genes for different traits are independently transmitted to the gamete. Under this law, the inheritance of each gene is an independent, random event, with the outcome of the segregation of one gene pair having no influence on the segregation of any other gene pair. Returning to the garden pea, let us now consider pea color and a second trait, pea shape. Mendel worked with peas that had two types of shapes, round and wrinkled. In the same way that yellow was dominant over green for pea color, round is dominant over wrinkled for pea shape. If two true-breeding (i.e., homozygous) strains of peas, one with round yellow peas and one with wrinkled green peas, are crossed, the expected hybrid offspring will produce round yellow peas, but be heterozygous for both shape and color genes. What will we expect to get if we cross two of these hybrid offspring? These hybrids can produce four types of gametes: round-yellow, round-green, wrinkled-yellow, and wrinkled-green. Figure 1.4 details the results of this cross: $9/16$ round–yellow, $3/16$ round–

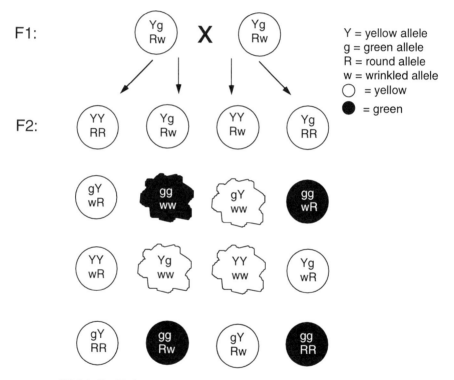

FIG. 1.4. Double heterozygote cross. Expected results of a cross between two pea plants heterozygous for both color and shape genes. The F$_1$ plants produce round, yellow peas, as both the yellow (Y) and round (R) alleles are dominant to green (g) and wrinkled (w) alleles. The F$_2$ offspring represent all possible combinations of parental alleles, written with the alleles from the first parent on the left and the alleles from the second parent on the right. Although only four phenotypes are visible (yellow–round, yellow–wrinkled, green–round, green–wrinkled), 9 genotypes are present (YY–RR, YY–Rw, YY–ww, Yg–RR, Yg–Rw, Yg–ww, gg–RR, gg–Rw, gg–ww). The shape of a pea has no influence on its color; twenty-five percent of both the round and the wrinkled peas are green, and 75% of each shape category are yellow. This illustrates Mendel's law of independent assortment.

green, $\frac{3}{16}$ wrinkled–yellow, and $\frac{1}{16}$ wrinkled–green producing plants. If we closely examine these numbers, we can see that the genes for color and shape are independently assorting. If we consider only the round peas (12 in number), we can see that 75% are yellow ($\frac{9}{12}$) and 25% ($\frac{3}{12}$) are green. Likewise, among the wrinkled peas (4 in number), 75% ($\frac{3}{4}$) are yellow and 25% ($\frac{1}{4}$) are green. The 75% yellow and 25% green ratio we found from the original cross is unaffected by the characteristics of pea shape, demonstrating independent assortment.

Sometimes genes do not follow this law but rather demonstrate a tendency to segregate particular combinations together. Let us assume that instead of the above results, we found something very different when we crossed the round–yellow pea producing hybrid plants with themselves. Suppose we found that $\frac{3}{4}$ of the offspring produced round yellow peas, and $\frac{1}{4}$ produced wrinkled green peas. Now if we were to consider only the round peas, we would find that 100% of them are yellow. Likewise, if we examined only the wrinkled peas, we would find that 100% of them are green. In this cross, the genes for yellow color and round shape seem to be transmitted together, while the genes for green color and wrinkled shape also seem to be traveling together. When we observe this type of violation of the law of independent assortment, we say that the genes under consideration are linked.

Genes are linked when they are located close to one another on a particular chromosome. Under this situation, when the chromosome pairs separate to create the haploid gametes, the copies of the different genes that happen to be together on a particular chromosome will be carried off together. The closer two genes are physically to one another, the more frequently they will segregate together and the tighter they are linked. However, even tightly linked genes may be separated during meiosis by a normal process of chromosome breakage and exchange. In fact, the probability that such a crossover event break, or will occur between two genes is proportional to the physical distance between them (Fig. 1.5). Even though there may be great variation among individuals in the different copies that exist for the same gene, there is extreme consistency in the order in which the genes are located on the chromosomes. If two genes are found to be tightly linked in a certain group of individuals, ordinarily they will be tightly linked in all individuals. Maps of the chromosomes which reflect the probability that neighboring genes will cosegregate, called linkage maps, can thus be generated from a specific set of individuals and generally applied to the population as a whole.

Genetic Linkage Studies

The goal of linkage studies is to identify a chromosomal region or set of regions related to an inherited disease, with the aim of isolating the responsible genes, identifying the causative mutations, and, ultimately, understanding function in both the normal and pathologic state. While linkage studies are most commonly used to study diseases, any inherited characteristic that can be observed or measured may be studied. The general method of these studies is to compare,

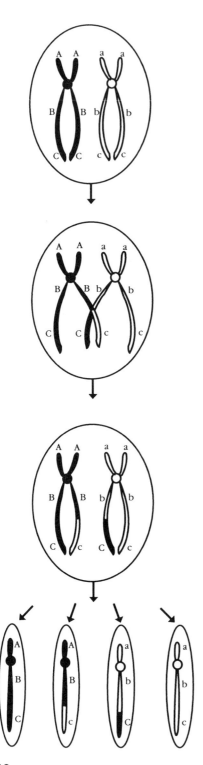

FIG. 1.5. Meiosis and recombination.

within families, the presence or absence of the characteristic with the pattern of inheritance of different chromosomal regions. If the characteristic is observed to consistently cosegregate with a particular region of a chromosome, then it is likely that the gene responsible for the characteristic is located in that chromosomal region. Readers interested in a more in-depth discussion of the mathematical principles of linkage analysis are referred to Ott (1991).

For this approach to work, a way to follow the inheritance of particular chromosomal regions through families is critical. Fortunately, there are many variations in the DNA sequences of individuals that may be used for this purpose. It is estimated that approximately 1 out of every 250 basepairs varies between two haploid copies of the human genome (Cooper, Smith, & Cooke, 1985). The vast majority of these variations occur outside of gene sequences, as the DNA in these noncoding regions is under much less evolutionary pressure to remain unchanged. Some of these alterations can be detected by bacterial enzymes called restriction endonucleases, which recognize specific 4 or 6 base pair sequences and cleave DNA containing that pattern. The variations in DNA sequence between individuals may produce variations in the places where these enzymes cut, producing DNA fragments of differing length. This variation in length is called *restriction fragment length polymorphism* (RFLP; Fig. 1.6). Sometimes the length of a region of DNA varies because it contains a DNA sequence that repeats a variable number of times. These elements can be short (two base pairs) or long (1,000 base pairs) but in either case again can produce polymorphisms that can be followed through families (see Thompson, McInnes, & Willard, 1991, for further discussion of DNA polymorphism and laboratory techniques of analysis).

Table 1.2 summarizes some of the important events in the history of linkage studies. While the phenomenon of linked genes was not included in Mendel's description of his work, it has been in the genetics literature since the beginning of this century. Long before the chemical nature of DNA was understood, the geneticist Morgan (1911) noticed that certain physical traits in Drosophila tended to cosegregate, and he began to generate linkage maps based entirely on these observable characteristics. Nearly 50 years ago, Haldane and Smith (1947) examined the cosegregation of colorblindness and hemophilia in humans, two diseases caused by defects in genes located on the X chromosome. Linkage studies in humans progressed very slowly for the next 30 years, stymied by the lack of a way to reliably follow the inheritance of most chromosomal regions. Then, in 1980, Botstein, White, Skolnick, and Davis proposed the use of RFLPs to generate a human linkage map. This provided the much-needed way to track the inheritance of a multitude of different chromosomal regions. Within 3 years, the Huntington's disease gene was localized using this method. Since the launch of the Human Genome Project in 1990, the number of human genes placed on linkage maps has been increasing at an ever accelerating pace (Fig. 1.7).

Over 100 disease genes have been mapped through linkage studies (Lander & Schork, 1994). Most of these successful studies have involved "simple" Men-a.

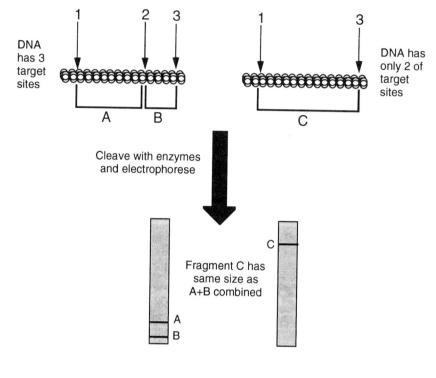

b.

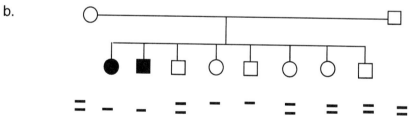

FIG. 1.6: Restriction fragment length polymorphism.

delian disorders, diseases with clear diagnoses and patterns of inheritance, such as Huntington's disease (autosomal dominant), Duchenne muscular dystrophy (X-linked), and cystic fibrosis (autosomal recessive).

Complicating Genetic Factors

Even among disorders that are caused by a single gene or clearly show a classical Mendelian pattern of inheritance, a variety of genetic mechanisms can present obstacles to a linkage study. Sometimes not all individuals inheriting a disease

TABLE 1.2
History of Linkage Studies

1911:	Morgan reports the phenomena of linked genes in Drosophila.
1915:	Eighty-five Drosophila genes are mapped by linkage.
1947:	Haldane and Smith describe likelihood analysis of 17 families with colorblindness and hemophilia.
1968:	The first gene assignment to a specific autosome is made, Duffy blood group to chromosome 1.
1971:	Four autosomal gene assignments have been made.
1979:	Linkage studies are very limited, based on a few biochemical and blood group markers.
1980:	Botstein et al. propose the use of restriction fragment length polymorphisms (RFLPs) to generate a human linkage map.
1983:	Huntington's disease is localized to Chromosome 4 by RFLP analysis.
1987:	Approximately 400 RFLPs have been described.
1989:	Dinuelcotide repeat polymorphisms are described. They are highly polymorphic, abundant, and utilize technology that requires very little DNA for analysis.
1990:	The Human Genome Project is launched.

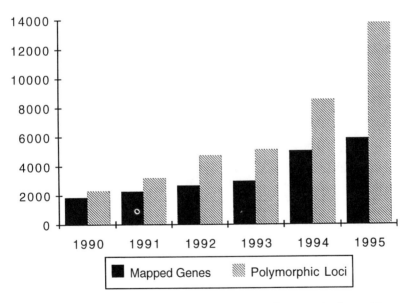

FIG. 1.7. Growth of the human gene map. Numbers of genes and polymorphic loci placed on the human gene map, since the start of the human genome project (data from McKusick, 1990 and GDB™).

gene show signs of the illness. This is called reduced penetrance, and is suggested when a disorder appears to be genetic but appears in fewer than the expected number of family members, such as in only one of two identical twins. Split-hand deformity or lobster-claw malformation is an autosomal dominant disorder of hand and foot formation. It is estimated that only approximately 70% of individuals inheriting the gene for this disorder will actually exhibit a deformity. Reduced penetrance can disguise the true mode of inheritance of the responsible gene by producing a pedigree with apparent skipped generations, when nonpenetrant individuals who have the defective gene produce children with the deformity.

Sometimes a single gene defect can produce a variety of different symptoms, so that even within the same family, affected members do not seem to have quite the same disease. This is called variable expressivity, and calls for good diagnostic criteria that will include all of the possible different forms. Marfan syndrome, an autosomal dominant disorder of fibrous connective tissue, illustrates this point. People with Marfan syndrome typically have abnormalities in skeletal, ocular, and cardiovascular systems, but specific individuals may have abnormalities limited to only one or two of these systems, and the severity of symptoms can vary widely.

Nongenetic cases of disease that look like an inherited illness are called phenocopies. Returning to the example of split-hand deformity, similar-looking defects may be caused by intrauterine trauma due to congenital constriction rings ("amniotic bands"). The presence of many phenocopies in the population can, again, disguise the mode of inheritance of the genetic cases of illness.

Sometimes different defects in the same gene may exist within the population. These different defects may produce similar but nonidentical diseases that seem to "breed true" within any specific family. This situation is termed *allelic genetic heterogeneity*. In two different forms of muscular dystrophy, Duchenne and Becker, the same gene coding for the protein dystrophin is involved. In Becker muscular dystrophy, the damaged gene produces a partially functional product, and so the disease is relatively mild. In Duchenne muscular dystrophy, the gene product is badly damaged or totally absent, and the resulting disease is very severe.

Multiple genes may be involved in causing a single disease, with the defects working singly or together. Sometimes defects in entirely separate genes can produce clinically identical disease. This possibility is called *nonallelic genetic heterogeneity*. For example, retinitis pigmentosa refers to a group of inherited diseases that cause loss of vision by retinal degeneration. Defects in any of several genes, located on different chromosomes, are able to cause this disorder. In addition to this situation, in which only one gene defect is present in any given family, diseases can be produced by the interaction of defects in more than one gene, producing more complex patterns of inheritance.

Complex Disorders

In addition to the complicating factors that may confound studies of "simple" Mendelian disorders, complex disease studies usually have to account for several additional problems (Lander & Schork, 1994). Often the relative importance of genetic and environmental factors in the development of illness is unknown. The mode of inheritance may be unclear or entirely unknown. Multiple genes, acting either individually or in combination, may be involved. There can often be difficulty in locating sufficient suitable families for a linkage study. Very frequently, there is some question as to the best way to define the diagnostic categories of "affected" and "unaffected" (Ott, 1990).

To understand why these problems pose difficulties in applying the general approach of linkage to the study of complex disorders, it is helpful to review the parts of a linkage study and consider how the differences between complex and simple disorders can affect each step. Linkage studies can be roughly broken into three stages: (a) defining the disorder and selection of suitable families for study; (b) genotyping of the individual family members through molecular genetic techniques; and (c) statistically analyzing the data to establish the presence or absence of linkage. The complications of complex disorders primarily affect the first and third steps.

As mentioned, the exact disease boundaries of complex disorders are usually not well defined. Hypertension is a useful example. While most clinicians would agree on the diagnosis of a severely hypertensive individual, there is considerable debate as to where the exact blood pressure cutoff should be. Also consider that blood presssure measurements vary from visit to visit, are clearly affected by environmental factors such as diet and anxiety, and that elevated blood pressure can be a secondary symptom of a number of medical illnesses. It is easy to see why many complex disease studies become stalled at this stage of phenotype definition.

Once a suitable disease definition is established, it may not be easy to recruit sufficient suitable families. Remember that the main principle behind linkage studies is to study the cosegregation of a disease with specific chromosomal region within families. To do this, families with multiple affected individuals are used. Roughly speaking, as the number of affected individuals increases, there will be more chances to observe cosegregation, and more statistical power to demonstrate linkage. However, for some complex disorders that appear to be genetic, families with multiple affected individuals may still be rare. Autism is such an example. While the risk of being autistic for a sibling of an autistic child is 50 to 100 times the risk for members of the general population, the low incidence of this disorder (about 2–4 in 10,000 births) still makes families with multiple autistic children rare (Pearson et al., 1994).

The news is much better for the genotyping of individuals. The molecular genetic advances described earlier are just as helpful in the study of complex and simple disorders. In fact, the increased amounts of genetic information that now can be obtained are making some difficult complex disorder studies feasible for the first time.

While the details of the different statistical approaches that may be used in a linkage analysis are well beyond the scope of this chapter, one principle does demonstrate how some of the difficulties with complex disorders can hinder a linkage study. In general, linkage studies are statistically more powerful if they are conducted with some ideas about how the disorder is inherited. Nonparametric methods that require far fewer assumptions about a genetic model have become increasingly popular, but they are much less powerful than the more commonly used likelihood-based methods. These lower power analytic methods may be compensated for by increased numbers of subjects, but this may not always be so easy for certain complex disorders.

Each of these three areas of linkage studies hold challenges and potential advances for the study of complex disorders. More information about the recent laboratory advances can be found in the program reports of the Human Genome Project. Further details about linkage analysis and recent advance in those techniques can be found in Terwilliger and Ott (1994), and Lander and Schork (1994). The remainder of this chapter will focus on issues of disease definition.

STRATEGIES FOR DISEASE DEFINITIONS

Are All Diagnostic Criteria Appropriate for Genetic Studies?

There are many ways to approach the development of diagnostic crtieria. It is helpful to have an understanding of some of the historical forces that have shaped the methods used in the past. While the field of medical genetics is not new, the emphasis on the genetic component of many diseases is recent, particularly when compared to how long these diseases have been described and studied. There may well be fundamental differences in how a disease is conceptualized if it is initially considered to be genetic or not. For genetic disorders, disease definitions can be developed by considering families affected by an illness, and then determining what traits seem to be shared by affected members. This type of approach will be sensitive to issues such as variable expressivity that might otherwise lead to the diagnosis of different disorders in different family members. Diseases that are not thought of as genetic will typically be defined by considering a series of unrelated individuals that seem to have some features in common. The reason for the diagnostic classification, such as to aid in treatment decisions or in making a prognosis, will color the dividing lines that are used to "lump" or "split" the individual cases into diagnostic groups. There is no a priori reason to assume that disease definitions that were developed for other purposes will correspond to discrete genetic entities. The two general types of errors that can exist in disease definitions are that they can be too broad or too narrow, when compared to the underlying genetic mechanisms. These problems will be considered from the epidemiologic perspective of specificity and sensitivity.

Specificity and Sensitivity

Sensitivity is a measure of the rate at which true cases are correctly identified by a certain procedure; specificity is the number of true noncases that are accurately classified. In terms of complex disease definitions, the problems are not diagnosing all of the true genetic cases (sensitivity) or diagnosing nongenetic cases along with the genetic cases (specificity). Diagnostic criteria that are too strict will not be sensitive enough and criteria that are too inclusive will not be specific enough. Different approaches may be used to try to compensate for the diagnostic shortcomings of each of these situations.

Low Specificity Phenotypes

Phenotype definitions that are of low specificity will diagnose nongenetic cases (phenocopies), along with true genetic cases. This is the situation encountered when a disease shows familial transmission in some cases but also appears to occur as sporadic cases. Sometimes, the majority of cases in the population can be nongenetic, despite the presence of a true genetic etiology for some.

For the linkage analysis of complex disorders, it may be desirable to begin the study with a subset of families with the disorder, a select group with clearer genetics and phenotypic expression than the general population. The advantage of careful phenotype definition is illustrated by the recent success in isolating BRAC1, the gene for one form of hereditary breast cancer (Futreal et al., 1994).

A number of studies have examined the patterns of breast cancer within families. Newman, Austin, and King (1988) described a model with a rare, autosomal dominant, susceptibility allele, occurring in the population at a frequency of 0.0033. Based on this, about one in 150 women would inherit a susceptibility to breast cancer. The lifetime risk of breast cancer for women with this genetic susceptibility is 90%; lifetime risk for those without it is only 10%. For an individual woman, inheriting this susceptibility allele vastly increases her risk for breast cancer (by nine times). However, the impact on the population as a whole is much less. Remember that only one woman in 150 inherits this allele. Ten percent of the other 149 (or about 15 women) will still develop breast cancer due to other causes. So the estimated contribution to the population breast cancer rate by this susceptibility allele is only 6% (about 1 out of 16) of all cases. The vast majority of cases (94%) are accounted for by the large number of women at relatively low risk.

Under these circumstances, searching for a gene responsible for such a minority of cases in the population may seem like a hopeless cause. Indeed, if a linkage study were set up to include as probands any woman with breast cancer, the 6% of cases with a genetic cause would be extremely difficult to detect. This would be the problem expected with a low-specificity disease definition. If, instead, a subgroup of women with breast cancer that seemed more likely to have a genetic susceptibility could be defined, and only those families were studied, the odds of finding linkage might be enhanced.

This was the approach taken by Hall et al. (1990). They used previously noted epidemiological features characteristic of familial breast cancer, namely younger age at diagnosis, frequent bilateral disease, and more frequent occurrence of disease among men, to identify 23 extended families with 146 cases of breast cancer. Within these families, 60% of women at risk and 20% of men at risk were affected. Linkage to the q21 region of chromosome 17 was detected in 40% of study families. Families with mean age of breast cancer diagnosis greater than 45 were unlinked to this location. This linkage finding provided the foundation for the recent isolation of the cancer gene BRCA1 (Futreal et al., 1994).

Low Sensitivity Phenotypes

Genetic disease definitions often allow for a wide range of expression of disease, from incidental findings that do not interfere with function to severe impairment. A good example of this is neurofibromatosis, a disease characterized by spots of increased pigmentation of the skin and the formation of variable numbers of soft tumors, arising from the cells surrounding the peripheral nerves (neurofibromas). One third of patients with neurofibromatosis come to medical attention for neurologic complaints (due to nerve compression by tumors), another third out of cosmetic concerns, with the final third discovered on routine physical exam, having no complaints (Adams & DeLong, 1984). If a disease definition of neurofibromatosis included a requirement of severe disease with some functional impairment, it would not be sensitive enough to identify these individuals who are mildly affected.

A strategy to address this type of problem is to select families on the basis of a well-diagnosed proband, and then examine all family members for a subclinical trait associated with the disorder. The presence of this subclinical trait, rather than the full disease, is then used as the affected phenotype for linkage analysis. Linkage to juvenile myoclonic epilepsy (JME) was found using this approach (Greenberg et al., 1988).

Historically, JME has been identified as a subtype of epilepsy twice, by Janz in 1957 and Delgado-Escueta in 1978. JME can be characterized by typical muscle jerks, age of onset, and EEG abnormalities. The hallmark early morning myoclonic jerks are the initial complaint, however, in only 5% of cases. In addition to the myoclonic jerks, patients may also have tonic clonic or absence seizures.

Significantly, some clinically normal relatives have been found to exhibit the characteristic EEG abnormalities.

Greenberg et al. (1988) identified families for their linkage study of JME through a well-diagnosed proband. EEGs were then obtained on all family members. Of the 30 families studied, only four had more than one family member with frank epilepsy, while seven had multiple individuals with abnormal EEGs, and in 19 only the proband was affected. If the study had been limited to the four families with more than one member with epilepsy, the power to detect linkage would have been severely limited. By using the abnormal EEG as the phenotype, several additional families could be used for the linkage study, and linkage was detected to the HLA region of chromosome 6.

Implications for the Study of Language Impairment

Can either of these strategies be helpful in studies of language impairment? It is quite possible that the diagnoses used to define language impairment, as well as many other behavioral disorders, suffer from the problems of both low specificity and low sensitivity. For any behavioral disorder, it is likely that there will be environmental as well as genetic causes of disability. But, as in the case of breast cancer, if a group can be defined by its greater homogeneity and the symptoms or diseases of the group by their greater heritability, there is a chance that the genetic cause of that disease or disability can be identified, even if it represents a minority of all affected cases. On the other side, it is also likely that current disease definitions are not identifying all affected cases. As we saw with the examples of neurofibromatosis and JME, disease definitions that require significant functional impairment may not correctly identify all individuals who share the same underlying biological deficit. Environmental factors, the effects of other genes, and even chance may influence the differential expression of the identical gene defect in two separate individuals.

To construct a more behavioral example, let us consider three children of different intellectual ability, each with an identical, mild perceptual deficit. To a child with average intelligence, this deficit might present only a mild obstacle to learning certain tasks, until some compensatory strategy is figured out. In contrast, a child with fewer intellectual resources might find the deficit difficult to overcome. To the third child, with above average intelligence, an alternative strategy might be figured out so quickly that no performance difficulty is ever outwardly observed. While a test of the underlying perceptual ability might be identical in all three children, a performance-based test of the dependent skills would give widely disparate results.

One of the difficulties in studying a behavior as complex as language is that it requires the processing and integration of so many types of information. Every facet of every step might be under the influence of a different gene, with the possibility for multiple, different deficits that would not result in a total inability to process language but rather some variable degree of impairment. How, then, might we ferret out a defect in a single one of these genes? First, we need to realize that our clinical definitions of *language impairment* likely represent a heterogenous group of underlying problems, with language deficits acting as a final common pathway. The more specifically we can subdivide and test the different perceptual, processing, and representational tasks necessary for language comprehension and production, the more likely we will be moving closer to the actions of single (or a small group of) genes. Second, we need to understand that there may be individuals who have clinically normal language skills yet demonstrate a deficit in some component process. Clinically, we like to have tests that are highly sensitive and specific for the disorder we are trying to diagnose. However, if the diagnosis does not share a one-to-one correspondence with a specific genetic defect, then that sort of test will not be helpful for genetic studies. The type of test we are looking for may not identify all individuals clinically diagnosed with the disorder and may identify as impaired some people who seem clinically normal. The critical feature necessary to determine whether a test may be helpful in locating a specific genetic defect would be the heritability of poor test performance.

Genetics of Auditory Processing

A number of lines of evidence exist that suggest that there may be a heritable component to specific language impairment. A number of single families with several language impaired members have been described (Arnold, 1961; Borges-Osorio & Salzano, 1985; Hurst, Baraitser, Auger, Grahm, & Norell, 1990; McReady, 1926; Samples & Lane, 1985). The most recent of these (Hurst et al., 1990) describes a three-generation pedigree with 30 members, half of whom are affected with speech and language disorders, suggestive of an autosomal dominant mode of inheritance. A number of studies of the incidence of language impairment among the relatives of language impaired probands (Bishop & Edmunsond, 1986; Byrne, Willerman, & Ashmore, 1974; Hier & Rosenberger, 1980; Ingram, 1959; Luchsinger, 1970) suggest elevated rates of language impairment (from 24–63% of probands having a positive family history), but are limited by sketchy disease descriptions in some cases and lack of comparison to matched control populations in all cases. Several recent family studies (Neils & Aram, 1986; Tallal, Ross, & Curtiss, 1989; Tomblin, 1989) have utilized matched control families, and have reported significantly higher rates (2–7-fold increase) of language impairment among the first-degree relatives of the language impaired probands. Rates of impairment among first-degree relatives of the probands ranged from 17% to 43% across the different studies and different classes of relatives (mothers, fathers, siblings). While the high rates of affected parents in these families would not be expected in an autosomal recessive disease, these rates are, on average, less than would be expected for a fully penetrant autosomal dominant disorder. One possible way to reconcile the results of these family studies with the autosomal dominant pattern suggested by some of the single pedigree studies is to consider the possibility that there is an autosomal dominant form of specific language impairment which represents only a percentage of all of the cases of language impairment identified by these family studies.

If only some cases of language impairment are genetic, can we find some diagnostic feature that will separate the genetic and nongenetic cases, as was done in the study of breast cancer? Performance on nonverbal auditory processing tasks may identify a genetically relevant subgroup of individuals with specific language impairment. Tallal et al. (1989) have reported, on the basis of parental questionnaire data, that family history of language impairment tends to exhibit a bimodal distribution, with most families reporting either absolutely no family history or greater than 32% of family members with a positive history of impairment. Families with low rates (1%–32%) of impairment were very rare. Examination of a group of longitudinally followed children with persistent language impairment revealed that individuals with a family history of language impairment perform significantly more poorly on tasks of rapid perception than those language-impaired children with a negative family history (Tallal, Townsend, & Curtiss, 1991). These results suggest that poor performance on these tasks could be used to define a more specific category of individuals with language impairment, like the group of women with early-onset breast cancer. Currently, a family study and a twin study, each using the Tallal Repetition Test, are exploring the possible heritability of these auditory processing deficits. Instead of collecting questionnaire data on family history, family members of language impaired probands are being directly tested for processing deficits. If it appears to be inherited, this deficit, and not clinical language skills, could be used as the phenotype for a gene linkage study of language impairment.

The Future of Complex Disorder Studies

Recent advances in molecular technology are providing a highly polymorphic, fine genetic map of the entire genome. Techniques developed to isolate genes, once linkage is found, and then to identify causative mutations have also been advancing. Improved understanding of nontraditional modes of inheritance and factors complicating Mendelian genetics, together with improved analytic methods and faster, more powerful computers are enhancing the power and accuracy of linkage analysis. The careful and appropriate definition of disease phenotypes clearly remains the critical issue in the study of more complex and challenging disorders.

REFERENCES

Adams, R. D., & DeLong, G. R. (1984). Developmental and other congenital abnormalities of the nervous system. In R. G. Petersdorf, R. D. Adams, E. Braunwald, K. J. Isselbacker, J. B. Martin, & J. D. Wilson (Eds.), *Harrison's Principles of Internal Medicine* (10th ed.). New York: McGraw-Hill.

Arnold, G. E. (1961). The genetic background of developmental language disorders. *Folia Phoniatrica, 13*, 246–254.

Bishop, D. V. M., & Edmundson, A. (1986). Is otitis media a major cause of specific developmental language disorders? *British Journal of Disorders of Communication, 21*, 321–338.

Borges-Osorio, M. R., & Salzano, F. M. (1985). Language disabilities in 3 twin pairs and their relatives. *Acta Geneticae Medicae et Gemellologiae* (Roma), *34*, 95–100.

Botstein, D., White, R. L., Skolnick, M., & Davis, R. W. (1980). Construction of a genetic linkage map using restriction fragment length polymophisms. *American Journal of Human Genetics, 32*, 314–331.

Byrne, B., Willerman, L., & Ashmore, L. (1974). Severe and moderate language impairment: Evidence for distinctive etiologies. *Behavioral Genetics, 4*, 331–345.

Cooper, D. N., Smith, B. A., Cooke, H. J., Niemann, S., & Schmidtke, J. (1985). An estimate of unique DNA sequence heterozygosity in the human genome. *Human Genetics, 69*, 201–205.

Delgado-Escueta, A., Trieman, D. M., & Enrile-Bascal, F. (1982). Phenotypic variations of seizures in adolescents and adults. In V. E. Anderson, W. A. Huser, J. K. Penry, & C. F. Sing (Eds.), *Genetic basis of the epilepsies*. New York: Raven Press.

Futreal, P. A., Liu, Q., Shattuck-Eidens, D., Cochran, C., Harshman, K., Tavtigian, S., Bennett, L. M., Haugen-Strano, A., Swensen, J., Miki, Y., Eddington, K., McClure, M., Frye, C., Weaver-Feldhaus, J., Ding, W., Gholami, A., Söderkvist, P., Terry, L., Jhanwar, S., Berchuck, A., Iglehart, J. D., Marks, J., Ballinger, D. G., Barrett, J. C., Skolnick, M. H., Kamb, A., & Wiseman, R. (1994). BRCA1 mutations in primary breast and ovarian carcinomas. *Science, 266*, 120–122.

GDB™ Genome Database [database online]. (1990). Baltimore, MD: Johns Hopkins University Press. Available from Internet: <URL:http://gdbwww.gdb.org/

Greenberg, D. A., Delgado-Escueta, A. V., Widelits, H., Sparkes, R. S., Treiman, L., Maldonado, H. M., Park, M. S., & Terasaki, P. I. (1988). Juvenile myoclonic epilepsy (JME) may be linked to the BF and HLA loci on human chromosome 6. *American Journal of Medical Genetics, 31*, 185–192.

Haldane, J. B. S., & Smith, C. A. B. (1947). A new estimate of the linkage between the genes for color-blindness and hemophilia in man. *Annals of Eugenics, 14*, 10–31.

Hall, J. M., Lee, M. K., Newman, B., Morrow, J. E., Anderson, L. A., Huey, B., & King, M-C. (1990). Linkage of early-onset familial breast cancer to chromosome 17q21. *Science, 250*, 1684–1689.

Hier, D. B., & Rosenberger, P. B. (1980). Focal left temporal lobe lesions and delayed speech acquisition. *Journal of Developmental and Behavioral Pediatrics, 1*, 54–57.

Hurst, J. A., Baraitser, M., Auger, E., Grahm, F., & Norell, S. (1990). An extended family with a dominantly inherited speech disorder. *Developmental Medicine and Child Neurology, 32*, 352–355.

Ingram, T. T. S. (1959). Specific developmental disorders of speech in childhood. *Brain, 82*, 450–454.

Janz, D., & Christian, W. (1957). Impulsiv-petit mal. *Zschr Nervenheilk, 19*, 155–182.

Khoury, M. J., Beaty, T. H., & Cohen, B. H. (1993). *Fundamentals of genetic epidemiology*. New York: Oxford University Press.

Lander, E. S., & Schork, N. J. (1994). Genetic dissection of complex traits. *Science, 265*, 2037–2048.

Luchsinger, R. (1970). Inheritance of speech deficits. *Folia Phoniatrica, 22*, 216–230.

McKusick, V. A. (1990). *Mendelian inheritance in man: Catalogs of autosomal dominant, autosomal recessive, and X-linked phenotypes*. Baltimore: Johns Hopkins University Press.

McReady, E. B. (1926). Defects in the zone of language (word-deafness and word-blindness) and their influence in education and behavior. *American Journal of Psychiatry, 6*, 267.

Morgan, T. H. (1911). Random segregation versus coupling in Medelian inheritance. *Science, 34*, 284.

Newman, B., Austin, M. A., & King, M-C. (1988). Inheritance of human breast cancer: Evidence for autosomal dominant transmission in high-risk families. *Proceedings of the National Academy of Sciences, USA, 85*, 3044–3048.

Ott, J. (1990). Cutting the Gordian knot in the linkage analysis of complex human traits. *American Journal of Human Genetics, 46*, 219–221.

Ott, J. (1991). *Analysis of human genetic linkage* (Rev. ed.). Baltimore: Johns Hopkins University Press.

Pearson, P., Francomano, C., Foster, P., Bocchini, C., Li, P., & McKusick, V. (1994). The status of online Mendelian inheritance in man (OMIM) medio 1994. *Nucleic Acids Research, 22*, 3470–3473.

Samples, J. M., & Lane, V. W. (1985). Genetic possibilities in six siblings with specific language learning disorders. *ASHA, 27*, 27–32.

Tallal, P., Ross, R., & Curtiss, S. (1989). Familial aggregation in specific language impairment. *Journal of Speech and Hearing Disorders, 54*, 167–173.

Tallal, P., Townsend, J., & Curtiss, S. (1991). Phenotypic profiles of language impaired children based on genetic/family history. *Brain and Language, 41*, 81–94.

Terwilliger, J. D., & Ott, J. (1994). *Handbook of human genetic linkage.* Baltimore, MD: Johns Hopkins University Press.

Thompson, M. W., McInnes, R. R., & Willard, H. F. (1991). *Thompson & Thompson genetics in medicine.* Philadelphia: Saunders.

COMMENTARY ON CHAPTER 1

Jon F. Miller
University of Wisconsin-Madison

This chapter makes clear that there will be no easy or fast solutions to the problem of identifying genetic causes of specific language impairment (SLI). Working from the behavioral perspective, it is tempting to consider the laboratory sciences as having better tools for quantifying causal constructs explaining disordered language performance than behavioral science. It is tempting to see the grass as greener in molecular genetics. The technologies used in sophisticated genetic studies hold great promise for identifying the causes of many language disorders in children. Dr. Brzustowicz considers the problem from the molecular genetics perspective, relating the opportunities as well as the limitations of current research methodologies in human genetics and how these methods can consider the problem of the genetics of SLI. Her chapter suggests that we will not find the solution to the genetics of the SLI problem solely through molecular genetics research. Progress in identifying the genotype can only follow progress in defining the phenotype. This chapter is a blueprint for a partnership between behavioral science, whose task is to define SLI and provide measurement tools to identify all and only the children who have SLI, and molecular genetics, whose task is to document the genotype-phenotype correlations in SLI.

The chapter outlines the process of human genetics research on complex disorders. Research to date has documented increased rates of SLI in the relatives of a proband, suggesting the problem is familial. Current research focusing on twin studies seeks to determine if SLI is genetic. The final step in the process will be to determine the chromosomal region or regions on specific chromosomes related to SLI. The steps in this process are: (a) definition of the disorder and selection of the families, (b) genotyping the probands, and (c) statistical analysis. Two types of error in definition, specificity and sensitivity, are discussed from an epidemiological perspective. These constructs refer to the rate at which true cases are correctly identified by a particular measure (sensitivity), that is, not diagnosing all of the true genetic cases. Specificity refers to the number of true noncases accurately classified or diagnosing nongenetic cases along with the genetic cases. The measurement problem is defined as identifying diagnostic criteria for SLI that is not too inclusive, losing specificity, and not too narrow, losing sensitivity.

A related example of work on genotype–phenotype correlations can be found in the molecular genetics of Down syndrome. The focus of this research parallels that of the human genome project but focusing only on Chromosome 21. Investigation of the molecular structure of Chromosome 21 employs a variety of methods to document the minimal regions involved in producing a particular physical or behavioral feature of the syndrome (Korenberg, Pulst, & Gerwehr, 1992). Early work suggested that only part of the long arm of Chromosome 21, the q22 band, was the cause of Down syndrome. More recent work has shown that genes in other areas significantly contribute to the phenotype. Phenotypic maps of Down syndrome are beginning to appear in the literature, documenting the specific regions of Chromosome 21 associated with particular features of the syndrome. These maps reveal that physical features have been the first to be identified and mapped. Behavioral features, such as mental retardation, are defined by large regions between p13 and q22.3. The definition of the behavioral features of mental retardation is extremely difficult and is viewed as a major roadblock by geneticists working on this problem (Korenberg, 1994). The parallel to the study of the genetics of SLI is unmistakable. Progress can only move as fast as the phenotype can be described explicitly and predictive features identified.

The final sentence of this chapter provides a sharp focus for researchers investigating the genetics of SLI. "The careful and appropriate definition of [SLI] phenotypes clearly remains the critical issue in the study of more complex and challenging disorders" (p. 23, this volume). The problem for child language researchers is to describe disordered language performance through the developmental period. Such descriptions should identify performance features that meet both specificity and sensitivity criteria. This chapter suggests that genetic studies may identify more than a single genotype associated with SLI. Our assumptions about the nature of SLI must be re-examined if we are to develop new descriptive models capable of sufficient precision to advance genetic studies of SLI.

REFERENCES

Korenberg, J. (1994, April). *Advances in the molecular genetics of Down syndrome*. Paper presented at the tenth anniversary meeting of the National Down Syndrome Society Science Seminar. Charleston, SC.

Korenberg, J., Pulst, S., & Gerwehr, S. (1992). Advances in the understanding of chromosome 21 and Down syndrome. In I. Lott & E. McCoy (Eds.), *Down syndrome: Advances in medical care* (pp. 3–12). New York: Wiley-Liss.

LINKAGE ANALYSIS WITH COMPLEX BEHAVIORAL TRAITS

Shelley D. Smith
Boys Town National Research Hospital

Bruce F. Pennington
University of Denver

John C. DeFries
Institute of Behavioral Genetics
University of Colorado

Behavioral phenotypes are observable, and sometimes measurable, characteristics presumably mediated by the central nervous system. The genetic influences on phenotypes have traditionally been classified as either single gene or polygenic (mediated by just one gene versus many genes acting together), but, as the study of genetic disorders has progressed, it has been found that the inheritance patterns of some phenotypes do not fit neatly into these classifications. The phenotypes may not be discrete, and the inheritance patterns may be a blend between single gene and polygenic influences. These complex traits show a quantitative variation in their presentation and do not show obvious Mendelian inheritance. There may be a single major gene involved in such a trait, but it may show decreased penetrance (i.e., not all individuals with the susceptible genes have the disorder) or variable expression (differences in the phenotype), due to the modifying influences of other genes or environmental factors. A major gene may have many possible mutations (alleles) that may have different phenotypes as well. These alleles may affect differences that are within normal variation as well as more extreme cases, so the concept of a "disease gene" may also be too narrow. There may be two genes involved simultaneously in the production of a phenotype, or there may be several quantitative trait loci (QTLs), each with a measurable effect on the phenotype. There may be heterogeneity, in that different genes or environmental mechanisms may be involved in producing the same phenotype in different families. Finding these types of

29

genes is a logistical and computational challenge. Despite this, we, the authors of this chapter, have been trying to localize genes for one type of language disorder, specific reading disability (RD), which appears to follow an autosomal dominant mode of inheritance in some families. Fortunately, the technical advances in the laboratory have greatly increased the interest in the development of statistical methods to tackle these problems, and these should be directly applicable to the study of other language disorders.

The following factors should be considered in launching a search for individual genes: (a) phenotype definition, (b) evidence for genetic transmission of the phenotype and its apparent mode of inheritance, and (c) the selection of subjects.

SELECTION OF A PHENOTYPE

With any genetic study, it is most important to define a phenotype that is reproducible, reliable, and appears to have a genetic basis. This may not be the same as a phenotype that has educational utility or fits a theoretical construct. Because the expression of a gene may be variable, the definition must be broad enough to encompass its spectrum of effects, but not so broad as to include other genetic or nongenetic phenocopies. Dr. Bruce Pennington has given some guidelines on how a phenotype can be defined that has a biological, etiological basis (Pennington, 1986; Pennington, Gilger, Olson, & DeFries, 1992).

In our studies of reading disability, we have looked at qualitative and quantitative phenotypes. The qualitative phenotype is a clinical diagnosis based on test results and clinical history, which have been described in detail in Pennington, Smith, McCabe, Kimberling, & Lubs, 1984. All available family members were tested. An IQ-discrepancy diagnosis (i.e., a significant disability in reading and spelling, without demonstrable intellectual or neurological handicap, and with an IQ greater than 90) was used. The diagnoses were all made by Pennington, who was blind to the genotyping data. Each individual in the family was classified as affected, unaffected, compensated, or unknown. For the linkage analysis, *compensated* individuals, adults who have largely overcome childhood reading problems, were coded as *affected* because they are presumed to be gene carriers. The phenomenon of compensation has been well documented (Felton, Naylor, & Wood, 1990), and it appears to be related to the initial severity of the reading problems, with less severely affected children being more likely to compensate as adults.

The quantitative phenotype was a discriminant score using the same test results, based on a function derived from a similar series of tests by the Institute for Behavioral Genetics (DeFries, Fulker, & LaBuda, 1987). In this procedure, a function is derived that weighs and combines the individual test scores into a single score that maximally separates individuals known to be in two groups (in this case, reading disabled and normal readers). The function can then be applied to the scores of individuals of unknown classification to produce a single score that reflects the severity of their reading problems. Because this score is based on current test scores, it may not be as sensitive to adults who have a history of reading problems in childhood but who have since largely compensated for those problems and test within the normal range.

EVIDENCE FOR GENETIC TRANSMISSION

There is considerable evidence that the RD phenotype is genetically influenced (Pennington, 1990). There are a number of family and twin studies demonstrating that relatives of disabled readers are at increased risk for reading disability. Twin studies by DeFries, Fulker, and LaBuda (1987) have found that the discriminant score phenotype has a heritability of about 0.64, meaning that 64% of the variation in the phenotype can be accounted for by genetic factors. Although this shows that genetic factors have an influence on reading performance, it does not indicate how it is inherited or how many genes are involved.

MODE OF INHERITANCE

Segregation analysis uses data from families to test hypotheses of modes of inheritance, comparing them against the null hypothesis of no genetic transmission of the phenotype. These methods can determine if there is a single gene, how it is inherited (i.e., dominant or recessive and the level of penetrance), and whether there is polygenic influence. A phenotype that shows the influence of a major gene with high penetrance is going to be the most useful for gene localization studies. The most recent and sophisticated segregation analysis of reading disability was done by Pennington et al. (1991). Families from four different studies were used, each with somewhat different phenotypic definitions. Three of the populations of families with reading-disabled participants (including the kindreds described earlier) gave evidence for single gene dominant inheritance, but the fourth population, with a more general language disability, showed polygenic inheritance. This difference in results points out the importance of the definition of the phenotype. The segregation analysis cannot determine if there is only one dominant gene in the families showing single gene inheritance or if there is genetic heterogeneity, and there may be many or a few polygenes in the fourth population. Still, these results give support to efforts to localize genes for reading disability.

POPULATION

Selection of the subject population can be critical, depending upon the type of genetic study. For segregation analysis, every effort must be made to ascertain families without regard to the inheritance pattern; a family with just one affected individual should be just as likely to be included as one with many affected relatives. Conversely, for linkage analysis, researchers should concentrate on large multigenerational families to maximize the probability of finding a major genetic effect. This will tend to enhance homogeneity and increase the power of the analysis per sample. Families with more severely affected probands or multiplex families may also be more likely to be genetic, or at least to be easier to diagnose. Once a gene is localized, a random sample of families (preferably from different ethnic backgrounds) can determine if genetic heterogeneity exists.

In our studies of reading disability there are two populations: 19 extended kindreds with apparent autosomal dominant transmission of reading disability, and the families of dyzygotic (fraternal) twins, at least one of which has reading disability. The twins are part of a larger study of specific reading disability through a research center grant based at the University of Colorado directed by DeFries. They are ascertained in as unbiased a method as possible through cooperating school systems in Colorado. We are continuing to get samples on the twin families with a goal of 100 families in all, but the preliminary analyses are being done on a set of 39 completed nuclear families.

LINKAGE ANALYSIS

Linkage analysis refers to the methods used to determine the location of genes on chromosomes. Genes are arranged in linear fashion along the chromosomes, and alleles of genes that are located close together on the same chromosome tend to be inherited together as the chromosome is transmitted from one generation to the next. In contrast, genes that are on separate chromosomes, and genes that are far apart on the same chromosome, are inherited independently of each other.

In humans, there are 23 homologous pairs of chromosomes, with one chromosome from each pair inherited from the mother and the other from the father. As a result, each gene is represented by a pair of alleles. In meiosis, the chromosomes are copied and separated for inclusion in the germ cells, which contain only one homologue from each pair to be transmitted to the offspring. During meiosis, the two homologues line up side-by-side and exchange sections of DNA, resulting in recombination of the original chromosomes. In other words, a chromosome from one parent will exchange DNA with the chromosome from the other parent, and the resulting recombined chromosome is then transmitted to the germ cell. As a result, a child not only gets one chromosome from each parent, but each chromosome contains genes from both grandparents. Exactly where an exchange, or crossover event, takes place along the length of a chromosome varies with each meiosis, and the chance that recombination will occur between two genes is a function of the distance between them. If two genes are far apart, there will be more instances of recombination. The percent of offspring showing recombination becomes a measure of distance between genes, with 1% recombination termed one *centimorgan* (cM). Unlinked genes (whether on the same or different chromosomes) are by definition at least 50 centimorgans apart, meaning that 50% of the time they show recombination, and 50% of the time both alleles from a given grandparent are transmitted together.

There are several approaches to gene localization, but all are based on discovering the position of a gene by inference, by comparing the transmission of the alleles at a trait locus (such as a QTL) with the inheritance of alleles at a known marker locus. If the alleles for the trait (e.g., the alleles affecting a behavioral phenotype) show decreased recombination with the alleles at the marker locus, it is concluded that the two genes are close together on the same chromosome. It is necessary, then, to be able to determine which alleles came from each parent, which can only be done if at least one parent has different alleles at both the trait and marker loci. Ideally, all four grandparental alleles should be different, so that each instance of recombination can be spotted in the offspring. Also, in order to find linkage, one must have chosen a marker locus that happens to be very near the trait locus. The incredible explosion in gene localization has occurred because of the discovery of types of marker loci that are quite variable and quite numerous. This has made the localization of genes for complex traits conceivable because the effectiveness of the analytical methods can be increased when recombination can be decreased and the contribution of each parent can be traced. The markers we are using now are based on variations in DNA sequence. These variations are presumed to have no clinical significance, but show consistent inheritance through families. They are highly polymorphic, meaning that they have many possible alleles, so that the probability is high that a set of parents will have different alleles. In this way, each allele can be tracked through the family.

We use the following three methods for our reading disability study, each with its own advantages: the lod score method, sib pair analysis, and a newly developed regression analysis.

Lod Score Method of Linkage Analysis

The classical method of gene localization is the lod score, family study method. In the lod score method of linkage analysis, the transmission of alleles for a given set (usually a pair) of genes is followed through the family to see if the genes are transmitted randomly or if they show decreased recombination. As described earlier, a recombination fraction of 0.50 (50%) is equal to random transmission; the hypothesis to be tested in this type of linkage analysis is whether the observed recombination fraction is significantly less than 0.50. The likelihood of the pedigree data given a particular recombination fraction is computed and compared with the likelihood given free recombination, producing an odds ratio. Because individual families are generally too small to determine if an apparent decrease in recombination is statistically significant, the $\log_{10}$ of the ratio is taken so that the results of different families can be added together. This statistic is termed the *lod score* (the *L*og of the *Od*ds of linkage). A lod score over 3 is generally accepted as showing evidence of linkage, and a lod score less than −2 rejects linkage (Morton, 1955). Any linkage that is found must be confirmed in a separate population.

The phenotypes used in this type of linkage analysis can be qualitative or quantitative, and estimates of the penetrance of the phenotype for each genotype (i.e., the mode of inheritance) and allele frequency need to be included. We use the LINKAGE program developed by Lathrop, Lalouel, Julier, and Ott (1985).

The lod score method of linkage analysis is the most powerful; it is responsible for the localization of a number of genes. Power is a consideration in determining the number of families needed to get significant results. Because the ascertainment of large numbers of families and the costs of genotyping can consume much time and money, efficiency is particularly important. With the lod score method, one large family with at least 10 informative individuals and a highly polymorphic marker can be enough to obtain a lod of 3.0. However, if there is any suggestion that there may be genetic heterogeneity (and this possibility must always be considered), more families are needed to be sure that the linkage is true in all of them.

Recently, it has been possible to test the hypothesis that two disease loci are present, linked to different markers, and that they simultaneously produce the disease. This has been demonstrated in a study of multiple sclerosis. Lod scores had shown linkage with the myelin basic protein gene on Chromosome 18 with lod scores between 2.49 and 3.42, depending on the penetrances used. At the same time, association and linkage disequilibrium with HLA DQA1 on Chromosome 6 had been observed. When both loci were considered together, a lod score of around 10 resulted, greater than that obtained with either locus separately (Tienari, Terwilliger, Palo, Ott, & Peltonen, 1993). Unfortunately, it took 5 days for the program to run. It is not clear what the level of significance should be to accept linkage in this type of analysis, although the traditional level of three may still be appropriate.

Heterogeneity and Admixture Testing

Even when the phenotype is well defined, there is a great possibility of genetic heterogeneity. Gene localization studies have the advantage of being able to demonstrate heterogeneity by showing that some families are linked to a given marker and others are not. But, if the heterogeneity is too great, it may obscure linkage to either marker. The ability to reliably detect a linkage is dependent upon the size of the population, the percent of that population having a particular linkage, and the closeness of the linked markers. In the lod score method, programs are available to detect heterogeneity in the lod scores between separate families and to estimate the proportion of linked families if heterogeneity is detected in the population. This is sometimes referred to as an *admixture test* for linkage because, in addition to the alternate hypothesis of linkage at a given value of theta, there is a second alternate hypothesis that the population is actually made up of two populations, one linked and one unlinked. The parameters for this hypothesis are theta, and the proportion of linked families, alpha. We have used the program HOMOG by Ott (1985) but others are also available.

Sib Pair Linkage Analysis

Another type of linkage study is termed *sib pair analysis*. Rather than following the transmission of trait and marker alleles through several generations, only one pair of relatives (usually siblings) is assessed for both phenotypes. If a pair of siblings have inherited the same allele for a given gene from the same parent (i.e., the alleles are identical by descent, or IBD), they should also tend to share the alleles that are linked to that gene. The siblings can be identical by descent for 0, 1, or both alleles at a locus. Thus, if both siblings are affected (i.e., have the same alleles for the trait locus), they should also share the same linked alleles; if the siblings are discordant for the trait, they should be discordant for linked alleles as well. There are several methods for this type of analysis: they can use affected sibs only, affected and unaffected sibs, or other pairs of relatives such as grandparent-grandchild. If a disorder is common enough, the analysis will be most effective when only affected sib pairs are used because this gets by the problem of nonpenetrance of a gene in a supposedly unaffected sibling. Many of the analyses are based on the method of Haseman and Elston (1972), where the squared difference in the siblings' phenotypes (e.g., quantitative measures or, in a categorical variable, in the case of a qualitative trait) is regressed on the computed identity by descent at a given marker locus (between 0.0 and 1.0). If there is linkage of the trait locus to that marker, there should be a significant regression, with a smaller difference in trait phenotypes associated with higher identity by descent values for the marker locus.

The estimate of the identity by descent for the marker loci is most accurate when the marker alleles of the parents and grandparents are known. This ideal case may not be practical, however, either logistically or financially. If the frequencies of the alleles are known, researchers can compute the probability that they are identical by descent; that is, if a given allele is rare in the general population but is shared by siblings, chances are both siblings inherited it from the same parent. If the allele is quite common, both parents may have had it, so the identity may just be identity by state, not by descent. In many of the newly developed PCR markers, however, it is difficult to be sure that the allele designations are consistent across families, especially in different laboratories. As a result, population frequencies may not be available or may not be reliable and it is, therefore, worthwhile to try to sample additional relatives.

The sib pair method of linkage analysis is less powerful than the lod score method, but it does not require specification of the mode of inheritance, which can be of great value in complex traits where the mode of inheritance may not be clear. However, it may take many more pairs of sibs to find a close linkage, depending on how close the marker is to the gene.

There are several variations on this methodology. Goldgar (1990) extended this technique to analyze segments of chromosome between markers, either individually or simultaneously. This method would detect several genes working together to produce a phenotype. Fulker et al. (1991) combined sib pair analysis with a regression analysis approach, which was initially developed to measure heritability.

DeFries–Fulker Sibling Regression Analysis

DeFries and Fulker (1985) realized that, if one twin (the proband) has an extreme score on a phenotypic measure, the score of the cotwin will tend to regress toward the mean for that phenotype. However, if there is a genetic influence on that phenotype, the score of an identical cotwin should show less regression toward the mean than the score of a fraternal cotwin. This can be observed in a regression equation such as $C = A + B_1P + B_2R$ where C = cotwin score and P = proband score, which is weighted for the usual twin resemblance by B_1. R is the coefficient of relationship (1.0 for identical twins, 0.5 for fraternals), and B_2 is twice the difference between twins based on relationship (i.e., the extent of difference in regression between identical and fraternal twins). An augmented model was also developed in which the product of the proband score and the coefficient of relationship is included; its coefficient is an estimate of heritability. In linkage studies, the coefficient of relationship between twins becomes the identity by descent (0, .5, or 1) for a given marker between sibling pairs (Fulker et al., 1991). If the IBD value for a given marker has a significant effect on the regression between siblings, it can be assumed that the marker is linked to a trait locus. Furthermore, the heritability of the trait is estimated. The closer the marker is to the trait, the better the estimate of the heritability of the trait.

By further expanding the regression equation, the effect of the trait locus on the other aspects of the phenotype can be assessed, taking into account other measured variables and their interactions. Fulker also found that the power of this technique appears to increase greatly as the population is limited to the most severely affected individuals, which would mean that one would not have to sample as many individuals (Fulker et al., 1991).

Selection of Marker Loci

There are basically three approaches to the selection of markers to be typed: candidate loci, a chromosomal abnormality, or a full genome search.

Candidate genes are those whose function is known to be related to the disorder, or those that are active in relevant tissues. Variants in the gene itself as well as markers in the immediate vicinity of the candidate can be tested for linkage. For example, in our work with deafness disorders, genes known to be causes of deafness in mice are used as candidates, as are genes that are expressed in cochlear tissues. Unfortunately, for most behavioral disorders the biochemical pathways are completely unknown, animal models are not helpful, and the number of candidates in the organ of choice, the brain, is at least 30,000 (Sutcliffe, 1988).

A visible abnormality in a chromosome can also give a clue to the location of a gene. A portion of a chromosome including the gene may be deleted, duplicated, disrupted, or moved (translocated) to another chromosome. For example, genetic analysis of a child with a chromosomal translocation and features of Waardenburg syndrome rapidly led to the localization of one gene for that condition in the region of the translocation (Farrer et al., 1992; Ishikiri-yama et al., 1989).

The full genome search is applied when one does not have a clue as to the location of a gene. The researcher must go systematically through a battery of markers spaced across the chromosomes, hoping that one is nearby the trait locus. If any of the markers shows interesting results, usually a lod score over 1.0 or 1.5 (or an equivalent significance level in sib pair analysis), other markers in the same region are typed. These should also show positive results, it is hoped with lower values of theta or higher significance values indicating that one is getting closer to the gene.

ASSOCIATION STUDIES AND SUSCEPTIBILITY LOCI

The location of a major gene with an influence on a trait may be suggested by association, that is, the finding that a particular allele of a candidate gene or marker locus is found with increased frequency in individuals with the disorder. This is not the same as linkage, in which the particular allele of the marker that is transmitted with the disorder will generally be different in different families. However, both linkage and association may be seen between families if the loci are very tightly linked and if the families are actually derived from the same common ancestor. This is termed *linkage disequilibrium*. In other cases, however, association may be the result of the influence of the allele on the phenotype.

If a gene is neither sufficient to cause the disorder, nor absolutely necessary for it (i.e., there are unrelated phenocopies), the gene can be referred to as a *susceptibility locus* (Greenberg, 1993). Greenberg noted that HLA loci, in particular, may confer increased susceptibility to certain diseases, particularly those that may have an autoimmune component (such as multiple sclerosis), as seen in association studies. The genes in the HLA cluster are involved in the interaction of cells of the immune system and are quite variable between individuals.

A susceptibility locus, such as a particular HLA type, can be difficult to detect, because it is not necessarily present in all affected, nor absent in all unaffected, individuals. The resulting heterogeneity would cancel out linkage, but may still show up on a population level as an association of the disease with the mutant allele; that is, more affected individuals in a population would have the allele than those unaffected from that same population. This type of effect can be detected by linkage analysis if the sample of families is selected to enhance it for that allele. For example, if the susceptibility allele confers greater risk for the disorder, multiplex families may be enriched for the gene, or, if the allele confers greater severity, those families showing more severe symptoms may enhance the linkage. It should be also remembered that association studies themselves must be carefully done, as they are quite susceptible to artifact, especially if affected and unaffected populations are not carefully matched so that allele frequencies are the same in the populations from which the two groups are derived.

LINKAGE ANALYSIS OF SPECIFIC READING DISABILITY: RESULTS

As noted earlier, we have used two populations (extended dominant kindreds and twin families), three methods of analysis (lod score, sib pair, and sib regression to the mean), and two phenotypes (qualitative and quantitative). Because of the advantages of the sib pair and regression methods in disorders where the mode of inheritance is uncertain, our approach is to use these methods as screens for any loci of interest, defined as those with a significance over 0.05. Any significant results are followed up with additional markers from the same region. If significant results are still found, we will test the kindreds with the lod score method to see if the linkage is confirmed and if heterogeneity exists.

The siblings, parents, and in some cases even grandparents, are tested for the phenotypic measures and are also typed for the marker loci. This is necessary for the lod score approach and enhances the sib pair approaches by increasing the accuracy of the IBD estimate without as much reliance on allele frequencies. Including the entire sibship in the sib pair analysis also increases the efficiency because one can make pairwise comparisons of all of the siblings. Even though the pairs are not as independent as unrelated sib pairs, this does not affect the analysis.

When we first started this study, our approach was to use the lod score method in a full genome search in the population of apparent dominant families. This initially gave us some very positive results with the cytogenetic heteromorphisms of the short arm of Chromosome 15. This lod score increased, particularly in one family, until that family had a lod score of almost 3. However, as has happened in so many studies of behavioral phenotypes, an individual with an apparent crossover between the reading disability and the heteromorphic region was finally found, so that the linkage results were no longer significant. In addition, studies by our group and by Rabin et al. (1993) were unable to confirm the linkage with other markers in the region. As we turned to other chromosomes, the BF (properdin factor B) and GLO (glyoxylase 1) markers on Chromosome 6 gave significant results with the sib pair analysis. BF appears to be involved in the activation of complement in the immune response, but the enzyme coded by GLO is not involved in the immune system. Because together they pointed to a "candidate region," namely the HLA region involved in immune function, we concentrate on this area. We have typed a number of DNA markers from that region in both the families and the twin samples (Fig. 2.1). These markers are much more polymorphic and thus more informative than the BF and GLO markers. Interestingly, the twin and kindred samples have both shown positive results for some of the markers, most consistently in those around the HLA region. Phenotype had some effect on exactly which markers were signifi-

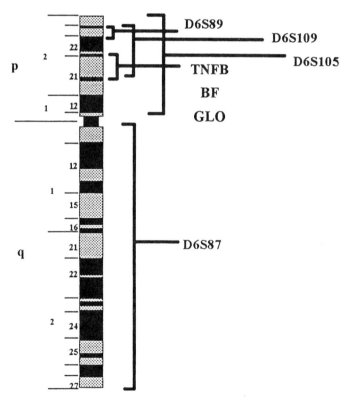

FIG. 2.1. Position of genetic markers on Chromosome 6. The approximate cytogenetic locations of the markers are indicated along the chromosome. TNFB and BF are within the HLA region (between HLA-B and HLA-DRA), but GLO is proximal to it.

cant, although for two markers, TNFB (tumor necrosis factor beta) and D6S87, there were no differences between the discriminant score and the reading recognition phenotype. Overall, the discriminant score phenotype shows the most separation between affected and unaffected family members. The results of the sib pair analysis using the S.A.G.E. programs are shown in Tables 2.1 through 2.4. Cytogenetic locations of the markers are approximate, although the order and distances between them are known.

Sib pair analysis, using the regression approach of DeFries and Fulker (1985), gave similar results. Subsequently, Cardon et al. (1994, 1995) modified the regression analysis approach to allow interval mapping. In this analysis, two adjacent markers can be considered simultaneously to figure the probability that the gene could be at varying distances between them. Using the discriminant phenotype, results highly suggestive of linkage were found in the DZ twin families ($p < 0.0094$) for the region between D6S105 and TNFB, and there was suggestion of linkage between the same two markers in the linkage kindreds ($p = 0.0667$).

We have looked at these markers using the lod score method in the kindreds. (For screening purposes, the twin families are too small for the lod score method.) If the QTL was a dominant with high penetrance, it should be detectable by this method. A gene frequency of 0.01 was used, with complete domi-

TABLE 2.1
Sib Pair Analysis: Twin Family Data, Discriminant Score Phenotype

Locus	Location	Degrees of Freedom	p Value
D6S89	6p23–p22.3	53	0.009**
D6S109	6p23–p22.3	49	0.22
D6S105	6p	43	0.032*
TNFB	6p21.3	54	0.021*
D6S87	6q	54	0.352

*$p < 0.05$. **$p < 0.01$.

TABLE 2.2
Sib Pair Analysis: Twin Family Data, Reading Recognition Phenotype

Locus	Location	Degrees of Freedom	p Value
D6S89	6p23–p22.3	55	0.016*
D6S109	6p23–p22.3	51	0.169
D6S105	6p	45	0.021*
TNFB	6p21.3	56	0.021*
D6S87	6q	56	0.352

*$p < 0.05$.

TABLE 2.3
Sib Pair Analysis: Kindred Data, Discriminant Score Phenotype

Locus	Location	Degrees of Freedom	p Value
D6S89	6p23–p22.3	157	0.052
D6S202	6p22–p21.3	131	0.263
D6S109	6p23–p22.3	175	0.373
D6S105	6p	166	0.316
TNFB	6p21.3	177	0.159
BF	6p21.3	147	0.0001***
GLO	6p21.3	92	0.023
D6S87	6q	176	0.835

***$p < 0.001$.

TABLE 2.4
Sib Pair Analysis: Kindred Data, Diagnosis Phenotype

Locus	Location	Degrees of Freedom	p Value
D6S89	6p23–p22.3	168	0.327
D6S202	6p22–p21.3	137	0.783
D6S109	6p23–p22.3	192	0.797
D6S105	6p	185	0.308
TNFB	6p21.3	193	0.030*
BF	6p21.3	175	0.672
GLO	6p21.3	169	0.020*
D6S87	6q	198	0.519

*$p < 0.05$.

nance and 1% sporadic cases. Using the diagnosis phenotype, some families have suggestive scores, but others show no indication of linkage. The discriminant score, on the other hand, gives lod scores around 0.00 for all families, so it does not appear to be as informative for this technique. Part of the problem is likely to be the misdiagnosis of compensated adults in the kindreds; because the discriminant score reflects only current functioning, they may appear to be unaffected. In the sib pair analysis of the twin population reported earlier, few of the subjects would have been old enough to show compensation.

We have also run the analyses with only the affected individuals, to reduce the effect of misdiagnosis of unaffected individuals. That is, if penetrance is low, unaffected individuals may still be gene carriers. The drawback to this procedure is that the sample size is decreased. This did not make an improvement in the lod scores although they were less negative. Decreasing the penetrance to 0.7 for females alone, and for both sexes, also did not improve the lod scores.

Finally, we looked at the results considering an autosomal recessive mode of inheritance, which also did not result in higher lod scores.

Multipoint analysis of two markers at a time can help localize a gene and may enhance a lod score if adjacent markers are consistent in their linkage. This was done with the diagnosis phenotype and did get some enhancement of linkage in some families. Given the results of the QTL analysis, it appears that the lod score method is less sensitive or there is mis-specification of the parameters.

Rabin et al. (1993) had been doing a search for RD genes using the lod score method in nine multigenerational kindreds. They initially had somewhat suggestive lod scores with Rh on Chromosome 1p32, and two other markers in the region gave positive lod scores, one of which has a lod of 2.3 at a theta of 0.20. In addition, Froster, Schulte-Korne, Hebebrand, and Remschmidt (1993) described a German pedigree in which RD and delayed speech development segregated with a translocation between Chromosomes 1p and 2q in a father and 2 of his 7 children. This will be another region for active analysis. We have looked at two markers from Chromosome 1p, D1S160 and D1S186, but all of the lod scores were less than 1.0 for the kindred data for both methods of analysis and for discriminant and diagnosis phenotypes. Additional markers between these two will be typed to more completely cover the region.

SUMMARY

Although the task is difficult, statistical and molecular genetic methods are available to characterize the genes that influence complex disorders. Segregation analysis can determine if one locus is involved, whether it is dominant or recessive, and what the penetrance is for each genotype. It can also determine if polygenic inheritance is a factor, either alone or in combination with a major gene influence. Linkage analysis can find the chromosomal location of single genes or QTLs with large effect on the phenotype. Genetic heterogeneity can also be determined, so that more precise genotype:phenotype correlations can be made. As the genetic analysis progresses, the definition of the phenotype can be refined to specific neurological or processing characteristics, and the identification of the responsible gene will permit the discovery of the underlying mechanism of the disorder.

It is an axiom of any type of linkage analysis that accurate phenotype definition is essential. Our results also demonstrate that particular phenotypes may be better suited to specific analyses. Quantitative scores can be quite powerful in sib pair analysis, especially if extreme scores are selected, but the lod score approach does not work as well if the quantitative score cannot adjust to phenotypic differences across generations.

Simulation studies are continually appearing that examine the reliability of single locus, two locus, and admixture tests for linkage with lod score and sib pair methods, and new methods of linkage analysis are being developed to detect QTLs. Analytical strategies are developing in response to the innovations in molecular genetic techniques, and this should have a profound effect on the feasibility of gene localization in language disorders. Our own results localizing a QTL for reading disability on Chromosome 6p, utilizing a novel means of analysis, is an example of development in this area.

ACKNOWLEDGMENTS

This work is supported by NIH grant P50-HD27802. Some of the results were obtained by using the program package S.A.G.E., which is supported by a U.S. Public Health Service Resource Grant (1 P41 RR03655) from the Division of Research Resources.

REFERENCES

Cardon, L. R., Smith, S. D., Fulker, D. W., Kimberling, W. J., Pennington, B. F., & DeFries, J. C. (1994). Quantitative trait locus for reading disability on chromosome 6. *Science, 266,* 276–279.

Cardon, L. R., Smith, S. D., Fulker, D. W., Kimberling, W. J., Pennington, B. F., & DeFries, J. C. (1995). Quantitative trait locus for reading disability: A correction. *Science, 268,* 5217.

DeFries, J. C., & Fulker, D. W. (1985). Multiple regression analysis of twin data. *Behavior Genetics, 15,* 467–473.

DeFries, J. C., Fulker, D. W., & LaBuda, M. C. (1987). Evidence for a genetic aetiology in reading disability of twins. *Nature, 329,* 537–539.

Farrer, L. A., Grundfast, K. M., Amos, J., Arnos, K., Asher, J. H., Beighton, P., Diehl, S. R., Fex, J., Foy, C., Friedman, T. B., Greenberg, J., Hoth, C., Marazita, M., Milunsky, A., Morello, R., Nance, W., Newton, V., Ramesar, R., San Agustin, T. B., Skare, J., Stevens, C. A., Wagner, R. G., Wilcox, E. R., Winship, I., & Read, A. P. (1992). Waardenburg Syndrome (WS) type I is caused by defects at multiple loci, one of which is near ALPP on chromosome 2: First report of the WS Consortium. *American Journal of Human Genetics, 50,* 902–913.

Felton, R. H., Naylor, C. E., & Wood, F. B. (1990). Neuropsychological profile of adult dyslexics. *Brain and Language, 39,* 485–497.

Froster, U., Schulte-Korne, G., Hebebrand, J., & Remschmidt, H. (1993). Cosegregation of balanced translocation (1;2) with retarded speech development and dyslexia [letter]. *Lancet, 342,* 178–179.

Fulker, D. W., Cardon, L. R., DeFries, J. C., Kimberling, W. J., Pennington, B. F., & Smith, S. D. (1991). Multiple regression analysis of sib pair data on reading to detect quantitative trait loci. *Reading and Writing: An Interdisciplinary Journal, 3,* 299–313.

Goldgar, D. E. (1990). Multipoint analysis of human quantitative genetic variation. *American Journal of Human Genetics, 47,* 957–967.

Greenberg, D. A. (1993). Linkage analysis of "necessary" disease loci versus "susceptibility" loci. *American Journal of Human Genetics, 52,* 135–143.

Haseman, J. K., & Elston, R. C. (1972). The investigation of linkage between a quantitative trait and a marker locus. *Behavior Genetics, 2,* 3–19.

Ishikiriyama, S., Tonoki, H., Shibuya, Y., Chin, S., Harada, N., Abe, K., & Niikawa, N. (1989). Waardenburg syndrome type I in a child with de novo inversion. *American Journal of Medical Genetics, 33,* 505–507.

Lathrop, G. M., Lalouel, J. M., Julier, C., & Ott, J. (1985). Multilocus linkage analysis in humans: Detection of linkage and estimation of recombination. *American Journal of Human Genetics, 37,* 482–498.

Morton, N. E. (1955). Sequential tests for the detection of linkage. *American Journal of Human Genetics, 7,* 277–328.

Ott, J. (1985). *Analysis of human genetic linkage.* Baltimore: Johns Hopkins.

Pennington, B. F. (1986). Issues in the diagnosis and phenotype analysis of dyslexia: Implications for family studies. In S. D. Smith (Ed.), *Genetics and learning disabilities* (pp. 69–96). San Diego: College-Hill Press.

Pennington, B. F. (1990). The genetics of dyslexia. *Journal of Child Psychology and Psychiatry, 31,* 193–201.

Pennington, B. F., Gilger, J. W., Olson, R. K., & DeFries, J. C. (1992). The external validity of age-versus IQ-discrepancy definitions of reading disability: Lessons from a twin study. *Journal of Learning Disabilities, 25,* 562–573.

Pennington, B. F., Gilger, J. W., Pauls, D. L., Smith, S. A., Smith, S. D., & DeFries, J. C. (1991). Evidence for major gene transmission of developmental dyslexia. *Journal of the American Medical Association, 266,* 1527–1534.

Pennington, B. F., Smith, S. D., McCabe, L. L., Kimberling, W. J., & Lubs, H. A. (1984). Developmental continuities and discontinuities in a form of familial dyslexia. In R. Emde & R. Harman (Eds.), *Continuities and discontinuities in development* (pp. 123–151). New York: Plenum.

Rabin, M., Wen, X. L., Hepburn, M., Lubs, H. A., Feldman, E., & Duara, R. (1993). Suggestive linkage of developmental dyslexia to chromosome 1p34-p36 [letter]. *Lancet, 342,* p. 178.

Sutcliffe, J. G. (1988). mRNA in the mammalian central nervous system. *Annual Review of Neuroscience, 11,* 157–198.

Tenari, P. J., Terwilliger, J. D., Palo, J., Ott, J., & Peltonen, L. (1993). Two-locus linkage analysis in multiple sclerosis. *American Journal of Human Genetics, 53,* Abs. 267.

COMMENTARY ON
CHAPTER 2

J. Bruce Tomblin
University of Iowa

This chapter by Smith, Pennington, and DeFries provides a glimpse into the future of research in language impairment. The work by this team of researchers has been at the forefront of research into the genetics of a complex developmental behavioral disorder—dyslexia. Two important points are demonstrated by the work described here. First, molecular genetics has advanced to the point that it can and will provide insight into the genetic contribution to the etiology of developmental communication disorders. The task of the interested researcher in language impairment will be to catch up and keep up with the opportunities provided by advances in molecular genetics. Second, despite the nearly certain long-term success of this research endeavor, we must anticipate that it will require a good deal of hard work. One only needs to read reviews of genetic research into schizophrenia, manic-depression, and dyslexia to realize that quick and easy answers are unlikely to come to those of us interested in exploring the genetics of language development and language disorders.

We can be both encouraged by the recent successes in genetics, but also humbled by the difficulty of studying complex traits, such as dyslexia and developmental language disorder. The revolution in human genetics brought on by advances in molecular genetics has thus far had its greatest impact on simple traits and diseases that are monogenetic and often have a straightforward relationship between the gene variant (allele) and an abnormal form of a protean. Thus, for instance, insight into the etiology of muscular dystrophy has resulted from an understanding of the relationship between a mutation in a gene coding for the dystrophin protean and the events leading to muscle atrophy.

Although there is no biological reason why complex behavioral disorders, such as schizophrenia, dyslexia, and language impairment, must have complex etiologies, and indeed some may be due to single genes, we must be prepared for the likelihood that these disorders are complex. Although the possibility of complex genetic etiologies has long been accepted, methods for the genetic analysis of these conditions have been limited until recently.

In the past few years, those working in molecular genetics and those concerned with complex behavioral phenotypes have begun to come together in efforts to extend the advances of simple diseases to these more complex disorders. Smith, Pennington, and DeFries in this chapter provide an excellent overview of how this enterprise has been initiated. Plomin (1994) recently noted that causal complexity for behavior disorders can be of at least two forms. One form of complexity that Plomin terms one gene, one disorder (OGOD) results when a constellation of genetically determined symptoms that are treated as a disease or syndrome is found to be comprised of several different simple genetic causes, each existing in different individuals. Thus, the etiology for the group is complex even though the etiology for each individual may be simple, though distinct from others in the group. This form of genetic complexity yields genetic heterogeneity that could be resolved if subgroups each with its own unique genetic etiology can be identified. In the case of OGOD complexity, the gene variants associated with disorder are often viewed as *disease genes* because their expression is usually a necessary and often a sufficient condition for impaired function.

The other form of complexity described by Plomin results from several genes—quantitative trait loci (QTL)—operating simultaneously to produce a continuously distributed trait. Therefore, these genes influence the full range of individual differences in a behavioral trait from what may be considered superior performance levels to deficit levels of performance. This form of complexity cannot be resolved by narrowing the phenotype to an etiologically homogeneous group as can be done with the OGOD condition. Instead, the phenotype of interest is etiologically and genetically complex even within the individual. Unlike single disease genes, the QTLs are not necessary or sufficient for impaired function, and because they can influence levels of function within what is considered "normal," they are only viewed as *susceptibility* or *liability* genes, not disease genes.

These different forms of etiologic complexity are recognized by Smith et al. as potential etiologic accounts of dyslexia. Until recently, the genetic basis of complex conditions like dyslexia could be demonstrated through twin studies demonstrating elevated concordance rates between monozygotic twins over dizygotic twins and family studies showing some form of familial transmission of the trait. These research designs provide important evidence that a trait is likely to have some form of genetic etiology and, hence, is heritable. The evidence of a strong heritability for dyslexia comes from numerous studies, several of them conducted by the authors of this chapter, and now we are beginning to accrue a similar base of evidence for specific language impairment (see chapter 7).

Once a sufficient amount of evidence for a genetic etiology has been accumulated through twin and family studies, the focus of research then must move on to identify the particular genetic mechanisms involved in the etiology. That is, we become interested in some form of linkage between phenotype and genotype. Much of Smith, Pennington, and DeFries' work has exemplified the OGOD strategy for linkage studies of a complex behavioral phenotype. This strategy requires a team of investigators who can examine different diagnostic systems in association with molecular genetic studies. The breadth of knowledge needed to associate traits involving behavior disorders and molecular genetics is so great that collaborations of this sort are necessary and can be expected to arise in the study of specific language impairment as well.

With the dramatic advances in molecular genetics, it is easy to assume that the "hard" work and the breakthroughs will occur in the molecular genetics laboratory. This, however, is a naive view. The authors of this chapter point out the importance in this type of research of the continued development and refinement of clinical phenotypes that is the study of our diagnostic systems. It is in this domain that the greatest amount of hard work is still ahead. We must keep in mind that what we are attempting to do is learn how genomic variations (individual differences in genotype) are associated with behavioral variations (individual differences in phenotype). When these individual differences in behavior, such as reading or language achievement, are cast into clinical categories, we confront the issue of what our clinical diagnostic categories are to represent and, in particular, the way causal agents like genes can lead to language impairment.

I have argued (Tomblin, 1991) that our view of the cause of language disorder will need multiple components. The first and central component is concerned with individual differences. If there were no individual differences in language performance, there would be no such thing as a language disorder. Individual differences by themselves do not provide us with an understanding of what constitutes disorder. Language disorder represents a particular domain of individual differences in language performance. Second, there is a domain of language variation that we designate as *disorder* because of associations (presumably causal) between these forms of *disordered* language function and undesirable consequences to the individual. It is society and the manner in which social values are applied to human function that determines which individual differences are undesirable and hence disordered.

Neither the first nor second aspect of a diagnostic system assumes anything about the causes for the individual differences, but we now have an account for why we want to explain (and take clinical action on) a particular domain of individual differences in language function. There is no reason to assume, however, that a diagnostic category formed in this manner will have any particular etiology or an etiology that is distinct from that which leads to individual differences among those we say have *normal* language. This means that as we consider different etiological types of language impairment there may or may not be a unique mapping between those causal agents (genes in this case) that contribute to individual differences labeled as language impairment and those that are labeled normal language status. When there is a unique mapping, as is expected in the OGOD type of complexity, we can view the gene variants associated with this type of function as disease or disorder genes.[1] When there is not a unique mapping, we may instead have genes that conspire to increase the likelihood of a person falling into the domain of language function labeled as language impairment. That is, the genes involved in language disorder may be QTLs. Thus, part of the difference between the OGOD and QTL forms of genetic etiology have to do with the coincidence of the way functions are valued and required by society and the genetic causes of the variance in these functions.

At this time, we have to be prepared for both types of genetic complexity contributing to genetically based forms of developmental language impairment, and we need to plan to use strategies that will allow us to discover linkages of both types. Fortunately, methods for the study of the full range of genetic influence on behavior are very rapidly becoming available. Today it is possible to consider exploring questions involved with the genetic basis of language development and disorders that a few years ago were only a dream.

REFERENCES

Plomin, R., Owen, M. J., & McGuffin, P. (1994). The genetic basis of complex human behaviors. *Science, 264*, 1733–1739.

Tomblin, J. B. (1991). Causes of specific language impairment. *Language, Speech, and Hearing Services in Schools, 22*, 69–74.

---

[1]We need to be careful here with our language. In biological terms, genes that produce altered protein are not disordered, except possibly those that make the organism inviable. Biological systems depend on variation for long-term survival. Genes are only disease genes because they cause functions that we disvalue.

3

LONGITUDINAL STUDY OF CHILDREN AT HIGH FAMILY RISK FOR DYSLEXIA: THE FIRST TWO YEARS

Dianne L. Lefly
Bruce F. Pennington
University of Denver

Written language is closely related to spoken language. Spoken language is a naturally occurring, species-specific behavior, but written language is not. Written language is an agreed-upon representation of spoken language (Liberman, 1991). Reading is the translation of written script into its spoken language equivalent (Henderson, 1982). Given this close relation, one would expect that the development of reading skill would depend on the development of spoken language. One would also expect that abnormal reading development (i.e., developmental dyslexia) to be related to abnormal language development.

Yet, surprisingly little is known about the relation between speech and language disorders and dyslexia, partly because these two disorders are studied at different ages by persons in largely different disciplines. A great deal has been learned in the last decade or so both about which spoken language processes are important in reading development and about the genetics of dyslexia. This information is highly relevant for efforts to understand the genetics of language. The study presented here helps provide some links between research on dyslexia and research on speech and language disorders.

In 1989, we, the authors of this chapter, began a longitudinal study of reading development at the University of Denver. The study is specifically concerned with which spoken language skills predict both normal and abnormal reading development. This chapter presents the results from the first two years of our longitudinal study when the children were five and six years old.

In our study, we included children considered both at high risk and low risk for developing dyslexia. Children were considered at high risk for developing dyslexia if they had a parent or sibling who reported a history of reading problems. Children were considered at low risk if they did not have a parent or sibling who reported a history of reading problems. Questions addressed by this initial report are (a) Do the risk groups differ in phonological skills at either age? (b) Is the factor structure of the phonological processing skills battery stable across these two ages? and (c) Are the predictors of early reading skill similar or different across the two risk groups? This study builds on previous work from (a) the genetics of dyslexia and (b) linguistic predictors of normal and abnormal reading development. Work in these areas is briefly reviewed in the following sections.

GENETICS OF DYSLEXIA

Recent work in behavioral genetics indicates that dyslexia is familial, heritable, and heterogeneous in its genetic influences (Pennington, 1990; Pennington & Smith, 1988). Across family studies, the magnitude of familial risk to first degree relatives has been found to be about 35% to 45%, which is a very considerable increase over the population risk, which is between 3% and 10%. Thus, family history alone is the best predictor we currently have of later dyslexia.

At least some forms of dyslexia appear to be due to major gene influence. In fact, a recent formal segregation analysis of four large familial samples found major gene transmission in three of them (Pennington et al., 1991). However, we still do not know if this putative major gene (or genes) acts in a Mendelian fashion or is a quantitative trait locus (QTL) with a major effect. Unlike a Mendelian locus, a QTL is neither necessary nor sufficient to produce the phenotype in question. So, the risk for dyslexia that is transmitted in families may be discrete and categorical, or it may be continuous. These two alternatives make different predictions about the reading development of children from high-risk families that do not become dyslexic. The discrete model predicts they should be similar to low-risk controls. The continuous model predicts they should be intermediate between dyslexics and low-risk controls.

The predictive value of family history was recently shown prospectively in the only longitudinal study of children at high familial risk for dyslexia conducted thus far (Scarborough, 1989). In this study, family history for dyslexia uniquely accounted for between 30% and 36% of the outcome variance in reading, with linguistic variables (sound and letter knowledge and discrete trial lexical retrieval; Boston Naming Test) accounting for an additional 7% to 8% of the outcome variance. In contrast, a host of other variables including sex, age, SES, preschool IQ, nonverbal measures, early education, reading and television exposure did not contribute significantly to reading outcome. Using a combination of family history and scores on the linguistic variables, Scarborough successfully predicted 82% of reading outcomes, which is approaching a level of accuracy acceptable for a population screening test (both false negatives and false positives were under 10%). Scarborough's study used only a rudimentary measure of phoneme awareness skill, which is the best behavioral predictor of reading outcome currently available, sometimes uniquely accounting for up to 50% of outcome variance (Wagner & Torgesen, 1987).

Thus, if family history could be combined with more sophisticated phoneme awareness (or possibly other linguistic) measures, an even higher rate of predictive success might well be reached. Further, if a reliable spoken language precursor to dyslexia could be found and used in combination with family history, it would make the early identification and treatment of dyslexia possible, thus revolutionizing current clinical and educational practice.

Scarborough's results also provide evidence pertinent to the discussion of discrete and catagorical models discussed earlier. While the at-risk children who became dyslexic were significantly different from normal controls on syntactic and phonological measures, the at-risk nondyslexic children were equal to or better than the control sample on all measures. Scarborough (1991) concluded from these results that "familial risk per se did not preclude normal developmental progress" (p. 39). These results are consistent with the discrete model, as described above.

LINGUISTIC PREDICTORS OF READING

Recent research has indicated that the written language deficit in dyslexia is in phonological coding; therefore, it is likely that the spoken language precursor involves phonological processes as well. Candidates for this underlying phonological processing deficit include deficiencies in phoneme perception (Brady, Shankweiler, & Mann, 1983), phoneme awareness (Bradley & Bryant, 1983; Liberman, Shankweiler, Fischer, & Carter, 1974), lexical retrieval of phonology (Denckla & Rudel, 1976), and verbal short-term memory (Jorm & Share, 1983).

In previous longitudinal studies of normal reading development, the strongest evidence for a predictive relation to reading outcome exists for phoneme awareness skills. A number of longitudinal studies (Liberman, Shankweiler, Fischer, & Carter, 1974; Wallach & Wallach, 1976) have found such a predictive relation, with at least 50% of the variance in reading outcome accounted for by phoneme awareness tasks in several studies (Calfee, Lindamood, & Lindamood, 1973; Lundberg, Olofsson, & Wall, 1980; Mann, 1984, 1993). However, Calfee et al. (1973) did not control for IQ, and the study by Lundberg et al. (1980) did not control for preexisting reading skill.

The study by Lundberg et al. (1980) tested the predictive relation between phoneme awareness skill and later reading skill using path analysis and studied a variety of phoneme and syllable manipulation tasks, which involved both synthesis and analysis. They found little predictive power for syllable or synthesis tasks, but impressive predictiveness for two phoneme analysis tasks:

phoneme segmentation and phoneme reversal. This is an important study, but a reanalysis of the data by Wagner and Torgesen (1987) revealed that when preexisting reading skill was partialled out, the predictiveness of the phoneme awareness tasks was considerably reduced.

Mann's (1984) study both controlled for IQ and screened early readers out of the sample, yet still found an impressive partial correlation of .75 between kindergarten phoneme reversal and first grade reading achievement. In a later study, Mann and Ditunno (1990) noted the potential impact of IQ, but did not specifically control for it or early reading ability, and found that 30% to 40% of first-grade reading ability could be predicted by kindergarten phoneme awareness skills.

Two studies, one by Bradley and Bryant (1983) and the other by Byrne and Fielding-Barnsley (1993) are interesting in that they combined longitudinal studies with training studies, thus allowing a test of both the direction and specificity of the causal relation between phoneme awareness skills and reading outcome. Their results support a specific, causal relation between preschool phoneme awareness and later reading skill.

It has long been noted that many children with speech and language impairments have a higher than normal incidence of reading disabilities. Hugh Catts (1993) conducted a longitudinal study of 56 kindergarten children who were speech and language impaired. He followed these children through the second grade. In this study about 50% of the subjects were reading within normal limits in first and second grade, but about the same percentage were not. This is a significantly higher percentage of reading disabilities than the population expectancy (3% to 10%).

Catts found that children who had only impaired articulation were less likely to be reading impaired than were those with other types of language impairments. Additionally, Catts reported that when reading was defined only as word recognition rather than comprehension, phonological awareness, and rapid automatized naming (lexical access) tasks were the best predictors of reading outcome. Interestingly, he also reported that phonological awareness and rapid automatized naming each explained large amounts of variance in word recognition with only a small amount of overlapping variance. This would seem to support the idea that phonological awareness and lexical access are related to different aspects of reading skill.

Other recent longitudinal studies have extended the predictive relation between phoneme awareness and later reading skill downward into the earlier preschool period. Maclean, Bryant, and Bradley (1987) found about a third of a broad sample of normal children showed evidence of some degree of phoneme awareness before age 4, and that early phoneme awareness skill was significantly related to later reading (but not arithmetic) skill, even when age, IQ, and maternal education were partialled out. In addition, early knowledge of nursery rhymes predicted later phoneme awareness skill. Similar results were obtained by Stuart and Coltheart (1988). Thus, it is possible that the behavioral precursor to dyslexia might be observable before age 4.

An early behavioral precursor to dyslexia was found in data from a longitudinal study of children at high familial risk for dyslexia that began when the children were 2½ years old (Scarborough, 1990). The future dyslexic subjects were significantly worse than both siblings and low-risk subjects in measures of natural language production (expressive syntax), but not on formal language measures. Nursery rhyme knowledge or other early measures of phoneme awareness were not evaluated in this study, so we do not know if a deficit in phoneme awareness is detectable this early in dyslexic development. Scarborough's results argued for a wider range of early language deficits in dyslexia. However, these early deficits in syntax were no longer detectable at ages 4 and 5.

Bryant, Maclean, Bradley, and Crossland (1990) completed a path analytic study of their longitudinal data. In this study, they concluded that rhyming skill is separate from phoneme awareness because it does not involve phonemes, but larger segments of words. They tested three models: (a) that rhyming is not related to phoneme detection or reading skill, (b) that rhyming leads to phoneme detection that then leads to reading skill, and (c) that rhyming skill contributes directly and independently to both phoneme detection and reading skill. They concluded that there is no evidence to support the first model; the data best fit some combination of models b and c. They found that rhyming skill is directly related to phoneme detection and is also directly related to reading skill.

Two confirmatory factor analytical studies have been completed recently that provide more convergent evidence that early linguistic skills are predictive of later reading ability. They also provide potential theoretical latent variable models of the phonological skills that underlie reading. The first study is by Sawyer (1992) and the second is by Wagner, Torgesen, Laughon, Simmons, and Rashotte (1993).

Sawyer's longitudinal study followed two cohorts of subjects ($n = 129$ and $n = 171$) from kindergarten through third grade. Her theoretical model was based on the hypothesis that there is a hierarchical relationship between early language processing skills specifically linked to reading and later language processing skills. In her view, basic language and auditory processing skills are related to early measures of reading achievement (letter naming, recognition of familiar words) and to later-developing language abilities, such as phoneme segmentation. The later-developing language abilities are then related to later measures of reading, such as naming unfamiliar words and reading comprehension.

From her results, she concluded that, consistent with her hypothesis, a composite of early kindergarten language processing skills (blending, sentence segmentation, PPVT score auditory discrimination, and auditory comprehension) accounted for a substantial amount of the variance in third-grade reading. However, she concluded that reading comprehension skills at Grades 2 and 3 lead to word recognition skills at those grade levels. This conclusion is inconsistent with previous research, but her methodology and this conclusion have been criticized by Torgesen and Wagner (1992). They argued that, technically, she was in error for using path analytical methods to compare tests given at the same age. They maintained that trying to establish paths from reading comprehension to word recognition at the same time is inappropriate for this type of analysis. Thus, her conclusion that reading comprehension leads to word recognition has been criticized, but Wagner and Torgesen did not challenge her overall conclusion that preschool language processing skills lead to third-grade reading skills.

The Wagner et al. study (1993) was a cross-sectional study of 95 kindergarten and 89 second-grade children. In this study, the investigators moved away from examining the predictive power of individual phonological processing skills. The purpose of their study was twofold: (a) to examine the relationship between phonological processing abilities and other cognitive abilities, and (b) to establish that phonological processing skills are coherent, enduring individual difference attributes that are related to, but separate from, general cognitive ability.

In their study, they evaluated the fit of the data to five potential models of the relationship between phonological processing abilities that resulted from their previous research (Wagner & Torgesen, 1987), and concluded that a single model fit the data acceptably well at both the kindergarten and second-grade levels. In the best fitting model, there are two underlying constructs: (a) the ability to be aware of individual phonemes and to manipulate those phonemes, and (b) the ability to recall phonological codes from long-term memory. In these results, phoneme awareness functioned in conjunction with other cognitive abilities (working memory and long-term memory), rather than by itself.

Wagner et al. contend that because the same model fit the data acceptably at both age levels, phonological processing abilities appear to be as stable in their structures as other cognitive abilities. This result is important, in their view, because only stable, enduring cognitive abilities can be evaluated for possible causal roles in important skills, such as reading. However, it is important to note that this study did not directly address the longitudinal stability of these factors.

In summary, longitudinal and factor analytical studies strongly suggest that phoneme awareness is a prerequisite, or at least a facilitator, of later reading skill. The training study by Bradley and Bryant (1983) helped to rule out the third variable or correlate relation, but a longitudinal study of all four phonological processing skills is still needed to completely address this possibility.

UNRESOLVED ISSUES IN THE DEVELOPMENT OF DYSLEXIA

What is not completely clear from any of these studies is whether phoneme awareness, with or without other cognitive abilities, has the same predictive relation for developmental dyslexia that it does for later reading skill. Do the same processes underlie the development of both dyslexic and normal reading?

Are dyslexics at the low end of the same continuum as normal readers or are different cognitive processes involved in reading success (Bryant & Bradley, 1985; Stanovich, 1986)?

Genetically, is the risk factor in dyslexia discrete or continuous? Do unaffected children in dyslexic families show any type of subtle phonological processing problems, or do the phonological processing difficulties affect only those who are diagnosably dyslexic? Regarding the broader language domain, is the linguistic phenotype in dyslexia restricted to written language and its developmental onset restricted to the early school years? The existing research data do not allow clear answers to any of these questions.

RATIONALE FOR THE CURRENT STUDY

There is no doubt that recent longitudinal and cross-sectional studies of reading ability and disability have considerably narrowed the search for behavioral precursors of reading ability and disability; however, many important questions remain.

In an attempt to address some of these questions, we are conducting a three-year longitudinal study of a small number of potential precursors in children at both high- and low-familial risk for developing specific reading disability. For this study, children were considered to be at high risk for developing dyslexia if they had a parent or sibling who reported a history of reading problems, and at low risk if they did not have a parent or sibling who reported a history of reading problems.

As originally designed, this study had five main hypotheses:

1. After controlling for estimated IQ, most of the significant predictive variance for later word recognition skill will be captured by phoneme awareness, and little unique variance will be contributed by the other phonological processing skills.

2. Skill in word recognition will have different predictors than skill in reading comprehension. Specifically, verbal IQ and verbal STM are expected to predict reading comprehension skill, but not word recognition skill independently of phoneme awareness. Conversely, phoneme awareness is expected to predict word recognition skill, but not have a direct predictive relation to reading comprehension.

3. The precursors of reading skill and difficulty will be similar in both the high-risk and normal-risk cohorts.

4. Reading experience is also expected to facilitate development of some or all of these phonological processing skills, so the relationship between phoneme awareness and reading is expected to be one of reciprocal causation.

5. Any differences between dyslexics and normals in self-esteem, depression or attention problems will only emerge after the onset of dyslexia, indicating that these clinical correlates of dyslexia are secondary rather than primary symptoms of the disorder. Some, but not all, dyslexics will have early articulation problems, indicating a correlate rather than a causal relation.

To address these hypotheses, measures of intelligence, reading ability, phonological processing abilities, and measures of emotional functioning were administered to both risk groups each year for three years. In addition, medical and developmental history, parental reports of the child's emotional and behavioral functioning, and parental reports of the parents' reading history were obtained from the parents. For the entire study, data were to be collected at four points: (a) prior to entry into kindergarten, (b) prior to first grade, (c) prior to second grade, and (d) at the end of second grade. The current report is based on the data collected from the first 2 years of the study.

METHODS

Subjects

Subjects originally entered the study just before beginning kindergarten (Year 1) and were tested again just before entering first grade (Year 2). The high-risk group consisted of 73 children considered to be at high risk for developing dyslexia because each had a first-degree relative who reported a history of dyslexia. To evaluate the precursors of early reading skill in the absence of family risk for dyslexia, a low-risk group was included in the study. The low-risk group consisted of 57 children, similar in IQ, sex, and SES distribution to the high-risk group. *Low risk* was defined as the absence of family history for dyslexia in first-degree relatives. Therefore, the risk for dyslexia in the low-risk subjects is somewhat lower than the population risk and considerably lower than the risk for the high-risk subjects. Research has shown that the risk to first-degree relatives of dyslexics is about 35% to 45%, which is much greater than the population risk (3% to 10%) (Pennington et al., 1991; Vogler, Decker, & DeFries, 1985).

Subjects for this study were recruited over a 3-year period from three sources: (a) 10 Denver-area preschools, (b) volunteers from the Colorado Branch of the Orton Dyslexia Society, and (c) from families with a student already in a special education placement either in the Denver or Jefferson County Public Schools. Most of the participants in the study were from the 10 Denver-area preschools.

In all cases, parents received an open letter discussing the rationale and specific details of the study and the potential advantages of participation (e.g., feedback regarding child's progress). If they were interested in participating in the study, they were asked to contact the study director.

When a parent contacted the study director, a brief telephone interview was conducted to determine whether there was a family history of reading problems. Children from families in which at least one biological parent or sibling reported a history of reading problems were classified as high risk, and those who had no family history of reading problems were classified as low risk.

To further document the telephone interview, a written reading history questionnaire (RHQ) was sent to each parent that required them to answer several questions about their reading history, educational history, and current reading habits. This questionnaire has been used in our familial dyslexia studies for several years. It was originally designed by Finucci, Isaacs, Whitehouse, and Childs (1982). We modified the original questionnaire by simplifying the questions and by focusing more on questions about specific reading and spelling-related skills (difficulty learning letter names, number/letter reversal, difficulty with spelling, difficulty with foreign language, etc.). These changes were based on extensive research and clinical experience with the original questionnaire.

The score on the RHQ was the total number of points obtained by an individual divided by the total number of points possible on the questionnaire; thus, higher scores indicate a stronger history for reading problems. There was 86.3% agreement between the telephone interview and the RHQ. In a recent reliability and validity study, the RHQ had an internal consistency reliability coefficient of .94 (Cronbach's alpha) (Lefly & Pennington, in preparation), and it correlated .86 with the original version of the RHQ. In terms of validity, the new RHQ had significant correlations from .57 to .73 with three diagnostic test criteria used to diagnose adult dyslexics in our laboratory. Hence, we have found the RHQ to be a reliable and valid measure of reading history.

As a result of this interview/questionnaire process, 73 children were assigned to the high-risk group and 57 were assigned to the low-risk group. The characteristics of the sample are shown in Table 3.1. There were no significant differences at either Year 1 or Year 2 on estimated FSIQ, the Vocabulary or Block Design subtests, age, or SES (Hollingshead, 1975). Ten children (five high risk [6.8%] and five low risk [8.8%]) were lost to attrition between Year 1 and Year 2.

In addition, parents completed questionnaires regarding the medical and developmental history of each child. Parents were asked whether or not their child had any difficulties at birth, any neurologic history, specific sensory impairments, or if early speech therapy was required. In an attempt to limit the variables under study, children were excluded from the study if they had serious birth complications, neurological impairments, or serious sensory impairments. This is a study of reading development in children of at least average intelligence; therefore, children with an estimated level of intellectual functioning (as estimated by the vocabulary and block design subtests of the WPPSI) that is less then 85 were excluded from the sample. Children were not excluded from the study because of previous placement in speech therapy.

Because some bright young children learn to read before they enter kindergarten, it was important to know which participants in this study were readers

TABLE 3.1
Demographic Information

| | | Year 1 | | | | | | Year 2 | | | | |
| | | | WPPSI | | | | | | WISC-R | | | |
	N	Age	VOC	BD	FSIQ*	N	Age	VOC	BD	FSIQ*	SES
HIGH RISK	73	5.4 (.4)	11.8 (3.2)	11.9 (3.2)	112.1 (12.6)	68	6.4 (.4)	12.4 (3.0)	12.4 (2.8)	113.2 (14.5)	2.34 (.99)
Male	40	5.4 (.3)			111.4 (13.7)	37	6.4 (.4)			114.3 (15.5)	
Female	33	5.3 (.4)			112.9 (11.3)	31	6.3 (.4)			111.9 (13.4)	
LOW RISK	57	5.3 (.3)	12.1 (3.2)	12.6 (3.0)	113.5 (14.4)	52	6.4 (.4)	12.3 (2.9)	11.6 (2.7)	111.5 (13.5)	2.27 (.96)
Male	31	5.4 (.3)			114.5 (16.3)	26	6.4 (.4)			111.7 (13.4)	
Female	26	5.3 (.3)			112.3 (11.9)	26	6.5 (.3)			111.2 (13.8)	

*prorated

and which were nonreaders before they entered the study. To accomplish this, we developed and administered a single word reading task at the outset of Year 1 testing. This task consisted of 36 words taken from basal readers, common young children's books, and the lowest levels of standardized reading tests. Words like *go, to, in, the,* and *cat,* were included in this task. Analysis of these data revealed that most children in the low-risk sample could read fewer than five words. Using this criterion, 91.5% of our sample were nonreaders, and 8.5% could read more than five words. Interestingly, the proportion of early readers across both risk groups was similar: 9.6% of the high-risk sample and 7% of the low-risk sample. There were some analyses in which it made sense to use nonreaders only, but in others it made sense to include the whole sample. To ensure that there were no statistically significant differences between the sample as a whole and the nonreaders alone, separate parallel analyses were performed on the whole sample and on the nonreader subset.

Measures

Measures administered at Year 1 and Year 2 were very similar and are detailed in Table 3.2. Each child's estimated overall level of intellectual functioning was prorated from the child's scores on the vocabulary and block design subtests of the age appropriate Wechsler intelligence test (Sattler, 1992). A child's general level of intellectual functioning is expected to be highly correlated with his/her reading ability. Therefore, we felt it was important to assess each child's general level of intellectual functioning each year. However, in the interest of time, a child's general level of intellectual functioning was estimated by prorating the child's full scale intelligence quotient score from the vocabulary and block design subtests of the age appropriate Wechsler intelligence test (WPPSI at Year 1 and WISC-R at Year 2). Sattler (1992) stated, "These two subtests have excellent reliability, correlate highly (.91) with the full scale over a wide age range, and are good measures of *g*" (p. 137). Once a child's FSIQ was estimated, it was possible to statistically control that variable so that the possible confound between FSIQ and the phonological processing skills could be examined, if necessary.

Reading ability at Year 1 was assessed by the single word reading task, as described earlier, and an upper-case letter identification task; at Year 2, it was assessed by the Woodcock-Johnson letter-word identification subtest, and the upper-case letter identification task. Psychosocial variables were assessed at both years by use of a parent behavioral questionnaire (DICA ADHD subscale) and by use of Susan Harter's scales of self-worth and depression for preschool and primary children to obtain the child's own report of his/her self-esteem and depression. Phonological processing skills were assessed in each of the four domains that research indicates hold potential for containing the spoken language precursors of later-reading ability and disability: phoneme perception, lexical access (naming speed), verbal short-term memory, and phoneme awareness (a detailed explanation of the measures used is contained in the appendix).

TABLE 3.2
Measures

Psycho/Social Scales	Phonological Processing Tasks
1. DICA ADD scale	1. Perception
2. Harter's self esteem & depression scales	phoneme perception task
IQ and achievement tests:	2. Lexical access
1. WPPSI, WISC-R	rapid automatized naming (Denkla & Rudel,
2. Letter identification	1976)
3. Single word reading	Rapid alternating stimulus test (Wolf, 1986)
4. Woodcock-Johnson letter-word	3. VSTM
identification**	Real words
	Rhyming words
	Pseudo words
	4. Phoneme awareness
	Initial consonant different
	Supply initial consonant**
	Strip initial consonant** (Stanovich, 1984)
	Roswell-Chall blending (1986)
	Bradley & Bryant sound categorization test
	(1983)
	Treiman's onset-rime task (1986)
	Syllable tapping (Liberman, 1974)*
	Lindamood test of auditory conceptualiza-
	tion (1973)**

*Year 1 only. **Year 2 only.

Some phoneme awareness tasks were well beyond the phoneme awareness skills of most five-year-old children (supply initial consonant, strip initial consonant, Lindamood test of auditory conceptualization). Conversely, some other phoneme awareness tasks administered at the 5-year-old level (syllable tapping) produce ceiling effects at the 6-year-old level and were not given then. Because recent research (Yopp, 1988) indicates that most tests of phonemic awareness are significantly and positively correlated, we can feel confident that we are assessing the same underlying skill across ages, even though some of the phoneme awareness measures differ at different ages. A list of the measures administered at each year is included in Table 3.2. The results discussed in this chapter are for only those measures that were given at both Year 1 and Year 2.

RESULTS

It should be reiterated at this point that the results reported here are based on risk status alone. No formal diagnosis of reading disability was made during the first 2 years of the study. This was because currently available diagnostic measures are not adequately normed to provide reliable diagnosis prior to about 8 years of age. Consequently, the high-risk group is a heterogeneous group made up of both those who will eventually become dyslexic (35% to 45%) and those who will not (55% to 65%). The low-risk group is a more homogeneous group and has a lower-than-population risk (less than 3% to 10%) of becoming dyslexic. Because the majority of the high-risk subjects will not become dyslexic, it is likely that the risk group comparisons are conservative. On the other hand, there is a possibility that even those high-risk subjects who are not eventually diagnosed as dyslexic may have some subtle phonological processing deficits. A clear test of these competing possibilities will be possible at later years in this study when we can divide the high-risk group into dyslexic and nondyslexic subgroups.

The data were analyzed in three different ways: (a) a developmental analysis accomplished by means of a two-way, between-subjects ANOVA with age and risk group as the two factors on all variables, (b) an exploratory factor analysis performed on each year's data with a multiple analysis of variance on the resulting factor scores by risk group each year, and (c) a regression analysis to analyze which Year 1 variables best predicted the Year 2 outcome measure (Woodcock-Johnson letter-word identification) in each risk group.

Developmental Analyses

The results of the developmental analysis are contained in Table 3.3. The phoneme perception tasks showed no main effect of either age or risk group and no interaction effect between risk and age. Most other tasks showed a significant main effect of age, but no main effect of risk group, and no interaction effect.

Four tasks showed both main effects of risk group and age, but no interaction effect, as follows: (a) letter knowledge—risk, $F(1, 117) = 5.66$, $p < .05$; age, $F(1, 117) = 79.1$, $p < .001$; (b) initial consonant different—risk, $F(1, 116) = 7.99$, $p < .01$; age, $F(1, 116) = 71.61$, $p < .001$; (c) rhyming as measured by the Bradley and Bryant sound categorization test—risk, $F(1, 111) = 3.96$, $p < .05$; age, $F(1, 111) = 15.46$, $p < .001$; and (d) the average response time of the rapid alternating stimulus test—risk, $F(1, 63) = 4.27$, $p < .05$; age, $F(1, 63) = 90.96$, $p < .001$. The first two tasks are phoneme awareness measures, the third is a measure of lexical access, and the fourth is a measure of letter knowledge. On all four tasks, post hoc analyses show that the high-risk group performed more poorly than the low-risk group. Thus, family risk is associated with differences in phonological processing skills before formal literacy training begins.

In addition, as described earlier, parents were asked at the outset of the study whether or not their child required speech therapy at the preschool level. In the high-risk group, 11 children were in speech therapy before entering kindergarten, but only 2 low-risk children were involved in speech therapy, $\chi^2(1, N = 130) = 4.52$, $p < .05$. In sum, differences in spoken language proficiency in the

TABLE 3.3
Phonological Processing Variables by Risk and Age:
Mean and Standard Deviation

	Year 1		Year 2		Risk	Age
Letter knowledge (mean letters identified)						
high risk	17.4	(8.8)	24.3	(3.5)	*	***
low risk	20.7	(7.2)	25.5	(1.1)		
Phoneme perception (mean percent correct)						
Words/quiet						
high risk	.74	(.19)	.75	(.18)		
low risk	.76	(.19)	.77	(.13)		
Words/noise						
high risk	.20	(.15)	.25	(.16)		
low risk	.24	(.14)	.24	(.16)		
Nonwords/quiet						
high risk	.47	(.21)	.46	(.21)		
low risk	.47	(.19)	.48	(.18)		
Nonwords/noise						
high risk	.07	(.09)	.05	(.07)		
low risk	.08	(.08)	.04	(.06)		
VSTM (mean percent correct)						
Real words						
high risk	.14	(.12)	.64	(.13)		***
low risk	.14	(.16)	.67	(.13)		
Rhyming words						
high risk	.15	(.10)	.59	(.10)		***
low risk	.15	(.16)	.60	(.13)		
Pseudo words						
high risk	.12	(.07)	.30	(.17)		***
low risk	.13	(.11)	.33	(.17)		
Phoneme awareness (mean percent correct)						
Blending						
high risk	.37	(.29)	.63	(.30)		***
low risk	.40	(.30)	.74	(.25)		
Rhyming						
high risk	.51	(.20)	.59	(.21)	*	***
low risk	.59	(.22)	.65	(.21)		
Initial consonant different						
high risk	.37	(.22)	.58	(.25)	**	***
low risk	.47	(.25)	.71	(.27)		
Lexical access (average response time)						
RAN						
high risk	84.7	(24.7)	70.1	(18.8)		***
low risk	80.8	(19.1)	64.0	(17.3)		
RAS						
high risk	106.8	(43.4)	82.9	(34.3)	*	***
low risk	89.5	(27.9)	68.8	(25.6)		

*p < .05. **p < .01. ***p < .001.

high-risk group are apparent on both experimental tests and in rates of clinical referral.

Factor Analyses

Because all subjects entered the study at essentially the same age, there was little age variance between or within groups. Although there were no group differences in IQ at either Year 1 or 2, variability on the cognitive and behavioral measures of this study may reflect individual differences in general intelligence because IQ was significantly correlated with most study variables (13 of 19 variables; rs .10–.48: Year 1; 18 of 21 variables; rs .11–.40: Year 2). The purpose of the factor analysis was to examine specific linguistic abilities independent of the effects of intelligence. Consequently, all measures were adjusted for the linear effects of IQ through a multiple regression procedure. These adjusted residual scores are employed in all analyses unless otherwise noted.[1]

To examine both the Year 1 and Year 2 data in relation to the four domains of interest (phoneme perception, phoneme awareness, lexical access, and coding in verbal short-term memory), the data were subjected to separate principal components factor analysis with varimax rotation to identify the underlying structure of the data. Excluded from both analyses were tasks involving timed responses (RAN and RAS) because the different methods used in these tasks would likely result in a factor based on method variance.[2]

In the Year 1 exploratory factor analysis, five factors emerged from the data. These factors accounted for 66.9% of the total variance of the sample. This solution was both interpretable and parsimonious (Table 3.4).

The first factor, a rhyming-phoneme awareness factor, explained 25.5% of the variance and was defined by rhyming tasks and syllable knowledge. The second factor was a blending factor and accounted for a 12.3% additional variance. The third factor accounted for an additional 10.9% of the variance and appeared to be a phoneme perception factor. The fourth factor appeared to be a verbal short-term memory factor that accounted for 10.5% of the variance, and the fifth

---

[1]The adjusted data was examined for normality, skewness, and kurtosis. Variables that were skewed and could not be brought into normality were dropped from the analysis. No variables were dropped because of skewness in the Year 1 factor analysis, and only two (letter knowledge and nonwords in noise) were dropped from the Year 2 analysis because of skewness (> 1.0). All other variables were found to be within acceptable limits. In addition, it is important to note that the purpose of this analysis was to examine the simple orthogonal structure of the data. For this reason, variables that cross-loaded on more than one factor were dropped from the analysis. Future factor analyses are planned to examine the complex structure of the data.

[2]Variables that crossloaded in the Year 1 factor analysis were initial consonant different, and the neither condition of the onset-rime task. Variables that crossloaded in the Year 2 factor analysis were nonwords in quiet and supply initial consonant. As discussed earlier, cross-loaded variables were dropped from the factor analyses.

TABLE 3.4
Mean Factor Scores

	Year 1			Year 2	
	High Risk	Low Risk		High Risk	Low Risk*
Rhyming	−.11 (.93)	.14 (1.08)	Blending	−.16 (1.06)	.24 (.87)*
Blending	.03 (1.02)	−.04 (.98)	Rhyming	−.08 (.97)	.08 (1.03)
Phoneme percep	−.05 (.95)	.06 (1.06)	Phoneme aware	−.19 (.95)	.26 (1.03)*
VSTM	.02 (.86)	−.03 (1.16)	VSTM	−.00 (1.04)	.02 (.96)
Onset-rime	−.09 (1.03)	.12 (.96)	Phoneme percep	−.04 (1.01)	.04 (.99)

*$p < .05$.

factor appeared to be an onset-rime factor and accounted for an additional 7.6% of the variance.

The Year 2 data were subject to factor analyses in the same manner as the Year 1 data. A few additional age appropriate measures were added to the Year 2 study, and some measures that were approaching ceiling levels were removed.

In the Year 2 factor analysis five factors, again, emerged from the data. These factors accounted for 65.6% of the total variance of the Year 2 sample. The five-factor solution was interpretable and parsimonious (Table 3.4).

Three of the five factors were related to phoneme awareness. The first factor was a blending factor and explained 31.2% of the variance; it was defined by blending tasks. The second factor was a rhyming factor and accounted for 11.1% of additional variance; it was defined by rhyming tasks. The third factor accounted for an additional 8.7% of the variance and appeared to be a phoneme awareness factor. The fourth factor appeared to be a verbal short-term memory factor that accounted for 8.4% of the variance and was defined by verbal short-term memory tasks. The fifth factor appeared to be a phoneme perception factor that accounted for an additional 6.2% of the variance.

The factors at both Year 1 and Year 2 were very similar. Additionally, they are very similar to the factors identified by Wagner and Torgesen in their recent research. The rhyming and phoneme awareness factors at each year can be considered analytical factors, and the blending factors at each year can be considered synthesis factors. The specific tasks Wagner and Torgesen employed were different from those used in the present study, but measured similar constructs. Although acknowledging that their results were from cross-sectional data, Wagner et al. (1993) hypothesized that their results show that phoneme awareness factors (analysis and synthesis) are coherent and stable across the preschool and kindergarten ages. In our results, the correlations between the Year 1 and Year 2 analysis factors were .43, $p < .001$ (rhyming) and .26, $p < .01$ (phoneme awareness and onset-rime). The correlation between the Year 1 and Year 2 synthesis factors was .28, $p < .01$ (blending). Our results are consistent with Wagner's hypothesis.

Profile Analysis

In addition to the exploratory factor analyses previously discussed, two exploratory MANOVAs were used to examine the relationship between risk status and the identified five factors both at Year 1 and Year 2.

The Year 1 MANOVAs revealed no significant main effects of group or a group × factor interaction on the factors at the preschool stage. The means for the Year 1 five-factor scores by risk group are presented in Table 3.4. Although there are not significant group differences on factor scores, it is important to recall there were Year 1 risk group differences on individual tests, one of which was excluded from the factor analysis because of cross-loading (initial consonant different). Also, the rapid alternating stimulus task was not included in the factor analyses.

The Year 2 MANOVA involving the factors revealed a significant main effect for risk status ($F = 6.84$, $p < .05$), but no interaction between risk status and factor scores. These results indicate that there is a level of performance difference between the risk groups. The means for the five Year 2 factor scores by risk group are presented in Table 3.4. Subsequent post hoc analyses revealed significant differences between risk groups on the blending, $F(1, 113) = 4.42$, $p < .05$, and phoneme awareness factors, $F(1, 113) = 5.87$, $p < .05$.

Predictive Analyses

Regression analyses were used to predict the Year 2 reading outcome measure (Woodcock-Johnson letter-word identification) by risk group from Year 1 variables for nonreaders only. Children who were able to read at Year 1 (4 low risk, 7 high risk) were dropped from this analysis because they might have already developed the skills that would predict the nonreaders outcome at Year 2; thus, potentially, they could introduce misleading information into the analysis. This analysis included only children who were not reading at Year 1.

There were no significant differences between the risk groups on the letter-word identification subtest at Year 2 (entire sample: high risk: $M = 96.71$, $SD = 14.2$; low risk: $M = 100.63$, $SD = 17.2$; nonreaders only: high risk: $M = 93.8$, $SD = 12.2$; low risk: $M = 97.7$, $SD = 13.8$). To achieve these mean standard scores, a child at this age would have to correctly name the nine letters on the test and correctly name two or three words. Thus, reading proficiency at Year 2 (before first grade) was still quite rudimentary for both groups and the variance in actual reading skill is restricted. Therefore, these predictive analyses are preliminary.

Although level of reading performance was not different, predictors of reading performance could still differ. To examine this possibility, separate stepwise regression analyses were performed for high-risk and low-risk children, allowing all factors identified at Year 1 to enter the equation along with variables that

TABLE 3.5
Stepwise Multiple Regression Summary: Analysis 1, Nonreaders Only

Low Risk Group[a]

Dependent variable: Woodcock-Johnson letter-word identification subtest

Step & independent variable	R^2	Change in R^2	Significance of Change in R^2
1. RAS time	.25	.25	.007
2. Phoneme percep	.42	.17	.01
3. Blending factor	.54	.12	.02

High Risk Group[b]

Dependent variable: Woodcock-Johnson letter-word identification subtest
Step & independent variable: No variables entered equation before .05 limits reached.

[a]Year 1 variables not in equation: ran time, rhyming factor, onset-rime factor, initial consonant different, VSTM, EIQ1, letter knowledge.

[b]Year 1 variables not in the equation: RAN time, RAS time, factors 1–5, initial consonant different, estimated IQ, letter knowledge.

for one reason or another (skewness, crossloading, etc.) were excluded from the Year 1 factor analysis. The results of these analyses are contained in Table 3.5. There were three predictors (rapid alternating stimulus response time, phoneme perception, and the blending factor) of the low-risk outcome, but there were no significant predictors in the stepwise regression for the high-risk group.

A second predictive analysis including all subjects, readers and nonreaders, was also performed on the same data (Table 3.6). In this analysis, six Year 1 predictors (verbal short-term memory factor, blending factor, phoneme perception factor, rapid alternating stimulus response time, estimated Year 1 IQ, and letter knowledge) accounted for 81.3% of the variance in the Year 2 reading outcome measure for the low-risk sample. For the high-risk sample, there was only one significant predictor (initial consonant different) accounting for 18% of the variance.

The specific variables predicting the low-risk cohort's Year 2 reading outcome were nonoverlapping with those predicting the high-risk cohort's outcome, regardless of reading status. A single phoneme awareness variable (initial consonant different) predicted the high-risk outcome, but a variety of predictors predicted the low-risk outcome.

It is important to assess the stability of the predictor variables from Year 1 to Year 2. The correlation for the response time of the RAS (.73, $p < .001$); for EIQ (.50, $p < .001$); for letter knowledge (.59, $p < .001$); for the blending factor (.28, $p < .01$); for the phoneme perception factor (.02, NS); and for the VSTM factor (−.09, NS). All but two of the predictor variables had reasonable stability.

TABLE 3.6
Stepwise Multiple Regression Summary: Analysis 2, Entire Sample

Low Risk Group[a]

Dependent variable: Woodcock-Johnson letter-word identification subtest

Step & independent variable	R^2	Change in R^2	Significance of Change in R^2
1. VSTM	.38	.38	.0002
2. Blending factor	.58	.20	.001
3. Phoneme percep	.67	.09	.01
4. RAS time	.72	.05	.04
5. EIQ1	.78	.05	.02
6. Letter knowledge	.81	.04	.04

High Risk Group[b]

Dependent variable: Woodcock-Johnson letter-word identification subtest

Step & independent variable	R^2	Change in R^2	Significance of Change in R^2
1. Initial consonant different	.18	.18	.03

[a]Year 1 variables not in equation: ran time, rhyming factor, onset-rime factor, initial consonant different.

[b]Year 1 variables not in the equation: RAN time, RAS time, factors 1–5, estimated IQ, letter knowledge.

DISCUSSION

This chapter presents the results of the first 2 years of a 3-year longitudinal study. The specific questions addressed are:

1. Do the risk groups differ in phonological skills at either year?
2. Is the factor structure of the phonological processing skills battery stable across these 2 years?
3. Are the predictors of early reading skill similar or different across the two risk groups?

In addition to these specific questions, two of the more general hypotheses of the study as a whole can be addressed in a preliminary way:

Hypothesis 1: After controlling for estimated IQ, most of the significant predictive variance for later word recognition skill will be captured by phoneme awareness, and little unique variance will be contributed by the other phonological processing skills.

Hypothesis 2: The precursors of reading skill and difficulty will be similar in both the high-risk and normal-risk cohorts.

Question 1 was addressed at two levels: global and specific. The global measure of spoken language differences between groups was assessed by documenting the incidence of early speech delays that were serious enough to warrant speech therapy prior to kindergarten, and the specific measures were addressed by examination of differences between risk groups on the spoken language precursors to reading ability/disability (phoneme perception, lexical access, verbal short-term memory, and phoneme awareness). The results of these analyses indicate that there were spoken language differences between the two groups at both the global and specific levels. At the global level, there was a significantly different incidence of preschool speech therapy in the high-risk group. The results of the specific spoken language skills indicated that there were detectable differences between the high-risk and low-risk groups in phoneme awareness, letter knowledge, and speed of lexical access, but that there were no discernible differences between groups on phoneme perception and verbal short-term memory.

The results pertinent to question 2 provide evidence that the factor structure obtained was similar and stable across the preschool and kindergarten years of the study and are consistent with other similar research. At Year 2, the analysis of the mean factor scores for the two groups revealed significant differences on two factors: blending and a general phoneme awareness factor. Both of these factors are considered to measure phoneme awareness.

The results pertinent to question 3 indicate that the predictor variables for the outcome reading measure are different for the high-risk and low-risk groups regardless of reading status. In the analysis of the entire sample, there was only one Year 1 predictor of Year 2 reading outcome in the high-risk group, but the predictors for the low-risk group were from six different factors and/or variables. In the analysis of the nonreader subsample, there were no Year 1 predictors of Year 2 reading outcome in the high-risk group, but the predictors for the low-risk group were from three different groups of phonological processing skills (lexical access [RAS], phoneme perception, and phoneme awareness [blending]).

We can think of two hypotheses to explain why considerably less reading outcome variance at Year 2 was predicted for the larger high-risk group. One is simply that there was less outcome variance in that group. We can reject this possibility because the means and variances for the two risk groups on Year 2 reading scores were quite similar, and neither was skewed or kurtotic. The variance in both groups is close to that of the norming sample for the test (i.e., $SD = 15$). The second hypothesis is that the high-risk group is composed of two different subgroups, one predyslexic and one not, and that the predictors of reading skill differ between subgroups. Therefore, this limits overall prediction for that group. A clear test of this hypothesis will only be possible after Year 4, when we can subdivide the high-risk group into dyslexics and nondyslexics.

It is quite striking that our IQ measure did not account for much variance in reading outcome in either group and that the vocabulary measure was not a significant predictor in either group. This result suggests that phonological skill is much more closely related to early reading (i.e., word recognition) than is lexical or semantic development. Thus, not all of spoken language development is equally important for predicting at least this aspect of written language development.

The results of this initial study address two of our original five study hypotheses. The first hypothesis addressed was that after controlling for IQ, most of the significant predictive variance for later word recognition skill would be captured by phoneme awareness, with little unique variance contributed by the other phonological processing skills. This hypothesis was not supported in the low-risk group, but it was supported in one of the analyses in the high-risk group. Although Year 1 phoneme awareness tasks did not account for the majority of the variance in Year 2 word recognition, they did account for a significant amount. The Year 1 blending factor accounted for a significant amount of variance for the low-risk sample, regardless of reading status and, in the analysis involving both the readers and nonreaders, the initial consonant-different task was the only predictor variable for the high-risk sample. These results are from only the first 2 years of the study, and we cannot be certain that the analyses of the entire project will result in the same conclusions.

The second hypothesis addressed was that the precursors of reading skill and difficulty would be similar in both the high-risk and low-risk cohorts. This hypothesis is not supported by these data. So far, precursors appear to vary by risk group, regardless of preschool reading status.

Implications for Dyslexia and Related Syndromes

These results are consistent with Scarborough's (1989) in indicating that there are spoken language differences in children at family risk for dyslexia in the preschool period, before formal literacy training begins. It appears that the linguistic phenotype in dyslexia is not restricted to written language and its developmental onset is not restricted to the early school years. We do not know how much continuity there is in that linguistic phenotype across the preschool years because our study and Scarborough's (1989) used different measures. In particular, we do not know if the linguistic phenotype is first expressed in the domain of syntax and then in the domain of phonology (heterotypic continuity), or if there are phonological differences throughout the preschool period (homotypic continuity), with additional syntactic differences early on, which later disappear. Answering these questions would have important implications for how we think about the developmental relations among domains of language competence that are ordinarily thought of as quite separate.

Wagner and Torgesen (1992) contended that the factor structures underlying phonological processing abilities are as stable as other cognitive abilities across time. This is important because only stable, enduring cognitive abilities can be evaluated for possible causal roles in important skills like reading. The results of the present study are consistent with Wagner and Torgesen's results. Analyses are under way to examine the factor structure underlying each risk group separately. These analyses will reveal whether or not the same structure of cognitive abilities underlies both risk groups. This is important because it may help to answer the question whether or not dyslexia is a distinct disorder.

Other future analyses planned for the data from the remaining years of the study will provide us the opportunity to examine the whole data set in a more exhaustive manner. Importantly, at the end of the last year we will have made diagnostic assessments of all the children and will be able to identify which children have become dyslexic and which have not. Then we will be in a position to look at which factors or variables predict dyslexia, and which predict normal reading outcome, and whether the predictors over the entire developmental course are the same or different for each group. We will also be able to evaluate whether children from high-risk families who do not become dyslexic, nonetheless, have a subtle phonological processing problem. The results of these analyses have implications for our understanding of what is being transmitted in dyslexic families: Is it a discrete or a continuous risk factor? What is the correlation between the risk factor and dyslexia itself? In other developmental disorders (e.g., schizophrenia), the familial risk factor appears to be more prevalent than the disorder itself, and etiological models, therefore, hypothesize the necessity of an environmental second hit. In contrast, the risk factor in dyslexia could be discrete and highly correlated with the disorder itself (Pennington, 1995).

In addition to being in a position to examine the predictors of reading ability and disability, future reports from this study will include analyses involving the potential reciprocal interaction between phoneme awareness and reading.

Limitations

The sample used in this study is subject to the issues inherent in longitudinal analyses and self-selected samples. Because we wanted to keep our attrition rate as low as possible, we made every effort to recruit subjects who would be willing to remain in a longitudinal study over a 3-year period. The result is that the generalizability of the data presented in this chapter may be limited to similar samples.

In addition, two of the variables used in the predictive analyses (phoneme perception and VSTM) had low stability. In spite of these limitations, this study adds important information to our knowledge of reading development. It offers the opportunity to follow the development of all four phonological processing domains longitudinally in both high-risk and low-risk cohorts. In this way, we can contribute to understanding the links between abnormal speech and language development and abnormal reading.

ACKNOWLEDGMENTS

The research reported here was supported by National Institute of Mental Health (NIMH) Research Scientist Development Award MH00419, by NIMH MERIT Award MH38820, by the March of Dimes Birth Defects Foundation Grant 12-FY93-0725, and by a grant from the Orton Dyslexia Society.

An earlier version of this chapter was presented to the National Conference of the Orton Dyslexia Society, New Orleans, LA, November 6, 1993.

APPENDIX: PHONOLOGICAL MEASURES

Phoneme Perception

This task has been adapted from the phoneme perception task used by Brady, Shankweiler, and Mann (1983). In their study, high- and low-frequency words were presented in quiet and noise for repetition. This task has high frequency words ($N = 24$) and nonwords ($N = 24$). Subjects hear a mixed list of real and nonwords randomly presented in either noise or quiet. The subject's task is to repeat each word as accurately as possible.

Lexical Retrieval

The RAN as used by Wolf, Bally, and Morris (1986) consisted of four different stimulus sets presented on continuous naming charts of 50 stimulus items presented in 10 rows of 5 items each. The four different stimulus sets are high-frequency lowercase letters (a, d, o, s, p), digits (2, 4, 6, 7, 9), high-frequency colors (red, yellow, green, blue, black) and line drawings of common objects (hand, chair, dog, star, ball). The stimuli in each set are presented 10 times in random sequence on a chart. The subjects' task is to name all the stimuli on a chart as rapidly as possible in a left to right, top to bottom order. Before the task begins, each subject will be tested for basic recognition of each stimulus. It is anticipated that some subjects may not know all the letter or number names; these subjects will be excluded from analyses of those tasks. Both latency and accuracy data will be recorded.

The RAS consists of two different stimulus sets presented on continuous naming charts of 50 stimulus items presented in 10 rows of 5 items each. The two different stimulus sets are (a) colors, numbers and lower-case letters, and (b) numbers and lower-case letters. The administration of this task is identical to that of the RAN, seen earlier.

Verbal Short-Term Memory

Three kinds of word lists, with 10 trials in each list, are used: real words, nonwords, and rhyming real words. The latter two conditions permit an evaluation of phonetic coding in short-term memory in different ways. The performance decrement caused by rhyming words compared with nonrhyming real words has typically been used as a measure of the use of phonetic coding, at least in young children. Nonwords, because they are unfamiliar and lack semantic context, depend solely on phonetic coding for maintenance in working memory, so the performance difference between real words and nonwords also provides a measure of phonetic coding. Finally, the magnitude of the recency effect in all three list types also provides a measure of phonetic coding. All of the word lists have half the trials consisting of three items, and half four items, in length. Subjects hear the word lists on a tape recorder presented at a rate of one word per second and are asked to repeat the list in order.

Phoneme Awareness

The Roswell-Chall test of auditory blending (1986) requires the subject to listen to the experimenter present sounds (e.g., "s-ing, a-t, de-sk") and tell what the whole word is. The test is divided into parts: (a) two-phoneme words such as "a-t" and "i-f"; (b) words broken down into segments such as "st-ep" and "c-ake"; and (c) single syllable words broken down into three phonemes such as "c-a-t" and "s-a-d."

Bradley and Bryant's (1983) sound categorization test was an oddity task. After careful instruction in attending to the rhyming sounds in words, subjects were asked to identify which word does not sound like the others in a string of three or four depending on the form used (e.g., hat, man, fat or pin, win, sit, fin). There are 30 trials total, 10 each focusing on the initial, medial, and final sounds, respectively, in CVC words. The initial condition taps knowledge of alliteration, which does involve awareness of individual phonemes. Normative data for English children has already been collected by Bradley and Bryant (1983).

The Treiman and Zukowski onset-rime task requires a child to identify whether words have similar sounds in them. In the onset condition the beginning syllable sounds are alike (e.g., sleep-slow). In the rime condition the ending syllable sounds are alike (e.g., crunch-bunch). The subject is randomly presented with 10 items in the onset condition, 10 items in the rime condition, and 20 items in which no sounds are alike. Treiman and Zukowski's (1988) research suggested rime awareness precedes onset awareness. Since neither onsets nor rimes require awareness of individual phonemes, this task may be very helpful in measuring the early stages of phoneme awareness.

The initial consonant different task (Stanovich, Cunningham, & Cramer, 1984) was similar to Bradley and Bryant's oddity task except that it focused on the initial consonant only and the words do not rhyme (e.g., boot, barn, buy, ear). Thus, every item requires awareness of individual phonemes. The child is trained in listening to the first sound in a word and choosing in which one of four words the initial consonant is different.

The supply initial consonant task and the strip initial consonant task (Stanovich et al., 1984) have been added to the battery for the 6-year-old participants. The supply initial consonant task required the subject to listen to two words and to figure out the sound that is missing in the second that is present in the first (e.g., "cat-at": What is missing from "at" that you hear in "cat"?). Again, training is provided so that the child clearly understands the task. The strip initial consonant task required the child to segment off the first phoneme of a word presented by the examiner, and tell what word is left (e.g., Listen to the word "feet". If you take away the /f/ sound, what word is left?).

REFERENCES

Bradley, L., & Bryant, P. E. (1983). Categorizing sounds and learning to read—a causal connection. *Nature, 301*, 419–421.

Brady, S., Shankweiler, D., & Mann, V. A. (1983). Speech perception and memory coding in relation to reading ability. *Journal of Experimental Child Psychology, 35*, 345–367.

Bryant, P. E., & Bradley, L. (1985). *Children's reading problems*. New York: Basil Blackwell.

Bryant, P. E., Maclean, M., Bradley, L., & Crossland, J. (1990). Rhyme and alliteration, phoneme detection, and learning to read. *Developmental Psychology, 26*, 429–438.

Byrne, B., & Fielding-Barnsley, R. (1993). Evaluation of a program to teach phonemic awareness to young children: A 1-year follow-up. *Journal of Educational Psychology, 85*, 104–111.

Calfee, R. C., Lindamood, P., & Lindamood, C. (1973). Acoustic-phonetic skills and reading—Kindergarten through twelfth grade. *Journal of Educational Psychology, 64*, 293–298.

Catts, H. W. (1993). The relationship between speech-language impairments and reading disabilities. *Journal of Speech and Hearing Research, 36*, 948–958.

Denckla, M. B., & Rudel, R. (1976). Rapid "automatized" naming (R.A.N.): Dyslexia differentiated from other learning disabilities. *Neuropsychologia, 14*, 471–479.

Finucci, J. M., Isaacs, S., Whitehouse, C., & Childs, B. (1982). Empirical validation of reading and spelling quotients. *Developmental Medicine and Child Neurology, 24*, 733–744.

Henderson, L. (1982). *Orthography and word recognition in reading*. London: Academic Press.

Hollingshead, A. B. (1975). *Four factor index of social status*. New Haven: Yale University.

Jorm, A. F., & Share, D. L. (1983). Phonological coding and reading acquisition. *Applied Psycholinguistics, 4*, 103–147.

Lefly, D. L., & Pennington, B. F. (in preparation). *Reliability and validity study of an adult reading history questionnaire*.

Liberman, A. M. (1991). Observations from the sidelines. In B. F. Pennington (Ed.), *Reading disabilities: Genetic and neurological influences* (pp. 241–245). Dordrecht, The Netherlands: Kluwer Academic Publishers.

Liberman, I. Y., Shankweiler, D., Fischer, F. W., & Carter, B. (1974). Explicit syllable and phoneme segmentation in the young child. *Journal of Experimental Psychology, 18*, 201–212.

Lundberg, I., Olofsson, A., & Wall, S. (1980). Reading and spelling skills in the first school years predicted from phonemic awareness skills in kindergarten. *Scandinavian Journal of Psychology, 21*, 159–173.

Maclean, M., Bryant, P., & Bradley, L. (1987). Rhymes, nursery rhymes and reading in early childhood. *Merrill-Palmer Quarterly, 33,* 255–281.

Mann, V. A. (1984). Longitudinal predictions and prevention of early reading difficulty. *Annals of Dyslexia, 34,* 117–136.

Mann, V. A. (1993). Phoneme awareness and future reading ability. *Journal of Learning Disabilities, 26,* 259–269.

Mann, V. A., & Ditunno, R. (1990). Phonological deficiencies: Effective predictors of future reading problems. In G. Pavlides (Ed.), *Dyslexia: A neuropsychological and learning perspective* (pp. 105–131). New York: Wiley.

Pennington, B. F. (1990). The genetics of dyslexia. *Journal of Child Psychology and Psychiatry, 31,* 193–201.

Pennington, B. F. (1995). Genetics of learning disablities. *Journal of Child Neurology, 10,* 69–77.

Pennington, B. F., Gilger, J. W., Pauls, D., Smith, S. A., Smith, S. D., & DeFries, J. C. (1991). Evidence for major gene transmission of developmental dyslexia. *Journal of the American Medical Association, 266,* 1527–1534.

Pennington, B. F., & Smith, S. D. (1988). Genetic Influences on learning disabilities: An update. *Journal of Consulting and Clinical Psychology, 56,* 817–823.

Renick, M. J., & Harter, S. (1989). Impact of social comparisons on the developing self-perceptions on learning disabled students. *Journal of Educational Psychology, 81,* 631–657.

Roswell-Chall Test of Auditory Blending. (1986). La Jolla, CA: Essay Press.

Sattler, J. M. (1992). *Assessment of children.* San Diego: Jerome M. Sattler, Inc.

Sawyer, D. J. (1992). Language abilities, reading acquisition, and developmental dyslexia: A discussion of hypothetical and observed relationships. *Journal of Learning Disabilities, 25,* 82–95.

Scarborough, H. (1989). Prediction of reading disability from familial and individual differences. *Journal of Educational Psychology, 81,* 101–110.

Scarborough, H. (1990). Very early language deficits in dyslexic children. *Child Development, 61,* 1728–1743.

Scarborough, H. (1991). Antecendents to reading disability. In B. F. Pennington (Ed.), *Reading disabilities: Genetic and neurological influences* (pp. 31–45). Dordrecht, The Netherlands: Kluwer Academic Publishers.

Stanovich, K., Cunningham, A., & Cramer, B. (1984). Assessing phonological awareness in kindergarten children: Issues of task comparability. *Journal of Experimental Psychology, 38,* 175–190.

Stanovich, K. E. (1986). Cognitive processes and the reading problems of learning-disabled children: Evaluating the assumption of specificity. In J. K. Torgesen & B. Y. L. Wong (Eds.), *Psychological and educational perspectives of learning disabilities* (pp. 87–131). New York: Academic Press.

Stuart, M., & Coltheart, M. (1988). Does reading develop in a sequence of stages? *Cognition, 30,* 139–181.

Torgesen, J. K., & Wagner, R. K. (1992). Language abilities, reading acquisition, and developmental dyslexia: Limitations and alternative views. *Journal of Learning Disabilities, 25,* 577–581.

Treiman, R., & Zukowski, A. (1988). Levels of phonological awareness. In S. Brady & D. Shankweiler (Eds.), *Phonological processes in literacy* (pp. 67–83). Hillsdale, NJ: Lawrence Erlbaum Associates.

Vogler, G. P., DeFries, J. C., & Decker, S. (1985). Family history as an indicator of risk for reading disability. *Journal of Learning Disabilities, 18,* 419–421.

Wagner, R. K., & Torgesen, J. K. (1987). The nature of phonological processing and its causal role in the acquisition of reading skills. *Psychological Bulletin, 101,* 192–212.

Wagner, R. K., Torgesen, J. K., Laughon, P., Simmons, K., & Rashotte, C. A. (1993). Development of young readers' phonological processing abilities. *Journal of Educational Psychology, 85,* 83–103.

Wallach, M., & Wallach, L. (1976). *Teaching all children to read.* Chicago: University of Chicago Press.

Wolf, M., Bally, H., & Morris, R. (1986). Automaticity, retrieval processes, and reading: A longitudinal study in average and impaired readers. *Child Development, 57,* 988–1000.

Yopp, H. K. (1988). The validity and reliability of phonemic awareness tests. *Reading Research Quarterly, 23,* 159–177.

4

How Can Behavioral Genetic Research Help Us Understand Language Development and Disorders?

Jeffrey W. Gilger
University of Kansas

OVERVIEW

Language-related disorders, such as specific language impairment (SLI), stuttering, and reading disability (RD) combined, may affect 10% to 20%, or more, of school-aged children in this country (Leske, 1981; Pennington, 1991a). Therefore, our grasp of how genes and environment work together to lead to deviations in language development is critical. Admittedly, however, our knowledge about the specific role that genes play in most complex neurocognitive disorders like SLI is limited, short of some very basic information applicable to genetic counseling and genetically special populations (Plumridge, Bennett, Dinno, & Branson, 1993; Roberts & Pembrey, 1985; Siegel-Sadewitz & Shprintzen, 1982; Sparks, 1984).[1]

Our understanding of the genetic and nongenetic mechanisms in normal language development is also relevant to several key theoretical issues in this area that need to be resolved (e.g., Bates, Bretherton, Beeghly-Smith, & McNew, 1982; Pinker, 1991, 1994). What aspects, for example, of the normally developing language system are under environmental control or are more or less teachable?

---

[1]In terms of abnormalities in speech and language, this chapter is primarily concerned with the etiology of nonsyndromic or developmental disorders. Space will not be devoted to a discussion of clinical genetic syndromes and their concomitant communication problems (see Carey, Stevens, & Haskins, 1992; Plumridge et al., 1993; Siegel-Sadewitz & Shprintzen, 1982; Sparks, 1984; Stoel-Gammon, 1990).

Which aspects are species wide, and which are governed more or less by innate mechanisms?

Several authors from a number of perspectives come together in this volume to address the role that genetics may have in normal and abnormal language processes and to provide some suggestions for future research in the area of the genetics of language. Our knowledge about genes and how they express themselves in complex human traits has literally exploded over the last 10 years. For instance, of the 50,000 to 100,000 or so genes estimated to comprise the human genome, some 30% appear to be expressed primarily in the brain; the remaining portion is expressed throughout the central nervous system (CNS) and elsewhere (Adams et al., 1991). Consequently, there can be no doubt that genes play a major role in almost every behavior requiring CNS processes.

There are several general aims of this chapter. The first is to provide the reader with some important concepts about genetic processes in human development—concepts that the reader may find useful in the laboratory or clinic when thinking about how a child's language system develops. Many of the key words and definitions are summarized in the Appendix accompanying this chapter. The second aim is to illustrate how behavioral genetic research can be applied to issues related to language. Primarily because of the shortage of behavioral genetic research on language development, examples from studies on other complex human traits will be used to satisfy this aim, including data on general intelligence and reading disability.[2] Finally, this chapter aims to present my opinion about what needs to be done and what can be done in the study of language if the interest is in genetic and nongenetic mechanisms. This will include suggestions for future research questions, research designs, and collaborations. It is important to note, however, that this limited chapter cannot provide detailed statistical and methodological information. References that address these topics in more detail, are given throughout this chapter (e.g., Falconer, 1981; Morton, 1982; Neale & Cardon, 1992; Plomin, DeFries, & McClearn, 1990; Smith, 1992).

BEHAVIORAL GENETICS AS AN INTERDISCIPLINARY APPROACH TO THE STUDY OF LANGUAGE

Human behavioral genetics can be defined in a variety of ways (Plomin et al., 1990). As behavioral geneticists, we are simply interested in how genes and environment combine to give us human behavior (be it normal or abnormal).

---

[2]Conveniently, the neurocognitive systems subserving language are probably closely tied to those for general intelligence, and, in the case of abnormal language skills, reading disability as well, making the use of IQ and reading disability as language analogues reasonable (Catts, 1989a, 1989b; Pennington, 1991a).

Because a behavior can be most anything measurable (e.g., pulse, hair growth, nonverbal intelligence, delinquency, neurochemical production, fine motor coordination, language, Tay-Sachs disease, and so on), behavioral genetics can at times bridge the traditionally distinct domains of the behavioral sciences, as well as the medical, psychiatric, epidemiological, molecular, and quantitative genetics (e.g., Fuller & Simmel, 1983; Plomin et al., 1990).

Because of its interdisciplinary nature, human behavioral genetics may be the best scientific approach to studying some of the perplexing questions pertaining to complex human traits like language development. Unfortunately, there are relatively few formal studies that use behavioral genetic techniques with the specific goal being the elucidation of the genetic and nongenetic factors in the development of the human language system.[3] Although there are a growing number of family and twin studies on abnormal speech-language processes (e.g., SLI), there are only a limited number of very basic studies on the genetic and environmental parameters of traits related to normal language development (see Felsenfeld, 1994; Hardy-Brown, 1983; Hardy-Brown & Plomin, 1985; Hardy-Brown, Plomin, & DeFries, 1981; Julian, Plomin, Braungart, Fulker, & DeFries, 1992; Locke & Mather, 1989; Mather & Black, 1984; Reznick, in press). In part due to these studies, there are several pieces of information that support the idea that there are genetic effects on language, even beyond the implications of the obvious genetic and biological complexity of the CNS. One such piece of datum is the clear and relatively invariant maturational patterns that appear to occur in the course of language development (Chomsky, 1988; Pinker & Bloom, 1990; Studdert-Kennedy, 1990). Such patterns suggest a biologically (genetic) preprogrammed pathway to mature language. Second, twin and adoption studies have demonstrated significant heritability for a variety of cognitive traits related to speech and language development (DeFries & Plomin, 1983; Hardy-Brown & Plomin, 1985; Locke & Mather, 1989; Munsinger & Douglass, 1976; Nichols, 1978; Plomin et al., 1990; Scarr & Carter-Saltzman, 1983). For example, receptive and expressive vocabulary skills yield average heritability estimates of approximately .50 to .60, indicating that roughly 50% of the variance in vocabulary ability is due to genetics. Third, familiality for language-related problems like SLI has been shown (Gilger, 1992a; Gilger, Pennington, & DeFries, 1991; Lewis, 1992; Lewis, Ekelman, & Aram, 1989; Tallal, Ross, & Curtis, 1989; Tomblin, 1989, this volume). Although familiality for a trait does not, by itself, indicate the effects of genes, the consistency and patterns of familial aggregation for language disorders make genetic effects highly suspect. And finally, language impairments are known to occur in individuals with an assortment of known genetic and

---

[3]Ethological, cross-cultural, and observational studies that show evidence suggestive of a biological basis for some aspects of communication are not included under the rubric of behavioral genetic methodologies (see Chomsky, 1988; Pinker, 1991, 1994; Pinker & Bloom, 1990; Studdert-Kennedy, 1990).

chromosomal abnormalities, again suggesting that gene-influenced forms of SLI and language parameters in general may exist (Bender, Linden, & Robinson, 1991; Plumridge et al., 1993; Roberts & Pembrey, 1985; Siegel-Sadewitz & Shprintzen, 1982; Sparks, 1984; see also Miller, this volume).

IMPORTANT QUESTIONS IN LANGUAGE RESEARCH

Why conduct behavioral genetic research in the area of language? What are some of the unique questions that traditional experimental designs are unable to answer? These are legitimate questions, and I touch on them in this section. Five questions well suited to behavioral genetic methodologies are listed in Table 4.1.

Questions 1 and 2 (i.e., what are the etiologies of normal and abnormal language abilities?) fall into the category of *simple etiology*. Such questions in behavioral genetics are typically addressed by partitioning trait variances into those portions due to genetic effects and those portions due to environmental effects. We know, for example, that roughly 50% of the postadolescent variance in standard IQ tests is due to genetic differences among people (Plomin et al., 1990; Scarr & Carter-Saltzman, 1983). Similarly, as mentioned earlier, some 50% of the range in vocabulary ability is also due to the fact that people differ in their genes (Plomin et al., 1990; Scarr & Carter-Saltzman, 1983). The remaining 50% of variance in these two traits is reflective of environmental differences among people. This partitioning of variability into genetic (heritability or h^2) and nongenetic (environmentality or e^2) components is made possible by behavioral genetic studies that compare the resemblance among twins and families for these traits and, until recently, this has been the most common type of behavioral genetic research.

Behavioral genetics can also explore a variety of more complicated and theoretically driven questions relevant to complex human traits.[4] This idea of applying behavioral genetic methods to the study of specific, theoretically based hypotheses, rather than the simple question of the degree of heritability for a trait is, fortunately, increasing in frequency. However, this area is wide open still, especially for behavioral genetic tests of basic cognitive and linguistic theories. Questions 3, 4, and 5 of Table 4.1 are examples of complex types of questions in language-related research.

Question 3 concerns the etiological relationship between normal and abnormal language development. There has been some debate over whether or not

---

[4]For this chapter, complex traits refer to traits that vary in their degree or form, and are not easily fit into simple Mendelian models (Smith, 1992; Brzustowicz, this volume). These traits are often measured on a quantitative or continuous scale, and have traditionally been assumed to reflect multifactorial-polygenic effects.

TABLE 4.1
Examples of Research Questions Relevant to
Language and Language-Related Disorders

---

1. What is the etiology of language in the normal range?
2. What is the etiology of language outside the normal range?
3. How do the etiologies of normal and abnormal language relate?
4. What are the phenotypic boundaries of speech-language impairments?
5. What are the parameters of treatment, assessment, and prognosis for speech-language impairments?

SLI, for example, is part of the normal continuum of language skills, or whether it represents part of a separate language distribution with unique etiology (Johnston, 1991; Leonard, 1991; Tomblin, 1991). Similar questions have been raised regarding RD, schizophrenia, gifted and low IQ groups, and other traits (Achenbach, 1982; Cherny, Cardon, Fulker, & DeFries, 1992; Eaves, Eysenck, & Martin, 1989; Gilger, Borecki, Smith, DeFries, & Pennington, 1995; Penrose, 1963; Plomin & Rende, 1991). Behavioral geneticists have traditionally argued that the range of complex human traits like language is of multifactorial-polygenic origin (Eaves, Eysenck, & Martin, 1989; Falconer, 1981; Fulker & Eysenck, 1979; Plomin et al., 1990). That is, many small and additive genetic and nongenetic effects are responsible for individual differences in a trait like language, including the parts of the continuum containing the language disabled and superior. The idea that some major single gene effect (separate from the gene effects contributing to normal variation) could be contributing to abnormality in traits like language, or personality, originally seemed unlikely, although it was theoretically possible (see Falconer, 1981; Plomin, 1990; Plomin & Rende, 1991).

Genetic theory and its accompanying statistical models is particularly well suited to testing the unique etiology hypothesis (Gilger et al., 1995; Plomin & Rende, 1991). Although focused behavioral genetic research has not been applied to the issue of unique etiology for SLI and language (but see Tomblin, 1991, and Tomblin & Buckwalter, 1994), it has been to some extent for RD. In this case, the current data suggest that RD is not due to a single major gene separate from the normal genes effecting reading (i.e., a disease gene), nor is it solely due to many small and additive genetic and nongenetic factors. Rather, the largest contributors to the likelihood of expressing a reading disability may be a limited number of genes, each with differing, yet significant major contributions to variance in reading ability. These genes, in conjunction with intra- and extra-uterine environmental risk factors and other small and additive multifactorial-polygenic effects tend to push an individual towards, and then over, the threshold for expressing RD (DeFries & Gillis, 1993; Gilger et al., 1995). The genetic system in these types of situations is sometimes referred to as *oligogenic* or as *quantitative trait loci* (Gelderman, 1975; Greenberg, 1993).

It is noteworthy that genetic research into RD often assumed a single disease gene model, a position that met with great argument by classic behavioral geneticists because of their view that complex traits like reading are multifactorial-polygenic in origin. Thus, today's understanding of RD is, in a sense, a compromise between these two early schools of thought, made possible by advances in our understanding of genetic mechanisms for other disorders and other organisms. That the number of key RD genes may be limited makes the potential of molecular genetic research to identify these genes promising. So, if there are five major genes, that together account for 60% of the liability to RD, identifying these genes would significantly improve diagnosis, prevention, and treatment (see Gilger et al., 1995; Pennington, 1991a; Pennington & Smith, this volume).

Question 4 in Table 4.1 (i.e., what are the phenotypic boundaries of speech-language impairments?) pertains to a variety of issues that abound in the field of clinical speech-language pathology that can be well addressed through behavioral genetic research. In reference to diagnosis, for instance, it is quite important to learn how to best identify SLI children and understand what behaviors should legitimately be classified as SLI-related. Should a nonverbal IQ-discrepancy score be used when identifying SLI children? Are traits correlated with SLI, such as reading difficulties or behavioral problems, simply alternate expressions of the same underlying neurodevelopmental etiology as SLI, or are they independent, causative, or secondary effects of the language disorder (Bishop & Edmundson, 1987; Bishop, North, & Donlan, 1995; Hall & Tomblin, 1977; Pennington, 1991a; Stevenson, Pennington, Gilger, DeFries, & Gillis, 1993)? When the two or more correlated traits are diseases or disorders, the situation is commonly referred to as *comorbidity*.

Again using RD as an illustration, behavioral genetic research has helped clarify two long-standing, complex questions about etiology, phenotypic boundaries, and the basis of correlated symptoms (see also chapters by Pennington & Smith, this volume): (a) To what extent is an IQ-based definition of RD etiologically distinct from an age-based definition? and (b) Given that reading and spelling are complex processes, what are the component processes of reading most susceptible to genetic influence? That is, which of the basic reading processes seem to be the key, genetically mediated, neurocognitive deficits in children with RD? Both of these questions have analogues in the area of SLI.

Using data from the Colorado Twin Study of Reading Disabilities, the first question was addressed by obtaining estimates of the extent to which IQ- and age-based definitions were heritable, and the extent to which the genes affecting one definition were related to the genes affecting the other definition (Pennington, Gilger, Olson, & DeFries, 1992). First, each subject was assessed with a variety of reading-related and IQ tests. Second, each subject was given two scores: a continuously varying score representing how much their reading performance deviated from their IQ expectancy, and another score indicating how well they read relative to age norms. Finally, statistics were used to assess the degree to which within-pair twin resemblance reflected the effects of genes and/or environments. Extrapolating these data to the population at large, suggests that both definitions are heritable, with nearly 50% of the difference between normal and poor readers on either of these reading ability indices being due to the effects of genes.[5] Furthermore, performance on both indices were highly correlated ($r = .93$). Namely, subjects that scored better than expected by IQ, also scored better than predicted by age, and vice-versa. Further analyses indicated that this correlation was primarily due to genes affecting performance on both indices simultaneously. Specifically, 92% of the covariance between the two diagnostic indices was explained by a common genetic origin. Thus, it appears that the neuropsychological and neurodevelopmental substrates and mechanisms tapped by age- or IQ-diagnostic methods have a common genetic etiology. Although still more valid and accurate diagnostic methods may later develop (e.g., finer grained analysis of reading skills), this finding implies that research on the genetics of RD need not be overly concerned about whether an age- or IQ-discrepant diagnostic scheme is applied in families, as both diagnoses seem to tap the same underlying genetic components. In terms of clinical application, this finding suggests that both diagnoses are similarly sensitive to the biological and neurocognitive deficits responsible for RD, a conclusion now well supported by standard experimental research conducted over the past two decades (Fletcher, 1992).

The second question was initially addressed by Olson and colleagues (Olson, Wise, Conners, & Rack, 1989; see also Stevenson, 1991). Using the Colorado twins, they tested whether a visual processing or a verbal processing deficit was at the root of genetically mediated RD. In this initial study, the results were quite clear and fit nicely into a dual-processing framework of reading development. Specifically, they found significant differences in the heritability for phonological versus orthographic coding for single word reading: Phonological coding was significantly heritable (group heritability of .46; see footnote 5) and orthographic coding was not. Moreover, the contribution of phonological coding was greater than the contribution of orthographic coding to the genetic basis of word recognition deficits in RD twins. Thus in these twins, the primary impairment and genetic etiology in reading disability appeared to lie in the domain of phonological processing of words that was somewhat independent of orthographic proc-

---

[5]The term h^2 is used in this chapter as it pertains to the sum of all possible genetic effects, including those of an additive and interactive nature. This is typically referred to as heritability in the broad sense, and is often symbolized h_b^2 although the symbol h^2 is used here (Falconer, 1981; Plomin et al., 1990). This type of heritability estimate is applied to the normal range of trait variance in populations. Sometimes group heritability is used. Group heritability is different from h^2, in that it provides an estimate of the extent to which mean differences between normal and selected extreme populations are due to genetic factors (see DeFries & Fulker, 1985; DeFries & Gillis, 1990).

essing. Such a basic problem in phonology may, in part, account for the common comorbidity between SLI and dyslexia (Catts, 1989a, 1989b; Pennington, 1991a; Tallal, Ross, & Curtis, 1989). However, in a more recent report based on a larger sample of the Colorado twins and different types of analyses, Olson and colleagues (Olson, Forsberg, & Wise, in press) show less disparate heritabilities for orthographic and phonological abilities than they did at first. Furthermore, the orthographic and phonological skills may contribute more equally to the inheritance of deficits in word recognition (i.e., they may be less independent than predicted by a dual-processing theory of reading ability).

Clearly, more research is needed in this area, but it is still important to bear in mind that hypotheses about the SLI-RD correlation/comorbidity and the cognitive process involved in the disorders can be tested using a twin design similar to those used by Olson and others (e.g., Bishop, North, & Donlan, in press; Olson et al., 1989; Olson et al., in press; Pennington et al., 1992; Stevenson, 1991). For instance, in a twin sample, we could perform essentially the same analyses as Pennington et al. (1992), only substitute reading and language ability measures for IQ- and age-discrepant diagnostic measures. Given certain assumptions, these analyses would start to address the etiological relationship between RD and SLI. Although these types of twin studies allow for a separation between genetic and environmental effects on the comorbidity observed, similar etiological questions can also be studied using nontwin family data, although genetic and environmental effects are not so easily separated (For examples of the family cosegregation and correlation method see Gilger, Pennington, Green, Smith, & Smith, 1992; Neale & Cardon, 1992; Pauls, Towbin, Leckman, Zahner, & Cohen, 1986.).

With only one or two exceptions (e.g., Bishop et al., 1995; Reznick, in press), family and twin studies of speech-language disorders in the literature have not included complex analyses and designs, particularly as they apply to the problem of phenotype (i.e., Question 4 of Table 4.1). Accurate and sensitive diagnostic schemes are critical for work on the simple genetics of traits like SLI (e.g., Table 4.1, Question 2; Pennington, 1986), and several interesting research projects are aimed at establishing a diagnostic method that is valid for a variety of age groups so that it can be used in family work (see Tomblin & Rice, this volume). The lesson here is that work on dyslexia shows that certain designs in behavioral genetics, such as those described previously, are also well suited to getting at the best diagnosis problem. Family and twin studies can first be used to find out what the best phenotype is for genetic study, and then that phenotype can be reapplied to a different set of subjects with a better probability of success (i.e., successful resolution of what is genetic about SLI, how it is transmitted, and perhaps, where the genes are located). This series of steps seems counter intuitive to most, but in reality, it holds promise and may be most efficient.

The final question of Table 4.1 is fairly self-explanatory, and closely related to the other four (i.e., what are the parameters of treatment, assessment, and prognosis for speech-language impairments?). It is chiefly concerned with the benefits of going beyond simply detailing h^2 and e^2 for language-related traits, to testing complex hypotheses relevant to treatment, assessment, and prognosis. Although classic experimental designs can be used to address hypotheses relevant to Question 5, behavioral genetics provides other extremely powerful ways to study these issues as well, even if the hypothesis of interest is not immediately concerned with genetics.

The specific hypotheses to be tested will depend on the theoretical perspective of the investigator. As noted previously, an example in the assessment domain may be the hypothesis that IQ-discrepant SLI children are etiologically and neuropsychologically different from children that only meet an age-based criterion. In terms of prognosis and treatment, a hypothesis might be that certain characteristics of the child's home (e.g., presence of SLI parents, birth order, socioeconomic class, parental education level, etc.) ameliorate or exacerbate an SLI condition.

COMMON MISCONCEPTIONS ABOUT HOW GENES OPERATE IN DEVELOPMENT

Our understanding about how genes operate in the development of complex human traits has dramatically improved since the 1970s (see footnote 4). However, there are several common misconceptions about genes, such as those listed in Table 4.2. These misconceptions may promote errors in clinical and research endeavors. The purpose of this section is to address these common misunderstandings about how genes and environment operate in development. Although there is some overlap, each of these misconceptions is listed in Table 4.2 and discussed separately below.

Traits Are All or None in Terms of Genetic Effects

One interpretation of the statement traits are "all or none" is that genes alone, or environment alone, control the development of a trait. Such a belief is often implicit in the straw man arguments of those that are pro-environment or pro-genetics (see the Snow and deViliers commentaries and chapters, this volume).

TABLE 4.2
Five Misconceptions About the Role of Genes in Complex Human Behaviors

---

1. Genes operate on traits in an "all or none" fashion.
2. The path from a gene to behavior is specific, basic, and simple.
3. A trait with a genetic component is immutable and unchanging.
4. A trait with a genetic component will show up early in a person's development.
5. The most important type of nongenetic influence operates to make members of the same family similar.

Note. Table adapted from Gilger (1995).

It is, however, false. More accurately stated, genes and environment contribute more or less to the variability observed for a trait. In some cases, such as height or finger ridge count, genetic effects typically outweigh environmental effects, and in other situations, as in religiosity or political affiliation, the reverse may be true (Eaves et al., 1989; Plomin et al., 1990; Truett, Eaves, Meyer, Heath, & Martin, 1992). In fact, it is hard to imagine any psychological trait that is purely environmental or purely genetic in origin. After all, the very basis for how everyone perceives and acts in life requires biologically influenced foundations at some level and environmental input on another level.

The concepts of heritability (h^2) and environmentality (e^2) are relevant to trait variability and the "all or none" misconception. Figure 4.1 shows a normal distribution that could apply to any continuously measured trait, such as IQ, expressive language skills, or certain dimensions of personality. This normal curve represents a distribution of values for a trait, and the spread of the curve represents the variance in the trait among people, or, individual differences for the trait of study. This trait variability (sometimes called *phenotypic variance* or V_p) in a population can be measured and is itself assumed to be due to two major factors: (a) variance due to genetic differences among people (V_g), and (b) variance due to environmental differences among people (V_e) (Falconer, 1981). Thus, $V_p = V_g + V_e$.

Heritability is a statistically standardized form of V_g and environmentality is the standardized form of V_e (see footnote 5). Namely, h^2 is the proportion of variance in a trait attributable to genetic effects and e^2 is the proportion of trait variance attributable to nongenetic factors (Falconer, 1981). Conveniently, the

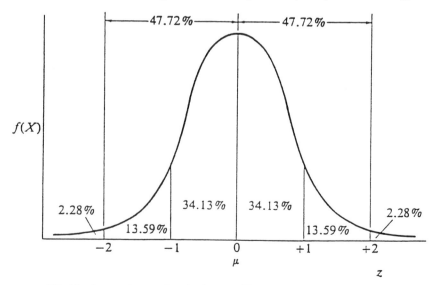

FIG. 4.1. A normal curve. Standard-scores (Z-scores) and the corresponding percent of curve area are shown.

sum of h^2 and e^2 (including gene and environment interactions) must account for all the phenotypic variance, and because they are standardized, estimates of h^2 and e^2 derived from the same sample and for the same trait will theoretically add up to 1.0 (Falconer, 1981). It is important to note that like any statistic, h^2 and e^2 are estimates of hypothetical true values, and they will tend to vary from sample to sample in response to measurement error and genetic and environmental differences in the samples (Falconer, 1981). Therefore, when h^2 of a trait is mentioned, it is not necessarily indicating a static all or none quantity or quality for a trait.

Two twin studies serve as good examples. Heritability estimates for similar Verbal Ability factors were obtained from a sample of 157 monozygotic (MZ) and 189 dizygotic (DZ) twin pairs by Bruun, Markkanen and Partanen (1966), and a sample of 337 MZ and 156 DZ pairs by Schoenfeldt (1968). In the former sample, an h^2 value of .50 was obtained, but the latter sample yielded an h^2 estimate of .64. On average, then, roughly 50% to 60% of the phenotypic variance for verbal ability reflected genetic differences among the people making up these samples, and the remaining 40% to 50% (i.e., 1.0–.50 or .60) of the phenotypic variance was because of different environments experienced by these people. Although it is obvious that people differ in their genes and environments, h^2 and e^2 apply only to those genes and environments directly or indirectly relevant to the development of the trait in question. If there are concerns about the basis for similarity and differences in verbal ability, for example, the fact that some people share the genes for brown hair is probably of little significance, but whether or not they experience the same teachers and teaching methods may be critical.

Although misconception 1 of Table 4.2 can be considered a general falsehood, in the case of some single gene medical disorders, the environmental effect may be so small that genes indeed appear to act in an all or none fashion (Roberts & Pembrey, 1985). For example, for the autosomal dominant disorder Huntington's chorea, simply having one copy of the deleterious gene is enough to produce the disorder 100% of the time (Gusella, 1991; Roberts & Pembrey, 1985). The gene for Huntington's chorea seems to be fully penetrant and most often expressed in mid-to-late adulthood as an ultimately fatal neurological disorder. It is likely that someday it will be fully understood how such disease genes work. By understanding the biochemical processes involved, environmental interventions can be applied that would modify the adverse effects of these genes, and thus, add environmental variance to a highly penetrant genetic trait. (This treatment effect also pertains to the immutability misconception discussed below.) Medicine has already had such successes, as in the case of hemophilia A and phenylketonuria (PKU), and work towards the treatment of many other genetic disorders is progressing (Martin, 1987; McKusick, 1990; Roberts & Pembrey, 1985). Classic PKU serves as a good illustration of this research.

PKU is an autosomal recessive disorder that can result in serious mental retardation and other physical afflictions due to the buildup of nonmetabolized phenylalanine and other metabolites in the blood and nervous system (Roberts

& Pembrey, 1985). There are actually several forms of PKU or related conditions, each due to different mutations in the phenylalanine hydroxylase gene (Roberts & Pembrey, 1985; Wood, Tyfield & Bidwell, 1993). By understanding how the defective gene modifies the normal metabolism of phenylalanine and how the buildup of nonmetabolized products could be reduced by manipulating the child's diet to lessen the intake of foods rich in phenylalanine, scientists added environmental variance to what was once thought to be a fairly penetrant genetic trait. Diet and chemical manipulations have become standard treatment for newborns bearing the genetic complement for PKU, and people with the PKU genetic complement are now leading relatively healthy and normal lives. (However, recent research suggests that the timing and maintenance of dietary constraints is more crucial than we once thought: It appears that if treatment is not initiated quite soon after birth and maintained indefinitely, reductions in IQ can result.) (Legido, Tonges, & Carter, 1993; Thompson et al., 1991).

It is noteworthy that the general principles about how genes and environment contribute to phenotypic variance in continuous or quantitative traits also apply to noncontinuous or qualitative traits (i.e., traits commonly measured in terms of distinct classes, such as a dichotomous yes/no, or presence/absence indicator for a disease; see footnote 5). In the qualitative case, an underlying trait continuity and a threshold for trait expression is assumed (see Falconer, 1981; Neale & Cardon, 1992; Plomin & Rende, 1991). Threshold or qualitative traits are common phenotypes when the concern is extremes in normal behaviors as with SLI, RD, and psychopathology.

The Path From Genes to a Behavior
Is Specific, Basic, and Simple

The heritability for a trait merely documents evidence for gene action and does not tell us how the genes are operating in development. Rarely do genes affect behavior in direct and simple ways. Understanding how genes work in development can lead to treatment advances in diseases.

The complexity of this process is captured by the concept of the gene-behavior pathway as illustrated in Fig. 4.2 (Plomin, 1981; Plomin et al., 1990). There are actually five different examples of the gene-behavior pathway in Fig. 4.2. The first of these models represents the simplest and most direct idea about how genes operate—an idea held by many today that genes lead deterministically to a completely expressed behavior. However, rarely does this situation ever occur, particularly for complex human traits, such as language (though some genetic traits are highly penetrant, such as Huntington's chorea).

As shown in other gene-behavior pathway models of Fig. 4.2, there is usually a cascade of multiple effects where gene action may be at the start, middle, and end, and where gene effects only culminate in a behavior after going through a variety of intermediary steps and interactions with other genes and environmental factors.

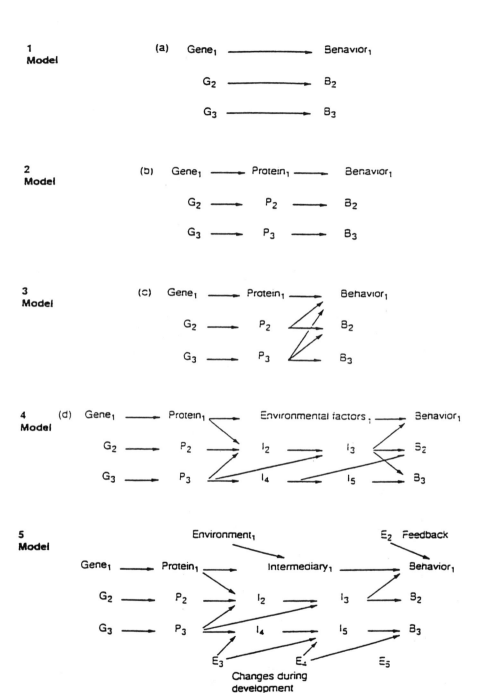

FIG. 4.2. Increasingly complex gene-behavior pathways. From Plomin (1981). Reprinted by permission.

Examining the paths in Model 4, for instance, shows that the set of behaviors (B1, B2, B3) are ultimately a function of G1-G3, several intermediary processes and interactions, and environmental effects. Model 4 can be made even more complicated by adding a consideration of genetic, behavioral, and environmental feedback loops, and developmental changes in gene expression, as illustrated in Model 5. Model 5 is probably the best representative of how genes work for complex human cognitive and language systems.

The gene-behavior pathway is a dynamic developmental system for complex human traits (Plomin, 1981, 1986). Different genes turn on and off at different times in our lives. Chromosomes contain genes that code for specific proteins (structural genes) that are necessary building blocks for life processes, as well as genes that control the activation of these structural genes sometimes called developmental genes, operators, or regulators (Hunkapiller, Huang, Hood, & Campbell, 1982; Plomin et al., 1990). Thus, developmental language or reading disorder in a child 8 years of age may reflect the distal (i.e., earlier and less direct) effects of a currently inactive gene that once coded for embryologic neuronal cell migration, death, and differentiation (Davidson, 1976; Freeman, 1985; Jacobson, 1978; Pennington, 1991b; see also Plante, this volume). On the other hand, the proximal effects of a currently expressed gene that regulates the levels of certain neurotransmitters used in learning could also be an influential factor in the disorder.

Differentiating between proximal and distal genetic effects is important. First, when looking for major gene effects for abnormal (or normal) language development, there is a need to remain cognizant about not searching for currently active genes. This is particularly true when behavioral genetic designs, such as linkage studies, use families that necessarily include some children and adults beyond the age of disease onset and periods of brain growth for areas critical to language.

Second, in terms of treatment, much of the neurological damage leading to a language disorder may have occurred during major developmental periods both before and soon after birth (Freeman, 1985; Pennington, 1991a, 1991b; see Plante, this volume). Thus, there is little hope of directly ameliorating such a genetic effect in school-aged children or adults, though there may be ways to help the SLI individual learn to use more intact brain systems. Standard remediation techniques do, in fact, operate through functional neurocognitive systems mediated by a variety of currently and once active genes, including the genes controlling the neurotransmitters needed for learning, motor control during speech, and auditory discrimination.

Third, because the expression of genes may not continue into adulthood, their effects may be hidden. In the case of RD, for example, some 22% of the adult subjects that were RD as children compensate for their disability, and no longer show reading deficits when given standardized reading tests as adults (Gilger, Hanebuth, DeFries, Smith, & Pennington, in review; Scarborough, 1989,

1990). How then can we access the effects of a putative RD gene when its expression has been modified or hidden? One approach for disorders like RD where expression occurs early in life and then may lessen with age, is to assess all family members when they are children. This, of course, would require a longitudinal and time-consuming study. Therefore, a more common approach has been to develop more sensitive assessment instruments that can be applied across generations and that are able to detect subtle differences between the yet to be affected (i.e., newborns), the once affected (i.e., compensated adults), and the never affected (i.e., normals; see Tomblin, this volume). Yet another approach is the high or at-risk study described in this volume by Lefly.

Finally, the type of molecular genetic work that relies on the current activity of genes may not be the best approach for language-related disorders. This is because such techniques depend on the presence of active gene transcription and translation to work. A brief description of this process will help make this point clear.

Each of the 46 chromosomes (23 pairs) that comprise the normal human complement is essentially a long strand of deoxyribonucleic acid (DNA) (Lewin, 1985; Watson & Crick, 1953). Different stretches of these DNA strands represent the 100,000 or so structural genes of the human genome. For a gene to be active or turned-on to produce its protein product, it goes through a number of biochemical steps, including the transcription of DNA into a form of ribonucleic acid (RNA) (Lewin, 1985; Watson & Crick, 1953). The RNA is, in turn, translated into the proteins that operate to produce behaviors (e.g., motor output, neuronal growth and migration, neurotransmitter production).

There are certain molecular techniques that rely on the presence of the protein product of RNA or on the RNA itself, and hence on genes that are active at the time of study (e.g., complementary DNA or cDNA cloning, RNA activity assays) (Lewin, 1985). For example, some (e.g., Adams et al., 1991) advocate identifying the 30,000 structural proteins active in the brain by establishing a complete cDNA library. In this way, researchers would be essentially characterizing all the genes actively involved in the everyday brain functions that are needed for complex human behavior. However, if this approach was used to identify major genes for intelligence or language, those genes not currently active in the brain when the cDNA library is established—genes that may have earlier effected brain areas important to cognitive skills—would be missed. Thus, a set of cDNA libraries should be established, each representing RNA activity at different ages (intra- and extrauterine). Similarly, these types of genetic techniques, as well as basic linkage or association methodologies (e.g., Plomin, 1993) are susceptible to missing the effects of genes that at the time of study have yet to be activated and expressed.

Figure 4.2 also illustrates two other concepts: *genetic heterogeneity* and *pleiotropic effects*. The former is the situation where the same or similar behavior can be produced by more than one (different) gene. This is shown in Model 3 where

G1, G2, and G3 can all lead to B1. Genetic heterogeneity probably exists for speech–language-related traits. For example, it is known that quite different chromosomal and genetic conditions can lead to related deficits in the language system (e.g., sex chromosome aneuploidies, Down's syndrome). In fact, a great many of the known genetic and chromosomal conditions have some form of cognitive or language deficit. This is consistent with the fact that a great many of our genes are directly responsible for brain development (Adams et al., 1991).

Given that many genes are involved in CNS development, and that human neurology and neuropsychology are so complex, it is quite unlikely that there would be a single unique gene for each specific language attribute observed (e.g., tense marking, number). However, not everyone would agree with this proposal. Gopnik and Crago (1991), for instance, originally interpret their case analysis of data from a single large pedigree as evidence for a genetically dominant form of *feature blind dysphasia*. In other words, they believe there is a domain specific, single gene mediated neural mechanism for the highly specialized deficit observed in the individuals in their family case study. On the other hand, Vargha-Khadem and Passingham (1990), studying the same family, reported a much broader range of language and cognitive deficits, and in a more recent chapter by Crago and Gopnik (1994) these greater deficits were also documented. Thus, resolution of the inheritance and specificity of the original Gopnik and Crago (1991) finding awaits further research, and the issue of the inheritance of modular and domain-specific aspects of language remains an interesting and debatable area—an area wide open for careful application of behavioral genetic methodologies (e.g., Pinker, 1991, 1994).

It is noteworthy that even if domain specific rules for language exist, each rule need not be mediated by the same single major gene in all people. The earlier discussion of genetic heterogeneity in reference to Fig. 4.2 is a simplified version of the case where separate single major genes can lead to similar behaviors (e.g., two different genes lead to SLI). However, multiple genes (and environmental factors) can act in concert to influence a single behavior like SLI, and this situation refers to the multifactorial-polygenic model (Lewin, 1985; Plomin, 1990; Plomin et al., 1990; Plomin & Rende, 1991; Roberts & Pembrey, 1985). The multifactorial-polygenic, oligogenic, and quantitative trait locus models were discussed earlier in this chapter as examples of how multiple genes acting in concert can modify a behavior. Such genetic systems are more likely to be operating for complex behaviors that involve multiple neurocognitive and neurological processes as is the case for performance on general IQ and verbal fluency tests. As the phenotype under study gets more narrowly defined, however, and more specific to the basic process or, in the case of language, domains and rules, it is possible that the number of genes involved may decrease, particularly for abnormal phenotypes. Thus, the issue of phenotype, how crucial it is for genetic studies, and how a person's theoretical slant can determine methodology, surfaces again (see Rice & Wexler in this volume; Hunt, 1983; Pinker, 1991, 1994).

Finally, a single gene can be pleiotropic, as occurs for the PKU gene. This gene may effect several different cognitive and physical phenotypes in a single individual (e.g., IQ, skin sensitivity, eye color, extrapyramidal symptoms). Pleiotropy is portrayed by the multiple behavioral effects of genes 1–3 in Model 4 of Fig. 4.2. The pleiotropic effects of genes may be partly responsible for the phenotypic correlations among specific cognitive abilities, and different personality traits, and correlations (comorbidity) between abnormal traits like RD and Attention Deficit Disorder, and Obsessive Compulsive Disorder and Tourettes (e.g., Baker, Vernon, & Ho, 1991; Pauls et al., 1986; Plomin & Rende, 1991; Stevenson et al., 1993).

If a Trait Has a Genetic Component, It Is Immutable and Unchanging, and Behaviors That Have a Genetic Component Show Themselves Early in a Person's Life

In this section, these two related misconceptions are jointly considered. These ideas have already been touched on: Specifically, some genes and their effects are expressed differently and at different times throughout the lifespan. For example, the same gene may be critical for very different behaviors at different times. In the case of some cancers, Alzheimer's disease, and even aging, for instance, it may be that genes that once coded for necessary and normal cell activity, or that were dormant, are responsible for deviant cell functioning later in life (e.g., Brown, 1985; Hatano, Roberts, Minden, Crist, & Korsemeyer, 1991; McClearn & Foch, 1985). On the other hand, some genes continue to have their effects on the same or similar behavior throughout the lifespan. In the case of intelligence, h^2 estimates generally increase from infancy (h^2 of roughly .15) to adolescence (h^2 of roughly .40) even though it appears that the genetic effects of early childhood are highly correlated with genetic effects in adults (DeFries, Plomin, & Labuda, 1987; Eaves, Hewitt, Meyer, & Neale, 1990; Eaves, Long, & Heath, 1986). That is, the genetic effects on intelligence early in life continue to have effects on individual differences in intelligence later in life.

It is also true that genetically mediated behaviors need not show up early in life. The onset of puberty and Huntington's chorea in midlife are striking examples of this point. Similarly, there are also data suggesting that some of the genes affecting cognitive and linguistic abilities may not be expressed until later in life (Hardy-Brown, 1983; Plomin et al., 1990). Again, because behavioral genetic research on language skills is in short supply, IQ is used as an illustration here.

As previously mentioned, the twin studies of IQ show increases in h^2 with age, as well as high, but imperfect, age-to-age genetic correlations for IQ. This suggests that some IQ genes turn on or turn off with development (otherwise age-to-age correlations would theoretically be perfect; DeFries et al., 1987; Eaves et al., 1990; Eaves et al., 1986). Given certain assumptions, adoption work has the unique ability to separate environmental and genetic effects (Plomin et al.,

1990). A few longitudinal adoption studies have been conducted, and they generally support the twin findings that the genetic contributions to IQ change with age. As an early and elegant example of this adoption work, a portion of the results from the Skodak and Skeels (1949) and Honzik (1957) studies are presented next.

The lines in Fig. 4.3 represent age changes in the correlations for IQ among pairings of fathers and sons of differing consanguinity. Because actual IQ scores were not available for all fathers, each father's educational level was substituted. Educational attainment is moderately correlated with IQ, and often serves as a stand-in for IQ when IQ tests are not possible (Honzik, 1957; Scarr & Carter-Saltzman, 1983). In Fig. 4.3, the adoptive father-child relationship is shown by the dotted line. The solid line shows the relationship between a child and his or her biological father, even though that child was adopted away in infancy and was not raised with that parent. Finally, the dashed line represents the relationship between biological fathers and their biological children that were not adopted away (i.e., that were raised by this father).

Children and their biological fathers share, on average, 50% of their genes, but children and their adoptive fathers are assumed to have none of their genes

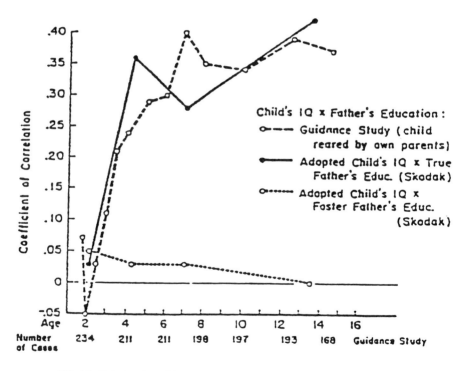

FIG. 4.3. Relationship of father's educational status and child's IQ. From M. Honzik (1957). Reprinted by permission.

in common, though they do share environments (Plomin et al., 1990). The interesting point about Fig. 4.3 is that it shows the line representing correlations between adoptive fathers and their adopted child to hover about zero across the ages of 2 to 15 years, but the lines tracking the age change in correlations among children and their biologically related fathers increase dramatically with age. In this way, these adoption data are in agreement with the twin data already mentioned and vividly suggest that genetic contributions to individual differences in IQ increase with age (Honzik, 1957; Plomin et al., 1990). The adoption data in Fig. 4.3 also make a very interesting point regarding the effects of the family environment on IQ. This is discussed in the next section.

The Most Important Type of Nongenetic Influence Is the Family Atmosphere. It has long been believed that being brought up in the same family will tend to make siblings and parents similar. This idea makes sense at an intuitive level, especially in terms of the development of personality, cognitive abilities, and psychopathology. Interestingly, however, research designed to look at this issue indicates something quite different (Eaves et al., 1989; Plomin & Bergeman, 1991; Plomin & Daniels, 1987; Plomin et al., 1990; Rowe, 1994): In general, as contributors to individual differences for most traits studied thus far, including personality and intelligence in older children, the effects of the family environment are typically small relative to the effects of environmental events not shared by family members and the effects of genes.

The environmental factors influencing a trait can be divided into two basic components—those shared by relatives and those not shared by relatives (Falconer, 1981; Plomin et al., 1990). Hence, in behavioral genetics, the summary concept of e^2 is often separated into common or shared familial environmental influences (symbolized E_s) and nonshared familial environmental influences (symbolized E_{ns}).[6] It is this latter type of environmental effect that seems to account for much of the nongenetic variance found for many complex human traits (Eaves et al., 1989; Plomin & Bergeman, 1991; Plomin & Daniels, 1987; Plomin et al., 1990; Rowe, 1994). However, E_s does appear to play a significant role for some traits, such as religiosity, political beliefs, and academic achievement, and the relative importance of E_s and E_{ns} can change with age and populations (Eaves et al., 1989; Rowe, 1994; Truett et al., 1992).

Figure 4.3 shows evidence for strong genetic effects and little or no E_s effect on IQ beyond 4 years of age. Specifically, the dotted line showing the relationship between a foster father and his adopted son hovers around zero. This line represents only the effects of E_s on the foster father–adopted son IQ correlation as these pairs of relatives do not share any genes. If the family environment, or E_s, were, in fact, a critical contributor to IQ, the correlations would be stronger. On

---

[6]The symbols used in this chapter are representatives of the variety of possible symbols used for shared and nonshared environmental and genetics effects in the literature (Falconer, 1981; Plomin et al., 1990).

the other hand, the two lines showing the father-son correlations when there are shared genes involved increase with age and then level-off at about .40. Here again is evidence that E_s effects on IQ are minimal: If E_s was important, the correlations between sons and biological fathers reared together (the dashed line) would be somewhat larger than the correlations between adopted-away sons and their biological fathers (the solid line), because in the former case shared genes (50% on average) and environment can contribute to the correlation, but in the latter, only shared genes (50% on average) can make a contribution. Note that the effects of environment and genes on IQ have been greatly simplified for this discussion, and there are several complexities not addressed, including the effects of gene-environment correlations and interactions, dominant gene effects, and so on (see Plomin et al., 1990; Rowe, 1994; Scarr & Carter-Saltzman, 1983).

Because many studies put an estimate of h^2 at approximately 50%, and, as was just demonstrated, there is little evidence for E_s, the environmental effects accounting for the remaining 50% of postpubertal IQ variance must logically be primarily of the E_{ns} type. If what is known about IQ and some other cognitive abilities is any indication, E_{ns} is also likely to be quite important, and perhaps more important than E_s, for language-related traits as well.

With regard to language disabilities, the general failure to find a common nongenetic event that triggers or substantially contributes to the liability for SLI is supportive of the proposition that E_s is not as important as E_{ns} in abnormal language development. Researchers trying to identify the actual environmental variables that tend to make persons alike for, say SLI or RD, have met with little success (e.g., poor communication models in the home), although significant E_s variance has been estimated for some reading-related traits in RD twins (Olson et al., in press). Attempts at identifying reliable sources of E_{ns} have not fared much better. The correlations between SLI or RD and variables thought to be examples of E_{ns} (e.g., anoxia, prematurity) have been inconsistently reported and are, at best, weak in their apparent contribution to the expression of the forms of developmental language disorders we see in 5% to 10% of the school-aged population. This does not apply to extreme cases, of course. By their very nature, E_{ns} events may not be easy to identify: Finding one event, anoxia, for example, that significantly and similarly contributes to an elevated risk for all people that experience it, has not been possible to date. In part, this elusiveness is because the E_{ns} event must interact with the individual's genotype. Different genotypes will not necessarily respond the same way to E_{ns} stressors.

The difficulty in finding the specific environmental factors involved in these disorders, though they are known to exist, suggests that a more fruitful approach towards identifying the etiologic factors making a substantial contribution to SLI or RD, is to focus more on the genetic side of the equation. The growing sophistication and power of genetic methodologies, and the greater ease with which genes can be controlled and identified, makes a "genetics of language" almost more realizable than an "environmentality of language" at this stage of understanding. The study of genetics is in a way easier than the study of environments, and, therefore, some of the initial great gains may lie in identifying the SLI and RD genes rather than specific environmental agents. Subsequently, studies where genes for SLI liability can be controlled and the adverse environmental effects, thereby, more easily identified and monitored, may be conducted.

In contrast to abnormal language development, normal language development may emphasize E_s rather than E_{ns}. It is not unreasonable to suspect that variations in normal language are much more susceptible to the shared family environment (see de Villiers, Rice, & Snow, this volume; Hardy-Brown, 1983).

Empirical support for an important E_s component in language can be found in the twin studies summarized by Nichols (1978). In this summary, Nichols compiled data on 211 twin studies of a variety of traits to obtain average MZ and DZ correlations. These data were used to calculate h^2, E_{ns}, and E_s for three composite variables having a clear language attribute (i.e., verbal comprehension, verbal fluency, language achievement). Each variable may have been based on studies that used different types of psychometric tests (see Nichols, 1987).

The estimates of E_s exceed those for E_{ns} for all three composite variables, and in all three cases explained over ⅓ of the phenotypic variance. Specifically, the verbal comprehension, verbal fluency, and language achievement composites yielded E_s estimates of .40, .37, and .35, respectively, and h^2 estimates of .38, .30, and .46, respectively. To precisely determine what these shared familial characteristics are that contribute significantly to variation in language requires careful study across a range of populations.[7] In particular, the best types of studies to access these characteristics may be of behavioral genetic design because such designs enable researchers to discriminate between the effects shared and nonshared genetic and environmental factors have on complex traits (Rowe, 1994).

Using standard, rather than behavioral genetic methodologies, researchers have, in fact, provided evidence that certain aspects of the normal language system are influenced by both the intra- and extrafamilial environment, and thus E_s as well as E_{ns} (Bates et al., 1982; Hardy-Brown, 1983; Leonard, Chapman, Rowan, & Weiss, 1983; Molfese, Molfese, & Carrell, 1982; Pinker, 1994; Rice & Woodsmall, 1988; Snow in this volume; Whitehurst, 1982). Examples include adult speech effects on infant phonological development and imitation effects in conversational style and pragmatics. Behavioral genetic studies could take this research a step further by providing a unique means of unambiguously separating environmental from genetic effects in families. Fortunately, careful behavioral ge-

---

[7]Because unique aspects of the twin environment may be operating, language-relevant environmental factors may lead to DZs being treated differently than nontwin siblings. This can lead to an elevated E_s estimate in these data. E_s may also be elevated due to the genetic effects of assortative mating by the parents for cognitive traits rather than environmental effects, per se (see Plomin et al., 1990).

netic studies of normal language development are starting to appear in the literature (e.g., Bishop et al., 1995; Cherny & Fulker, 1993; DeFries & Plomin, 1983; Felsenfeld, 1994; Hardy-Brown, 1983; Hardy-Brown & Plomin, 1985; Hardy-Brown et al., 1981; Julian, Plomin, Braungart, Fulker, & DeFries, 1992; Locke & Mather, 1989; Mather & Black, 1984; Munsinger & Douglass, 1976; Ooki & Asaka, 1993; Plomin, 1986; Plomin, DeFries, & Fulker, 1988; Reznick, in press).

One reason why behavioral genetic designs are important in the language area is that, in families, the environment created by the parents and passed on to their offspring (e.g., number of books in the home, amount and style of talking to the child, parental education, socioeconomic status), is in part a reflection of the parent's genes that are also passed on to their offspring. Family data confound genetic and nongenetic effects passed on from parent to child (Hardy-Brown, 1983; Plomin et al., 1990). Therefore, if a correlation between an environmental factor in the home and a child's behavior is observed, it may not be a phenomenon of learning, but instead, it may be a consequence of the genes transmitted from the parent to the child, the effects of which will manifest themselves in the child's development as well as the home environment the parent creates (Eaves et al., 1989; Hardy-Brown, 1983; Plomin et al., 1990; Plomin & Rende, 1991; Rowe, 1994). Thus, experiments that do not unambiguously separate genetic from environmental effects transmitted in families may be drawing erroneous conclusions from inherently confounded data. An example in SLI makes this concept clearer.

Tomblin, Hardy, and Hein (1991) identified several risk factors in families that were predictive of poor communication skills in young children. Among these factors was the father's educational achievement level. This level predicted the child's communicative competence such that fathers of children with lower than normal communication skills had lower educational achievements than fathers of normally developing children. This risk factor, among others, was interpreted in terms of how the environment created by the parent can adversely affect the child's acquisition of communication skills. The point here is that behavioral genetic studies strongly suggest that a large proportion of the variance shared between father's education and child's language may be due to the genes passed from the father to the child, rather than the environment the father creates (see the discussion of Fig. 4.3). These shared genes may influence IQ, general language competence, and educational attainment in both fathers and their children. So, the observed correlation between father's education and child's language does not tell us about the underlying causal relations, regardless of how plausible or intuitively appealing a given causal hypothesis may seem.

Even data showing that a parent's ability to adjust communication style to match that of a child, and, thereby, contribute to the child's language development (e.g., Furrow, Nelson, & Benedict, 1979; see also Snow, this volume), confound genetic and environmental effects on language acquisition. It is plausible that parents and their children genetically endowed with language skills are generally good communicators and can naturally better adjust their communication for whomever they are dealing with not only because the parents have provided a conducive environment, but also because they have received the good, language-relevant genes of their parents. Again, the only way to truly get at the cause of parent-child correlations is through careful behavioral genetic methodologies where the effects of genes and environment can be separated.

SUMMARY AND SUGGESTIONS
FOR FUTURE RESEARCH

This chapter has used illustrations primarily from behavioral genetic research on traits other than language. However, there are twin, family, and adoption studies of language-related traits, although they are limited in size, design, and analysis, and tend to focus on very broad phenotypes (e.g., tests of expressive vocabulary rather than specifics of grammar, phonological skills). In general, these preliminary studies suggest mild-to-moderate h^2 for global language-related skills, and significant E_s, at least in preschool children (see Cherny & Fulker, 1993; DeFries & Plomin, 1983; Felsenfeld, 1994; Hardy-Brown, 1983; Hardy-Brown & Plomin, 1985; Hardy-Brown et al., 1981; Hay, Prior, Collet, & Williams, 1987; Locke & Mather, 1989; Matheny & Bruggemann, 1973; Mather & Black, 1984; Munsinger & Douglass, 1976; Ooki & Asaka, 1993; Osborne, Gregor, & Miele, 1968; Plomin, 1986; Plomin et al., 1988; Scarr & Carter-Saltzman, 1983).

There has also been a growing interest in behavioral genetic work in SLI. Like research in normal language, studies of SLI have tended to focus on questions related to the simple etiology category of Table 4.1 (see Gilger, 1992a; Lewis, 1992; Tomblin, this volume). It is known for instance, that 20% to 30% or more of the first degree relatives of SLI individuals are themselves SLI, and that the h^2 for self report and other report of SLI is significant (see Bishop et al., 1995; Gopnik & Crago, 1991; Lewis, Cox, & Byard, 1993; Lewis & Thompson, 1992; Parlour & Broen, 1989; Tallal, Ross, & Curtiss, 1989; Tomblin, 1989). Still, much work needs to be done on SLI, particularly in the area of finer definitions of phenotype. The twin work of Dorothy Bishop and colleagues (1995) is a good start, and it has already yielded some interesting support that SLI profiles may breed-true, that is, there may be heritable subtypes of neurolinguistic deficits underlying SLI. Still, even in the area of simple etiology, more diverse research is needed.

Table 4.3 summarizes the conceptual steps in genetic analyses to determine the simple etiology of disorders like SLI. The questions addressed in each step are increasingly specific and concern familiality (the tendency to "run" in families), heritability (that genes are to some degree responsible for familiality), mode of transmission (that the familiality observed follows patterns in line with specific Mendelian genetic mechanisms, such as autosomal dominant major

TABLE 4.3
Four Basic Steps in an Analysis of Genetic Etiology for Complex Disorders

Steps and Associated Questions	Means to Address Steps
1. Is the disorder familial?	Family risk or aggregation
2. Is the familiality genetic in basis?	Twin or adoptions studies
3. What is the mode of genetic transmission?	Segregation analysis
4. Where are the genes located and what are the gene characteristics?	Linkage analysis, gene sequencing and cloning

gene; see Brzustowicz, this volume), and gene locations and characterizations (linkage studies) (see Smith & Brzustowicz, this volume). Within each step are a variety of study designs that permit the collection and analysis of data to address both simple and complex questions in speech-language science (e.g., co-segregation, correlation, concordance, modeling, regression, aggregation, and other analyses). Examples of some of these have been described previously, and space does not permit a more thorough discussion here (Falconer, 1981; Loehlin, 1987; Morton, 1982; Neale & Cardon, 1992; Plomin et al., 1990; Plomin & Rende, 1991). Because genetic studies of nonsyndromic SLI are relatively new, the bulk of the current data revolves around familiality (see Tomblin, this volume), and only a couple of studies attempt to formally address the specific mode of transmission (e.g., Lewis et al., 1993). There are no adoption or linkage studies specifically aimed at SLI in the literature to date. Therefore, the issues of heritability, mode of transmission, and gene location for even the simple etiology of SLI are difficult to address at this time.

In the future, the complex questions of interest will depend on the theoretical orientation of the experimenter: Current models of how the brain works or how the language system develops are perhaps not as fully specified as would be desirable for this endeavor, although certain models are available that can be tested (e.g., Gilger's commentary, this volume; Hunt, 1983; Pinker, 1991, 1994). For example, in this volume de Villiers makes the distinction between closed and open grammatical systems. Her use of the terms, along with genetic theory, suggests hypotheses that could be explored in future work: SLI deficits arising from the more closed aspects of language should be less variable, more canalized genetically, and perhaps largely explained by the influence of a limited number of major genes. In contrast, deficits due to more open systems, may show more variability in phenotype, may be less canalized genetically, and may better fit a multifactorial-polygenic model.

It is argued, in closing, that there are several areas where future behavioral genetic research can truly advance the field of language studies; illustrations and suggestions towards this goal have been provided. Behavioral genetic research on a variety of complex human phenotypes provides lessons applicable to language and language disorders. The contributors to this volume, with their different perspectives, offer additional lessons. Future researchers may also want to consider the following points:

1. More behavioral genetic work of all types is needed in the language area, including replications and expansions of findings already in the literature. In this way, the area is wide open. Future work, somehow combining family and twin research with neuroimaging, linguistics, and cognitive psychology theory will prove especially critical (e.g., see Hunt, 1983; Pinker, 1991, 1994; Plante, this volume).

2. Excessive time should not be spent simply handing out surveys and global tests that, on the surface, are relevant to language development, and then compute a variety of h^2 estimates based on the simple questions of etiology listed in Table 4.1. Rather, it would be better to focus more on finer-grained, theory-based phenotypes with research designs that can address more complex questions in speech-language.

3. Given the current state of the field, it would be prudent to conduct work in this area using a variety of samples and methods (e.g., twin and family samples, heritability and segregation analyses). To jump into, say, a pedigree study strictly for linkage analyses, is unlikely to lead to success. Although molecular techniques in genetics are developing quickly, there is much to learn about how the genes for SLI act and react, irrespective of their location, and how genetics modify treatment efficacy. Linkage studies, although able to answer the *where* question, are not so well suited to finding out how the genes for complex disorders work in development and how they interact with other genes and the environment. Nonmolecular behavioral genetic work, as well as more conventional experimental approaches, are required to address this concern.

4. Speech and language professionals should draw on the knowledge of other fields, including behavioral and molecular genetics, linguistics, cognitive psychology, and medicine. Working towards an understanding of the genetics of language will require a truly interdisciplinary approach.

ACKNOWLEDGMENTS

This chapter is based in part on a presentation at the 1993 annual American Speech-Language-Hearing Association (ASHA) convention in Anaheim, CA, and some of the ideas herein also appear in a *Journal of Speech and Hearing Research* (in press) tutorial.

The author wishes to thank Robert Plomin for his comments on an earlier draft.

Address inquiries to the author at the Department of Speech-Language-Hearing: Sciences and Disorders, 3031 Dole Human Development Center, Lawrence, KS, 66045-2181 (email: gilger@falcon.cc.ukans.edu).

APPENDIX

The following is a list of some key words and their definitions used in this chapter. See the text and citations for further explanation. Portions of this appendix were taken from J. W. Gilger (1995). Behavioral genetics: Concepts for research and practice in language development and disorders. *Journal of Speech and Hearing Research.*

Autosomes. A chromosome other than a sex chromosome (X or Y). Normal humans have 22 autosomal pairs of chromosomes and one pair of sex chromosomes.

Autosomal dominant disorders. Genes are represented as pairs in normal humans. Each member of a pair may interact in a dominant, recessive, or additive manner with the other. Dominance refers to the situation where one gene of the pair is expressed over the other. An example is Huntington's chorea, a disease that is expressed when an individual carries one or two copies of the mutant gene. Various factors are relevant to the expression of dominant genes, such as penetrance.

Autosomal recessive disorders. Recessivity refers to the situation where the effects of one gene are subordinate to those of another. An example is phenylketonuria (PKU), a disease that is expressed when an individual carries two copies of the mutant gene, but is not expressed when just one copy of the gene is carried (see *autosomal dominant disorders*).

Chromosomes. These are strands of deoxyribonucleic acid (DNA) and contain the coding sequences called genes. An individual's normal complement is 23 homologous pairs, or 46, chromosomes (one member of each pair from their father and one from their mother). In some conditions, such as Down's syndrome, the complement is not normal, and typically includes an extra chromosome number 21. Thus, Down's individuals have 47 chromosomes (see *gene*).

Deoxyribonucleic acid (DNA). The essential chemical compound or molecule of chromosomes and, therefore, genes. Specific coding properties of DNA specify genes and gene products, and allow for the transmission of genetic information across generations.

Dizygotic twins (DZ). Fraternal or nonidentical twins. Like nontwin siblings, DZ twins come from separate eggs and sperm, and share, on average, 50% of their genes. DZ twins may be of the same or opposite sex.

Distal effects. Genetic effects noted in the present that reflect gene action of the past. For example, much of the genetically mediated brain cell migration, connection, and death occurs prenatally. Thus, the brain structures used for learning and language were largely set before a child goes to school or is tested for SLI. This is to be contrasted with proximal effects.

Environmentality (e^2). The standardized proportion of phenotypic variance or individual differences in a population due to environmental effects. The proportion of variance in a trait remaining once the contribution of genes to trait variance has been controlled. In standardized form, it ranges from 0.0 to 1.0.

Familiality. The extent to which a trait runs in families. If a trait, such as SLI, is found to more often cluster in families of SLI children relative to families of non-SLI children, the trait may have a genetic basis.

Gene. A hereditary unit. A gene is a specific arrangement of chromosomal DNA that is required to produce a functional product (e.g., protein) necessary in development (see *chromosome*).

Gene-behavior pathway. The series of steps through which a gene is eventually expressed as a behavior. Some genes are quite direct in their expression, as in the case of Huntington's chorea, but others, particularly those relevant to complex traits, are modified by a long chain of environmental and biological events.

Genetic heterogeneity. The situation where different and distinct genes can give rise to the same or similar behavior. Genetic heterogeneity is suspected for many complex traits, including dyslexia, SLI, and psychiatric conditions. This can be a problem for studies attempting to localize the specific genes involved in complex human traits.

Heritability (h^2). The standardized proportion of phenotypic variance or individual differences in a population due to genetic effects. The proportion of variance in a trait remaining once the contribution of environment to trait variance has been controlled. In standardized form, it ranges from 0.0 to 1.0. The heritability statistic does not apply to the individual, but only to populations. Hence, if the h^2 estimate for IQ is .50, it does not mean that 50% of a person's IQ is due to genes. It does, however, reflect the extent to which differences in the IQs among people are due to genes (see *environmentality*).

Human behavioral genetics. The field of study combining behavioral sciences with genetics. It is essentially concerned with modeling how genes and environment act and interact to produce and modify behavior.

Mode of transmission. How a trait is transmitted in families and, more specifically, if this transmission fits a specific genetic model. To identify a mode of transmission, tests are made on data from sets of families to see if the familiality of a trait best fits single (e.g., autosomal recessive or dominant) or multiple gene models, with or without environmental effects.

Monozygotic twins (MZ). Identical twins. Unlike nontwin siblings or DZ twins, MZs come from the same sperm and fertilized egg, and share 100% of their genes. MZ twins are always of the same sex.

Nonshared familial environmental effects (E_{ns}). Nonshared effects refer to those environmental/experiential factors that contribute to behavioral differences among people belonging to the same family. It implies that siblings being raised by the same parents will tend to be different to the extent that nonshared effects are important for the behavior being studied (see *shared environmental effects*).

Nonsyndromic or developmental language-related disorders. In the current context, developmental disorders are those that are not part of a constellation of symptoms that would lead to the diagnosis of a specific medical condition (as in Down's syndrome), nor are they acquired (as in aphasia). Developmental disorders are thought to reflect underlying and subtle neurologic abnormalities that may have occurred during sensitive periods of normal development. These perturbations may have been a consequence of environmental or genetic events, or random developmental variation. In contrast to many of those with syndromic disorders, individuals afflicted with developmental dyslexia or language disorder do not have a unique physical appearance, and are by far the most common type of disordered child seen by school-based clinicians (see *syndromic language disabilities*).

Penetrance. The likelihood that the effects of a gene will be expressed in a person with the gene. A gene is said to be fully penetrant if 100% of those carrying the gene show the gene effects (e.g., Huntington's chorea). A gene is said to have incomplete penetrance if, say, only 60% of those with the gene manifest the trait (e.g., dyslexia) (see *variable expressivity*).

Phenotypic variance. Individual differences in the population for a trait. More technically, *phenotype* refers to the observable characteristics of gene expression.

Pleiotropy. The multiple phenotypic effects of a gene. One gene, or a cluster of genes, may effect multiple systems. In the case of PKU, the deleterious gene adversely effects neurology, skin pigmentation, behavior, and other systems.

Proteins. The essential chemical products of genes that regulate living processes. Different proteins and the systems they act on are coded and governed by different genes.

Proximal effects. Genetic effects noted in the present that reflect fairly immediate and direct gene action. For example, many everyday cognitive tasks are a function of the activation of various neurotransmitters in the brain. These neurochemical compounds are produced by certain genes. Thus, though the brain structures used for learning and language were largely set in utero by no longer active genes, the use of these brain areas is very much dependent upon the currently active, proximal effects of other genes (see *distal gene effects*).

Quantitative trait loci (QTL). Variation in stretches of DNA, whether they represent specific genes or not, that are correlated with variations in a quantitatively measured trait. For example, variations in a stretch of DNA on Chromosome 6 are significantly associated with variations in reading ability. Individuals with particular variations in this chromosomal region tend to have lower reading test scores than those with different variations.

Ribonucleic acid (RNA). These are a class of intermediate chemical compounds, some types of which are responsible for the translation of the DNA genetic codes into proteins.

Shared familial environmental effects (E_s). Shared effects refer to those environmental/experiential factors that contribute to behavioral similarities among people belonging to the same family. It implies that siblings being raised by the same parents will tend to be alike to the extent that shared effects are important for the behavior being studied (see *nonshared environmental effects*).

Syndromic language-related disability. In the current context, syndromic language disabilities refer to disabilities that occur as part of a collection of characteristics that together are sufficient for a clinical diagnoisis. A variety of genetic or chromosomal conditions are syndromes with concomitant speech-language problems (e.g., fragile X, Klinefelter Syndrome, Galactosemia, Apert Syndrome) (see *nonsyndromic or developmental disorders*).

Structural genes. A gene coding for any protein product. "Developmental genes" (e.g., regulators or operators) that code for proteins or RNA may act to regulate the expression of structural genes, such that not all genes are functional at all times.

Variable expressivity. The extent or manor in which a gene is expressed. If gene expressivity is variable, the trait may vary from mild to severe across people, from mild to severe within a person and across time, or as apparently different traits across people or across time. This situation complicates the phenotypic picture and can make genetic studies of disorders, such as SLI, quite difficult.

REFERENCES

Achenbach, T. M. (1982). *Developmental psychopathology* (2nd ed.). New York: Wiley.

Adams, M. D., Kelley, J. M., Gocayne, J. D., Dubnick, M., Polymeropoulos, M. H., Xiao, H., Merril, C. R., Wu, A., Olde, B., Moreno, R. F., Kerlavage, A. R., McCombie, W. R., & Venter, J. C. (1991). Complementary DNA sequencing: Expressed sequence tags and human genome project. *Science, 252,* 1651–1656.

Aram, D. (1991). Comments on specific language impairment as a clinical category. *Language, Speech, and Hearing Services in the Schools, 22,* 84–87.

Baker, L. A., Vernon, P. A., & Ho, H-Z. (1991). The genetic correlations between intelligence and speed of information processing. *Behavior Genetics, 14,* 351–367.

Bates, E., Bretherton, I., Beeghly-Smith, M., & McNew, S. (1982). Social bases on language development: A reassessment. In H. W. Reese & L. P. Lipsitt (Eds.), *Advances in child development and behavior* (Vol. 16). New York: Academic Press.

Bender, B. G., Linden, M., & Robinson, A. (1991). Cognitive and academic skills in children with sex chromosome abnormalities. In B. F. Pennington (Ed.), *Reading disabilities: Genetic and neurological influences.* The Netherlands: Kluwer Academic Press.

Bishop, D. V. M., & Edmundson, A. (1987). Language-impaired four-year olds: Distinguishing transient from persistent impairment. *Journal of Speech and Hearing Disorders, 52,* 156–173.

Bishop, D. V. M., North, T., & Donlan, C. (1995). Genetic basis of specific language impairment: Evidence from a twin study. *Developmental Medicine and Child Neurology, 37,* 56–71.

Brown, W. T. (1985). Genetics of aging. In M. P. Janick & H. M. Wisniewski (Eds.), *Aging and developmental disabilities: Issues and approaches.* Baltimore: Paul H. Brookes.

Bruun, K., Markkanen, T., & Partanen, J. (1966). *Inheritance of drinking behavior: A study of adult twins.* Helsinki: The Finnish Foundation for Alcohol Research.

Carey, J. C., Stevens, C. A., & Haskins, R. (1992). Craniofacial malformations and their syndromes: An overview for the speech and hearing practitioner. *Clinics in Communication Disorders, 2*(4), 59–72.

Catts, H. (1989a). Defining dyslexia as a developmental language disorder. *Annals of Dyslexia, 39,* 50–64.

Catts, H. (1989b). Phonological processing deficits and reading disabilities. In A. Kamhi & H. Catts (Eds.), *Reading disabilities: A developmental language perspective.* Austin, TX: Pro-ed.

Cherny, S. S., Cardon, L. R., Fulker, D. W., & DeFries, J. C. (1992). Differential heritability across levels of cognitive ability. *Behavior Genetics, 22,* 153–162.

Cherny, S. S., & Fulker, D. W. (1993, June). *Continuity and change in specific cognitive abilities from ages 3 through 9 years.* Paper presented at the annual Behavior Genetics Association Meeting, Sydney, Australia.

Chomsky, N. (1988). *Language and problems of knowledge.* Cambridge, MA: MIT Press.

Crago, M. B., & Gopnik, M. (1994). From families to phenotypes: Theoretical and clinical implications of research into the genetic basis of specific language impairment. In R. V. Watkins & M. L. Rice (Eds.), *Specific language impairments in children.* Baltimore: Brooks.

Davidson, E. H. (1976). *Gene activity in early development.* New York: Academic Press.

DeFries, J. C., & Fulker, D. W. (1985). Multiple regression analysis of twin data. *Behavior Genetics, 15,* 467–473.

DeFries, J. C., & Gillis, J. J. (1990). Etiology of reading deficits in learning disabilities: Quantitative genetic analysis. In J. E. Obrzut & G. W. Hynd (Eds.), *Advances in the neuropsychology of learning disabilities: Issues, methods, & practices.* Orlando, FL: Academic Press.

DeFries, J. C., & Gillis, J. J. (1993). Genetics of reading disability. In R. Plomin & G. McClearn (Eds.), *Nature, nurture and psychology.* Washington, DC: APA Press.

DeFries, J. C., & Plomin, R. (1983). Adoption designs for the study of complex behavioral characteristics. In C. L. Ludlow & J. A. Cooper (Eds.), *Genetic aspects of speech and language disorders.* New York: Academic Press.

DeFries, J. C., Plomin, R., & LaBuda, M. C. (1987). Genetic stability of cognitive development from childhood to adulthood. *Developmental Psychology, 23,* 4–12.

Eaves, L. J., Eysenck, H. J., & Martin, N. (1989). *Genes, culture, and personality.* New York: Academic Press.

Eaves, L. J., Hewitt, J. K., Meyer, J. M., & Neale, M. C. (1990). Approaches to quantitative genetic modelling of development and age-related changes. In M. E. Hahn, J. K. Hewitt, N. D. Henderson & R. Benno (Eds.), *Developmental behavior genetics: Neural, biometrical and evolutionary approaches.* New York: Oxford University Press.

Eaves, L. J., Long, J., & Heath, A. C. (1986). A theory of developmental change in quantitative phenotypes applied to cognitive development. *Behavior Genetics, 16,* 143–163.

Falconer, D. S. (1981). *Introduction to quantitative genetics* (2nd ed.). London: Longman House.

Felsenfeld, S. (1994). Developmental speech and language disorders. In J. C. DeFries, R. Plomin, & D. W. Fulker (Eds.), *Nature and nurture during middle childhood.* Cambridge, MA: Basil Blackwell.

Fletcher, J. M. (1992). The validity of distinguishing children with language and learning disabilities according to discrepancies with IQ: Introduction to the special series. *Journal of Learning Disabilities, 25*(9), 546–548.

Freeman, J. M. (Ed.). (1985). *Prenatal and perinatal factors associated with brain disorders.* Rockville, MD: National Institutes of Health.

Fulker, D. W., & Eysenck, H. J. (1979). Nature and nurture: Heredity. In H. J. Eysenck (Ed.), *The structure and measurement of intelligence* (pp. 102–132). Berlin: Springer.

Fuller, J. L., & Simmel, E. C. (Eds.). (1983). *Behavior genetics: Principles and applications.* Hillsdale, NJ: Lawrence Erlbaum Associates.

Furrow, D., Nelson, K., & Benedict, H. (1979). Mother's speech to children and syntactic development: Some simple relationships. *Journal of Child Language, 6*, 423–442.

Gelderman, H. (1975). Investigations on inheritance of quantitative characters in animals by gene markers. I. Methods. *Theoretical and Applied Genetics, 46*, 319–330.

Gilger, J. W. (1992). Genetics in disorders of language. *Clinics in Communication Disorders, 2*(4), 35–47.

Gilger, J. W. (1995). Behavioral genetics: Concepts for research and practice in language development and disorders. *Journal of Speech and Hearing Research, 38*, 1126–1142.

Gilger, J. W., Borecki, I., Smith, S. D., DeFries, J. C., & Pennington, B. F. (1996). The etiology of extreme scores for complex phenotypes: An illustration using reading performance. In C. Chase, G. Rosen, & G. Sherman (Eds.), *Neural and cognitive mechanisms underlying speech, language, and reading*. Tinmonium, MA: York Press.

Gilger, J. W., Pennington, B. F., & DeFries, J. C. (1991). Risk for reading disabilities as a function of parental history in three samples of families. *Reading and Writing, 3*, 205–217.

Gilger, J. W., Pennington, B. F., Green, P., Smith, S. A., & Smith, S. M. (1992). Dyslexia, immune disorders, and left-handedness: Twin and family studies of their relations. *Neuropsychologia, 30*(3), 209–227.

Gopnik, M., & Crago, M. (1991). Familial aggregation of a developmental language disorder. *Cognition, 39*, 1–50.

Greenberg, D. A. (1993). Linkage analysis of "Necessary" Disease Loci versus "Susceptibility" loci. *American Journal of Human Genetics, 52*, 135–143.

Gusella, J. F. (1991). Huntington's disease. In H. Harris & K. Hirschhorn (Eds.), *Advances in human genetics*. New York: Plenum.

Hall, P. K., & Tomblin, J. B. (1977). A follow-up study of children with articulation and language disorders. *Journal of Speech and Hearing Disorders, 43*, 227–241.

Hanebuth, E., Gilger, J. W., DeFries, J. C., Smith, S. D., & Pennington, B. F. (in review). *How does parental compensation modify the transmission of familial reading problems?*

Hardy-Brown, K. (1983). Universals and individual differences: Disentangling two approaches to the study of language acquisition. *Developmental Psychology, 19*, 610–624.

Hardy-Brown, K., & Plomin, R. (1985). Infant communicative development: Evidence from adoptive and biological families for genetic and environmental influences on rate differences. *Developmental Psychology, 21*, 378–385.

Hardy-Brown, K., Plomin, R., & DeFries, J. C. (1981). Genetic and environmental influences on the rate of communicative development in the first year of life. *Developmental Psychology, 17*, 704–717.

Hatano, M., Roberts, C. W. M., Minden, W. M., Crist, W. M., & Korsemeyer, S. J. (1991). Deregulation of homeobox gene HOX11 by the t(10;4) in T cell leukemia. *Science, 253*, 79–81.

Hay, D. A., Prior, M., Collett, S., & Williams, M. (1987). Speech and language development on twins. *Acta Geneticae Medicae et Gemellogiae, 36*, 213–223.

Honzik, M. P. (1957). Developmental studies of parent-child resemblance in intelligence. *Child Development, 28*, 215–228.

Hunkapiller, T., Huang, H., Hood, L., & Campbell, J. H. (1982). The impact of modern genetics on evolutionary theory. In R. Milkman (Ed.), *Perspectives on evolution*. Sunderland, MA: Sinaver Associates.

Hunt, E. B. (1983). On the nature of intelligence. *Science, 219*, 141–146.

Jacobson, M. (1978). *Developmental neurobiology*. New York: Plenum.

Johnston, J. R. (1991). The continuing relevance of cause: A reply to Leonard's "Specific Language Impairment as a clinical category". *Language, Speech, and Hearing Services in the Schools, 22*, 75–79.

Julian, A., Plomin, R., Braungart, J. M., Fulker, D., & DeFries, J. C. (1992, June). *Genetic influence on communication development: A sibling adoption study of two and three year olds*. Paper presented at the Behavior Genetics Association conference, Boulder, CO.

Legido, A., Tonges, L., & Carted, D. (1993). Treatment variables and intellectual outcome in children with phenylketonuria: A single center based study. *Clinical Pediatrics, 32*(7), 417–425.

Leonard, L. B. (1991). Specific language impairment as a clinical category. *Language, Speech, and Hearing Services in the Schools, 22,* 66–68.

Leonard, L. B., Chapman, K., Rowan, L. E., & Weiss, A. L. (1983). Three hypotheses concerning young children's imitations of lexical items. *Developmental Psychology, 19,* 591–601.

Leske, M. C. (1981). Prevalence estimates of communicative disorders in the U.S.: Speech disorders. *Journal of the American Speech and Hearing Association, 23,* 217–225.

Lewin, B. (1985). *Genes II* (2nd ed.). New York: Wiley.

Lewis, B. (1992). Genetics in speech disorders. *Clinics in Communication Disorders, 2*(4), 48–58.

Lewis, B. A., Cox, N. J., & Byard, P. J. (1993). Segregation analysis of speech and language disorders. *Behavior Genetics, 23,* 291–299.

Lewis, B. A., Ekelman, B. L., & Aram, D. M. (1989). A familial study of severe phonological disorders. *Journal of Speech and Hearing Research, 32,* 713–724.

Lewis, B. A., & Thompson, L. A. (1992). A study of developmental speech and language disorders in twins. *Journal of Speech and Hearing Research, 35,* 1086–1094.

Locke, J. L., & Mather, P. L. (1989). Genetic factors in the ontogeny of spoken language: Evidence from monozygotic and dizygotic twins. *Journal of Child Language, 16,* 553–559.

Loehlin, J. C. (1987). *Latent variable models: An introduction to factor, path and structural analysis.* Hillsdale NJ: Lawrence Erlbaum Associates.

Martin, J. B. (1987). Molecular genetics: Applications to the clinical neurosciences. *Science, 238,* 765–772.

Matheny, A. P., & Bruggemann, C. E. (1973). Children's speech: Heredity components and sex differences. *Folia Phoniatrica, 25,* 442–449.

Mather, P. L., & Black, K. N. (1984). Heredity and environmental influences on preschool twins' language skills. *Developmental Psychology, 20,* 303–308.

McClearn, G., & Foch, T. T. (1985). Behavioral genetics. In J. E. Birren & K. W. Schaie (Eds.), *Handbook of the psychology of aging* (2nd ed.). New York: Van Nostrand Reinhold.

McKusick, V. A. (1990). *Mendelian inheritance in man* (9th ed.). Baltimore: Johns Hopkins University Press.

Molfese, D. L., Molfese, V., & Carrell, P. L. (1982). Early language development. In B. B. Wolman (Ed.), *Handbook of developmental psychology.* Englewood Cliffs, NJ: Prentice-Hall.

Morton, N. E. (1982). *Outline of genetic epidemiology.* London: S. Karger.

Munsinger, H., & Douglass, A. (1976). The syntactic abilities of identical twins, fraternal twins, and their siblings. *Child Development, 47,* 40–50.

Neale, M., & Cardon, L. (1992). *Methodology for genetic studies of twins and families.* The Netherlands: Kluwer.

Nichols, R. (1978). Twin studies of ability, personality, and interests. *Homo, 29,* 158–173.

Olson, R., Forsberg, H., & Wise, B. (1994). Genes, environment, and the development of orthographic skills. In V. W. Berninger (Ed.), *The varieties of orthographic knowledge I: Theoretical and developmental issues* (pp. 27–71). The Netherlands: Kluwer.

Olson, R., Wise, B., Conners, F., & Rack, J. (1989). Specific deficits in component reading and language skills: Genetic and environmental influences. *Journal of Learning Disabilities, 22,* 339–348.

Ooki, S., & Asaka, A. (1993, June). *Genetic analysis of motor development, language development and some behavior characteristics in infancy.* Paper presented at the annual Behavior Genetics Association Meeting, Sydney, Australia.

Osborne, R. T., Gregor, A. J., & Miele, F. (1968). Heritability of factor V: Verbal Comprehension. *Perceptual and Motor Skills, 26,* 191–202.

Parlour, S. F., & Broen, P. A. (1989, June). *Familial risk for articulation disorders: A 25 year follow-up.* Paper presented at the annual Behavior Genetics Association meeting, Charlottesville, VA.

Pauls, D. L., Towbin, K. E., Leckman, J. F., Zahner, G. E. P., & Cohen, D. J. (1986). The inheritance of Gilles de la Tourette's syndrome and associated behaviors. *New England Journal of Medicine, 315,* 993–997.

Pennington, B. F. (1986). Issues in the diagnosis and phenotype analysis of dyslexia: Implications for family studies. In S. D. Smith (Ed.), *Genetics and learning disabilities.* San Diego: College Hill Press.

Pennington, B. F. (1991a). *Diagnosing learning disorders: A neuropsychological framework.* New York: Guilford.

Pennington, B. F. (Ed.). (1991b). *Reading disabilities: Genetic and neurological influences.* The Netherlands: Kluwer.

Pennington, B. F., Gilger, J. W., Olson, R., DeFries, J. (1992). External validity of age versus IQ discrepant diagnoses in reading disability: Lessons from a twin study. *Journal of Learning Disabilities, 25*(special), 562–573.

Penrose, L. S. (1963). *The biology of mental defect* (3rd ed.). London: Sidgwick & Jackson.

Pinker, S. (1991). Rules of language. *Science, 253,* 530–535.

Pinker, S. (1994). *The language instinct: How the mind creates language.* New York: Morrow.

Pinker, S., & Bloom, P. (1990). Natural language and natural selection. *Brain and Behavioral Sciences, 13,* 707–784.

Plomin, R. (1981). Ethological behavioral genetics and development. In K. Immelmann, G. Barlow, L. Petrinovich, & M. Main (Eds.), *Behavioral development: The Bielefeld interdisciplinary project.* New York: Cambridge University Press.

Plomin, R. (1986). *Development, genetics, and psychology.* Hillsdale, NJ: Lawrence Erlbaum Associates.

Plomin, R. (1990). The role of inheritance in behavior. *Science, 248,* 183–248.

Plomin, R. (1993, June). *Molecular genetic investigation of low and high cognitive ability in children.* Paper presented at the annual Behavior Genetics Association Meeting, Sydney, Australia.

Plomin, R., & Bergeman, C. S. (1991). The nature of nurture: Genetic influence on "environmental" measures. *Behavioral and Brain Sciences, 14,* 373–427.

Plomin, R., & Daniels, D. (1987). Why are children in the same family so different from each other? *Behavioral and Brain Sciences, 10,* 1–16.

Plomin, R., DeFries, J. C., & Fulker, D. W. (1988). *Nature and nurture during infancy and early childhood.* New York: Cambridge University Press.

Plomin, R., DeFries, J. C., & McClearn, G. E. (1990). *Behavioral genetics: A primer* (2nd ed.). New York: W. H. Freeman.

Plomin, R., & Rende, R. (1991). Human behavioral genetics. *Annual Review of Psychology, 42,* 1–66.

Plumridge, D., Bennett, R., Dinno, N., & Branson, C. (1993). *The student with a genetic disorder.* Springfield, IL: Thomas.

Reznick, J. S. (in press). Intelligence, language, nature and nurture in young twins. In R. Sternberg & E. Grigorenko (Eds.), *Intelligence, heredity, and environment.* New York: Cambridge University Press.

Rice, M. L., & Woodsmall, L. (1988). Lessons from television: Children's word learning when viewing. *Child Development, 59,* 420–429.

Roberts, J. A. F., & Pembrey, M. E. (1985). *An introduction to medical genetics.* New York: Oxford University Press.

Rowe, D. C. (1994). *The limits of family influence: Genes, experience, and behavior.* New York: Guilford.

Scarborough, H. S. (1989). Prediction of reading disability from familial and individual differences. *Journal of Educational Psychology, 81,* 101–108.

Scarborough, H. S. (1990). Development of children with early language delay. *Journal of Speech and Hearing Research, 33,* 70–83.

Scarr, S., & Carter-Saltzman, L. (1983). Genetics and intelligence. In J. Fuller & E. Simmel (Eds.), *Behavior genetics: Principles and applications*. Hillsdale, NJ: Lawrence Erlbaum Associates.

Schoenfeldt, L. F. (1968). The hereditary components of the Project Talent two-day test battery. *Measurement and Evaluation in Guidance, 1*, 130–140.

Siegel-Sadewitz, V., & Shprintzen, R. J. (1982). The relationship of communication disorders to syndrome identification. *Journal of Speech and Hearing Disorders, 47*, 338–354.

Skodak, M., & Skeels, H. M. (1949). A final follow-up of one hundred adopted children. *Journal of Genetic Psychology, 75*, 85–125.

Smith, S. D. (1992). Identification of genetic influences. *Clinics in Communication Disorders, 2*, 73–85.

Sparks, S. N. (1984). *Birth defects and speech-language disorders*. San Diego: College-Hill Press.

Stevenson, J. (1991). Which aspects of processing text mediate genetic effects? In B. Pennington (Ed.), *Reading disabilities: Genetic and neurological influences* (pp. 61–82). The Netherlands: Kluwer.

Stevenson, J., Pennington, B. F., Gilger, J. W., DeFries, J. C., & Gillis, J. (1993). ADHD and spelling disability: Testing for shared genetic aetiology. *Journal of Child Psychology, Psychiatry, & Allied Disciplines, 34*, 1137–1152.

Stoel-Gammon, C. (1990). Down's Syndrome: Effects on language development. *Journal of the American Speech and Hearing Association, 32*(9), 42–44.

Studdert-Kennedy, M. (1990). Language development from an evolutionary perspective. *Haskins Laboratory Status Report on Speech Research, 101–102*, 14–27.

Tallal, P., Ross, R., & Curtiss, S. (1989). Familial aggregation in specific language impairment. *Journal of Speech and Hearing Disorders, 54*, 167–173.

Thompson, M., McInnes, R., & Willard, H. (1991). *Genetics in medicine*. Philadelphia, PA: W. B. Saunders.

Tomblin, B. (1989). Familial concentration of developmental language impairment. *Journal of Speech and Hearing Disorders, 54*, 287–295.

Tomblin, B. (1991). Examining the cause of specific language impairment. *Language, Speech, and Hearing Services in the Schools, 22*, 69–74.

Tomblin, B., & Buckwalter, P. R. (1994). Studies of the genetics of specific language impairment. In R. Watkins & M. L. Rice (Eds.), *Language impairments in children* (pp. 17–34). Baltimore, MD: Brookes.

Tomblin, B., Hardy, J. C., & Hein, H. (1991). Predicting poor communication status in preschool children using risk factors present at birth. *Journal of Speech and Hearing Research, 34*, 1096–1105.

Truett, K. R., Eaves, L. J., Meyer, J. M., Heath, A. C., & Martin, N. G. (1992). Religion and education as mediators of attitudes: A multivariate analysis. *Behavior Genetics, 22*, 43–62.

Vargha-Khadem, F., & Passingham, R. (1990). Speech and language defects. *Nature, 346*, 226.

Watson, J. D., & Crick, F. H. C. (1953). Molecular structure of nucleic acids: A structure for deoxyribose nucleic acids. *Nature, 171*, 964–967.

Whitehurst, G. (1982). Language development. In B. Wolman (Ed.), *Handbook of developmental psychology*. Englewood Cliffs, NJ: Prentice-Hall.

Wood, N., Tyfield, L., & Bidwell, J. (1993). Rapid classification of phenylketonuria genotypes by analysis of heteroduplexes generated by PCR-amplifiable synthetic DNA. *Human Mutation, 2*(2), 131–137.

LINGUISTICS AND LANGUAGE ACQUISITION

5

THE DEVELOPMENT OF INFLECTION IN A BIOLOGICALLY BASED THEORY OF LANGUAGE ACQUISITION

Kenneth Wexler
Massachusetts Institute of Technology

The field of first language acquisition has made large theoretical and empirical strides in recent years. On the one hand, the amount and quality of empirical material that is investigated has shown a significant increase. On the other, the theoretical integration of this material has proceeded rapidly, so that a variety of empirical material, in a much more unified fashion than in the past, is understood. In this chapter, there is no way to review the range of work in language acquisition that shows these characteristics. What I do instead is present a few of the general ideas in the field and develop one of the areas of active interest in a bit more detail, namely the area of inflectional development. In this way, I hope to have developed enough of the ideas in the field to discuss how they might interact with the study of genetics.

In the following section, I discuss the modern view of language and its acquisition arising from linguistic theory, in particular the theory of principles and parameters and its general relation to genetic programs. The next section, on "Inflection and Syntax," illustrates the theory in one particular syntactic/morphological domain. The following section, on "Children, Inflection, and Verb Movement," presents evidence about the knowledge that children have of these principles and parameters, focusing on the Optional Infinitive (OI) stage and the relation of this stage to maturation and brain-based theories of linguistic knowledge. The following section develops the OI model in detail for English. The chapter ends with a short discussion of "Learnability and Maturation," returning to the theme of language as a piece of biology.

PRINCIPLES, PARAMETERS, UG, AND GENETICS

As with the study of any complex system, complete understanding of language acquisition requires that one understand many kinds of phenomena and levels of description. In language these include phonetics, phonology, morphology, syntax, semantics, pragmatics, and discourse. There is active investigation of language acquisition in each of these areas. The example I have chosen—inflection—relates to many of these levels. I concentrate on considerations that are primarily morphological, syntactic, semantic (and possibly ultimately pragmatic).

The fundamental problem of linguistic theory (Chomsky, 1965; Wexler, 1990b; Wexler & Culicover, 1980) is the problem of learnability: How is it possible for any normal child to learn any natural language? The study of language acquisition has as its goal the solution to this problem along with the solution to why the developmental course of language is as it is.

The most important thing, and possibly the most conceptually difficult thing, to understand about language acquisition is that it is not at all obvious how it works. What is it about human beings that allows them to learn language relatively quickly (i.e., for the most part within a few years, but in fact, much is known much earlier)? What possible kind of learning mechanism allows this learning to take place?

For somebody from outside the field of psycholinguistics or linguistics, it is crucial to understand that language is not just a list of words that must be memorized. This list, the *lexicon*, is an extremely small part of the linguistic competence of human beings. There is no way to describe in this space what is known about the competence that is studied in great detail in linguistic theory, with an enormous amount of evidence from a wide range of human languages. But the general conclusion is quite clear: Language is a highly specific and intricate system located in the human brain/mind. The detail of the structure in human language goes way beyond the information in the linguistic environment that would be obvious to an uninformed, general purpose learning device. In other words, there is ample reason to believe that a general purpose learning device could not learn natural (i.e., human) language. (For extensive mathematical work in this area, see works in *learnability theory*, i.e., Gibson & Wexler, 1993; Wexler & Culicover, 1980.) There are also large numbers of arguments from linguistic theory and from the study of language acquisition itself to show that the learner contributes much of the structure of human language.

In the vocabulary of contemporary studies of learning theory (Rumelhart & McClelland, 1986), language learning is not a case of supervised learning. Adults do not instruct children in the principles by which sentences are formed, nor do they even correct the utterances of children that are not well-formed according to the adult system, nor do they fail to understand and thus give indirect correction to non-well-formed utterances. If a child says "her go" at the age of

2, it is the rare adult who says "Say 'she goes.' " If the adult did, in fact, say this, it would be unlikely in general that the child understood that her grammar was being corrected. Thus, there is no negative evidence in language acquisition. (The term, I believe, was first introduced in Wexler and Hamburger, 1973, and the claim was made there that negative evidence did not exist and that this was a central problem for the field of language acquisition because learning would be more difficult without negative evidence. The empirical basis for the claim was results in Brown and Hanlon, 1970. Work since then has generally confirmed the results. See Marcus, 1993, for a recent review of the literature.) Even if it turns out that there is some negative evidence, it should be emphasized that the general problem of learnability remains; how can the language learner create a complex system on the basis of evidence that does not give most of the details of that system.

Thus, the conclusion is that the human brain contains specific knowledge about human language. In other words, language is a species-specific ability, analogous to species-specific abilities that every animal has. Language unfolds according to a genetic program, interacting with environmental events.

From the standpoint of any other animal such a conclusion, that it had genetically based abilities, would be unexceptional in the world of biology. The fact that the claim has not always been accepted in the study of language results, I believe, from the fact that language has been studied more in psychology than in biology; in other words, traditionally in psychology the assumption has been that complex abilities are learned. However, now that psychologists in many areas are attempting to integrate research more closely with biology, it is likely that the claim that there is a basis in the brain for the structure of human language will be viewed as the natural expectation, simply another case of an organism specializing in various ways through long years of evolution. For very important early work laying out the biological basis for locating the structure of language in the human brain, see Lenneberg (1967).

One central element in understanding how language develops, then, is the question of what constitutes knowledge of language. What *is* the structure of language? Linguistic theory gives the answer as *Universal Grammar* (UG), a set of principles that characterize knowledge of language, in fact, of the fundamental basis of all languages. (Note that *grammar* is used here in the broad sense to include, for example, phonetics and pragmatics, that is, any part of the human language system. *Grammar* simply means the underlying system in the brain that is the basis of human language.)

Linguistic theory has an extensive, detailed, and formal characterization of UG. Of course, UG is a research topic; linguists spend their time trying to characterize UG and the characterization, at any time, is a hypothesis about the nature of UG. There is no reason to expect that the characterization of UG is any more correct at any given time than the characterization of, say, any biological process. Thus, the characterization of UG changes over time, just as the characterization of the biological basis, for example, of vision changes. The fact that there are receptors for particular colors in the human eye was a fact unknown to researchers 40 years ago, though it was, perhaps, guessed at. The fact that there is more than one channel for vision, depending on the density of the scene, is a fact hardly guessed at years ago. There is no reason to think that these "facts" will not change in some fundamental way. The study of UG has exactly the same scientific status; namely, its details are an empirical hypothesis. As new evidence accumulates from all sorts of languages, as old phenomena are looked at in new ways that are, it is hoped, more integrative, as evidence comes in from language acquisition, language processing, studies of deviant language and related fields, it is expected that the description of UG will change. All this is only to say that the study of language is a living research field. If the described nature of UG were not to change, then linguists would no longer be dealing with a living science.

As soon as the hypothesis of UG is invoked, however, we must deal with the observation that there appears to be more than one language. English is different from French and from Chinese. Moreover, a speaker of one of these languages usually cannot speak the other one. Furthermore, which language a person speaks depends on which environment she was raised in. How do we account for this if all languages share one underlying UG?

The solution given to this problem in contemporary linguistic theory is the hypothesis that there are small, particular ways in which languages differ, and that these differences are learned through linguistic experience. UG allows certain particular kinds of variation. For example, there is nothing in UG that says that H_2O is pronounced *water* or *eau* or anything else. Thus, the phonetic form of a vocabulary item is not specified. And there are other differences allowed between languages. These places where there is some openness in UG, an openness that is closed by experience, have come to be called (linguistic) *parameters*. Some of them are more structural than the phonetic example given above. Parameters are set by experience. Examples of linguistic parameters will soon be mentioned. For now, imagine a (probably too simple) case, which would determine whether the (grammatical) object of a sentence came before or after the verb; call this the *object direction parameter*. Thus, English places the object *after* the verb (*she read the book*, where *the book* is the object), and Japanese places the object *before* the verb (the Japanese equivalent of *she the book read*). A child learning English or Japanese would set the object direction parameter as *object after verb* (English) or *object before verb* (Japanese) on the basis of experience, that is, by hearing and attempting to understand sentences, either English or Japanese.

It is not to be thought that any of this learning process is conscious; there is no evidence that it is. It is expected that it is implicit, as much learning is. Even more, we expect that the principles of UG are implicit; they are not conscious. They only become available to us consciously as scientific hypotheses when studying the nature of UG. To speakers of the language, they are generally implicit.

There are a number of hypotheses on the nature of parameters under investigation. One hypothesis is that parameters are varying properties of *morphological* (roughly, *lexical*, i.e., word-like) items. Even though languages appear to differ in more fundamental ways, this hypothesis holds that the appearance is wrong (Borer, 1983; Borer & Wexler, 1987; Fakui, 1988; Wexler & Chien, 1985). I will not review here any of the hypotheses about parameters; the point to be understood is that parameters will clearly delineate places where languages can differ and that learning through experience sets the parameters. Parameters by themselves cannot explain most of the structure of language; rather this core of linguistic abilities is characterized by UG.

On this *Principles and Parameters* view of the nature of human language, the problem of learning theory becomes the problem of *parameter-setting*. There is a good deal of discussion of parameter-setting in the literature (see, e.g., Gibson & Wexler, 1993; Hyams, 1986; Manzini & Wexler, 1987).

Researchers studying language development now have some particular questions to ask, including the following central ones:

1. Do children show linguistic systems that conform to UG? Do their underlying systems change over time?
2. Do children show that they have set the parameters, either correctly or incorrectly?
3. How do children set the parameters? That is, how does learning take place?

One of the major results of the study of language acquisition in recent years, I believe, is the demonstration that children's language conforms to UG in many essential respects. This is particularly intriguing because there is so much abstract structure in UG (i.e., structure that is not immediately obvious from the input) and, in fact, processes that are hardly the kinds of processes one thinks exist in language before it has been studied scientifically. (For an elementary example, see the discussion on *c-command* in Wexler, 1990a. For a nontechnical discussion of some linguistic properties, see Lightfoot, 1982; Pinker, 1994.)

At the same time, there has been evidence that certain aspects of UG mature (i.e., develop according to a general human program, as opposed to being guided in a detailed way by experience; Borer & Wexler, 1987, 1992; Wexler, 1990a). The sense of maturation I have in mind is, say, the maturation that underlies the development of a second set of teeth or of secondary sexual characteristics. These developments take place according to a biological program, with somewhat varying times in the population. Although the environment certainly can affect the maturation (e.g., nutrition might affect the development of secondary sexual characteristics), it is uncontroversial that the development is essentially guided by a biological, genetically determined program. There is reason to believe that some aspects of UG share this rather omnipresent aspect of biological phenomena. Biological structures and processes mature according to a biological program, either before or after birth.

The idea of genetically programmed maturation is so strong in the study of biology that a special term has been defined for exceptions. This term is *plasticity*. Plasticity means that there is experience-dependent variation in biological structures or processes. It is considered a major discovery in the study of the brain in neuroscience, for example, when it is demonstrated that a certain process is plastic. The reason this is considered a major discovery is because the general view is one of a biological, genetically based program guiding development (see Nadel & Wexler, 1984, for discussion).

Thus, it should not be surprising to find both brain-based, somewhat fixed, structures in language, as well as certain details depending on experience. Such an outcome would fit language into the kinds of structures otherwise known in the biological world and make language much less mysterious. Language can now be viewed simply as a biological phenomenon, for the most part, given current knowledge, unique to our species, though most likely related to homologues in the animal world (because most structures are related to homologues in other species).

INFLECTION AND SYNTAX

One basic structure of human language is the sentence. Here I am concerned only with simple sentences (i.e., sentences that do not have other sentences as subparts). There is a basic structure to the sentence in any human language. Some parts of this structure are obvious. There are nouns and verbs that combine into large noun phrases and verb phrases. The noun phrases (NPs) are related to the verbs (V) and verb phrases (VPs) in particular semantic ways. Thus, consider sentence (1):

(1) The woman kissed the child.

In (1), the *subject* of the sentence is the noun phrase *the woman*, and the *object* of the sentence is the noun phrase *the child*. These noun phrases also have semantic or thematic roles. For example, *the woman* is the *agent* of the sentence (i.e., the actor), and *the child* is the *patient* of the sentence (i.e., the acted-upon). The verb *kissed* denotes the kind of action that the agent performs on the patient. So much is obvious.

There is much more to the syntax of such a sentence, however, that is not so immediately obvious. The verb I chose in (1), *kissed*, shows past tense, indicating that the action took place before the moment at which the sentence was spoken. Had I uttered (2), however, the speaker would have been saying that the action takes place at the current moment (actually, the current moment or later):

(2) a. The woman kisses the child.
    b. The woman is kissing the child.

Sentence (2a) is in present tense. *Tense* in English is either *past* (sentence 1) or *present* (sentence 2a or 2b). Sentence (2b) is in present tense, but it is in *progressive* aspect, as indicated by the use of a form of *be* (*is*) and the *-ing* marking on the end of the verb.

Thus, *-ed* is the marking for past tense in English. *-(e)s* is the marking for present tense in English, when the subject of the sentence is third person singular (first person = speaker, second person = person spoken to, third person = person or thing spoken about, i.e., other than the speaker or person spoken to). *Singular* means that the subject is only one person, as opposed to more than one, in which case the *plural* is used (*the women kiss the child*, i.e., *kiss* instead of *kisses*).

Markings like past tense *-ed* and present tense *-s* are called *inflection*. Because these particular inflections occur on verbs, they are called *verbal inflections*. English has a fairly simple, undifferentiating, system of verbal inflection. Many languages have much more verbal inflection than English. But there is enough in English to illustrate one point: There is a process called *agreement*, which means that the verb has an inflection that can agree with the number (singular vs. plural) and person (first, second, third) features of the subject. (Thus *-s* occurs on verbs with third singular subjects, whereas nothing occurs on verbs with other than third singular subjects.) Agreement in English only occurs in the present tense. Thus, the only past tense inflection in English is *ed*, for all subjects of any person or number. There is no agreement in past tense for English verbs.

Although it looks as if verbal inflection systems are very simple, and purely the result of glueing little morphemes onto bigger words, systematic investigation shows that there is much more to the study of inflection than this. In particular, a major result of contemporary linguistic theory is that there is a deep relation between the morphological processes of inflection and the syntax of the sentence. I illustrate this with an example from French.

French is similar to English in that the basic word order is subject-verb-object; that is, it is an SVO language (in contrast to, say, Japanese, which is SOV). Like English, French marks tense and agreement on its verbs. Consider sentence (3):

(3) Marie embrasse Jean.
    Mary kisses John.

The verb *embrasse* shows the third singular, present tense marking.

Something very interesting happens when sentences are negated, however: the word order changes, depending on the kind of inflection on the verb. First, notice that the basic word for *not* in French is *pas* (we will neglect here the role of *ne*). Ordinarily, the verb precedes *pas*, as in (4a):

(4) a. V finite pas . . .
    b. pas V nonfinite
    c. *V nonfinite pas
    d. *pas V finite . . .

The first thing to note is that in (4b) the order of the verb and *pas* is exactly the opposite from that in (4a). In fact, the order given in (4b) is the only possible order in this case; if we try to use the order from (4a), with the form of the verb in (4b), we get sentence (4c), which is not part of French. (The technical term is that (4c) is *ungrammatical*; this means that French speakers consider (4c) to not be part of their language or to violate some property of French. As implicit throughout, this is a *descriptive*, not *prescriptive* finding; what we mean is that the brains of French speakers have a system that finds (4c) to be a deviant structure. The asterisk before (4c) is notation for *ungrammatical*, that is, deviant.) Similarly, (4d) indicates that with the verb inflection type in (4a), the order of the verb before *pas* is ungrammatical.

What is the difference between (4a) and (4b) such that the word order of the verb and *pas* has been reversed? (4a) involves the kind of verb I have discussed so far, a verb in a particular tense that indicates a tense marking directly, namely present tense. The verb in (4b) in nontensed; it does not indicate a tense marking. It is an infinitival form. One traditional term for tensed verbs is *finite* verb, and for nontensed verbs, *nonfinite* verb. The generalization that is true of French, as can be seen by detailed study of many sentences, is shown in (5):

(5) a. Finite verbs precede *pas* (in French).
    b. Nonfinite verbs follow *pas* (in French).

There is a large amount of well studied evidence in many languages that shows that there is much interaction between verb inflection and word order or other processes that have been thought of as *syntactic*. In other words, verbal inflection (morphology) and syntax are deeply related. Of course, I cannot review the evidence and theories that result from this point of view. I will give one conclusion that has been arrived at to account for example (5), without giving most of the evidence.

It is assumed that pieces of inflection (like past tense *-ed* in English, or third person singular present tense *-s*) correspond to *functional categories*. That is, there is a category Tense (TNS) and a category Agreement (AGR). These are real categories, finding their way into the structural representations of sentences. They are, thus, similar to categories like V (verb) or N (noun), which also appear in syntactic representations. They have some different properties than nouns (N) or verbs (V). Thus the inflectional categories like TNS and AGR are called *functional* categories, as opposed to the *lexical* categories like N and V. Notice that one important distinction is that lexical categories are usually fairly open-ended; there are lots of nouns or verbs, and they have to be listed in the lexicon (new ones can be created, etc.). Functional categories, on the other hand, usually only have a small number of particular members. For example, TNS is *past* or *present*. AGR shows values for the (usually six) person/number combinations. (Most of the phonetic forms for these are equivalent in English, but not in other languages. In English, the verb *be*, exceptionally for English, shows three different finite forms, *am*, *is*, and *are*.)

Now consider French. I show how the use of functional categories helps explain the word order in French, finite versus nonfinite sentences in example (4). The representation for sentence (4a) is:

(6) SUBJECT   TNS   NEG   V . . .

In example (6) I show that TNS exists as a category outside the VP. NEG (for negation) is also considered a functional category. The underlying structure of (4a) is given in (6). Note that the verb is marked +Present. Part of syntactic theory is that a verb that has a TNS marking (present, past) must check this marking against the functional category TNS to make sure that the correct TNS values are on both the TNS position and the verb itself. The only way that this checking can take place is if the verb is in a position (the *checking domain*) next to TNS. So the verb raises to TNS, so that its TNS value can be checked. The resulting output is in (7):

(7) SUBJECT V+TNS NEG . . .

(7) represents the surface structure of the sentence (4a), derived from the underlying representation in (6). The word order, the phonetic form more generally, of the sentence is read off from the surface structure (7). Thus, when the sentence is pronounced, the word order is that in (4a), with the verb before NEG *pas* because in (7) the verb precedes *pas*.

In other words, it is the need for checking the value on TNS in the underlying position of TNS against the value on the verb itself (i.e., the form of inflection chosen, past or present) that requires the verb to move up and thus changes the word order of the sentence. If the verb does not have to move, it will appear after *pas*.

Thus, consider (8), which is the underlying representation for sentence (4b), with a nonfinite verb.

(8) SUBJECT NEG   V . . .

The verb in (8) does not have a TNS-marking; it is nonfinite, an infinitival verb. Thus, it does not have to be checked against the TNS value in TNS (which, in fact, must not have a value of past or present, or else the derivation will crash for other reasons; all values of past or present must be checked, whether in TNS or on the verb).

It follows that the verb does not have to move up to TNS. I assume that there are economy considerations that insure that if a verb does not have to move, then it doesn't move; in other words, there is a principle of *procrastination: move only if necessary to check a grammatical feature.* Procrastination, therefore, insures that the verb does *not* move in the infinitival case (8). Thus, the order of (4b), *pas*, before the nonfinite verb, is derived. (For a detailed current presentation of the relevant theory, see Chomsky, 1995.)

It is important to observe that modern linguistic theory, in any of its implementations, for example, the one just sketched, finds it necessary to assume that there is a deep interaction between morphology and syntax. Inflectional forms are not simply little markers that go on other content words (e.g., verbs), perhaps adding meaning. Rather, the inflectional forms are signs of the existence of central grammatical categories in the internal representation of a sentence. The existence of these categories has large implications for the syntax of a sentence, as can be seen, for example, from the effect on the surface word order of a sentence.

In fact, this is a very satisfying view of inflection, because it brings it centrally into the representation of a sentence, and makes the ubiquity of inflection much less mysterious. These little forms that one adds onto words bear much grammatical and (sometimes) semantic weight. On this view, it should be obvious that the study of inflectional development might turn out to be a deep study on syntactic development, not simply a study of the memorization of some forms. In fact, this turns out to be the case.

I have illustrated one particular kind of movement that exists in natural language, verb movement. It is useful to think of verbs undergoing movement.[1] Verb movement is ubiquitous; it has been found and systematically studied in large numbers of languages.[2]

Thus, it is clear that to study the development of inflection we must ask not only what children know about morphology, but also what they know about syntax, because syntax and morphology are deeply interrelated in the domain

---

[1]There may be other ways to consider this relation than movement, that is, there may be *representational* rather than *movement* theories. The point is the same, however. Namely, there is an important discontinuous syntactic relation, one that exists between noncontiguous elements in the initial representation.

[2]Verb movement is a special kind of *head* movement, so-called because only the head of a phrase, not the whole phrase, moves. In the cases of verb movement, the verb itself is the head of the verb phrase; only the verb (the head) moves. The second kind of movement that is ubiquitous in natural language is whole phrase movement (as in most *wh*-questions, and passive raising). Current syntactic theory has provided evidence that head movement and whole phrase movement are essentially the only kinds of movement that UG allows.

of inflection. For example, to study whether children understand inflectional processes involving the verb, we have to ask what they know about verb movement. Moreover, because inflectional processes are deeply involved with functional categories, to study inflection in children, we have to study what they know about the existence, structural position, and meaning of functional categories. The study of inflection, thus, becomes a study of many of the foundational elements of sentence structure.

CHILDREN, INFLECTION, AND VERB MOVEMENT

The traditional study of inflectional development was a study of the inflectional forms that appeared on words. For example, at one point in a child's development did a TENSE morpheme appear to be attached to a verb stem? The major breakthrough in the study of inflectional development in recent years has occurred with the application of a very simple new idea, namely, the idea of correlating verbal inflection with syntactic position (word order) in children's utterances. The idea is extremely natural from the point of view of current syntactic theory, as I have just explained, because syntactic position (as exemplified by word order) and inflection (as exemplified by the morphological form of a verb) are strongly connected in the representation of a sentence. Thus, the idea to explore the connection between morphology and word order in children's utterances, as a sign of their knowledge of underlying grammar, is one which grows directly out of syntactic discoveries of the last 10 years. It turns out to be a very powerful idea.

An Example from French

Let me exemplify with the study of young children's knowledge of verbal inflection in French. I have already shown that French word order and verbal morphology are linked in a particular way; in particular, a tensed verb precedes negative *pas*, whereas a nonfinite verb follows negative *pas* (5). In the discussion that followed (5), I discussed how these facts were accounted for by the assumption that finite and nonfinite verbs occur in the same underlying position with respect to negation, but that finite verbs undergo a process of verb movement to a higher functional category, TENSE, which yields the different surface word orders.

Pierce (1989, 1992) and Weissenborn (1988) studied this phenomenon in children, with Pierce providing systematic counts, thus illustrating the method of investigation that has proven so powerful. I will, therefore, present some of her numbers here. (For further confirmation, see Friedemann, 1994.)

Pierce studied natural production data from three French children (two children originally recorded by Lightbown, 1977, and one child recorded by Suppes et al., 1973). She extracted all negative sentences (those that included *pas*) and noted two facts for each of the sentences: First, what was the morphological

TABLE 5.1
Distribution of Verbs

	[+finite]	*[−finite]*
pas verb	11	77
verb *pas*	185	2

Note. From Pierce (1989). Reprinted by permission.

form of the verb, finite or nonfinite? Second, did the verb precede *pas* or follow *pas*? The data from the total recordings of the three children lumped together are presented in Table 5.1.[3]

The results are striking. The three children provide many finite and nonfinite verbs used together with *pas*, and the verbs come both before and after *pas*. However, almost all of the finite verbs *precede pas* and almost all of the nonfinite verbs *follow pas*. The two other cells of Table 5.1 (*pas* following a nonfinite verb and *pas* preceding a finite verb) have entries close to zero, close enough that it is at least possible to consider such forms as performance errors.[4]

As Pierce argued, the systematic difference in the position of *pas* depending on whether the verb is finite or nonfinite, shows that these children know that finite verbs raise to the left of *pas*. In other words, these children know the existence of the functional category TENSE, the relation of the phonetics of inflectional items to morphological and syntactic features, the possibility of verb movement, and the morphosyntactic conditions that require verb movement when the verb is finite and disallow it when the verb is nonfinite. In the terms that I have previously outlined, the children know that the finite verb has a feature that must be checked against TENSE, and that economy conditions prevent a verb from raising on the surface if it does not have to.

In other words, the introduction of the method of correlating morphology and word order has shown that children at a very young age (before 2;0) know many of the principles of UG that are responsible for inflectional and clause structure and the particular values that the language they are learning has chosen for parametric differences. This is a remarkable demonstration. These facts were completely unknown 10 years ago, and researchers in the field of language acquisition had no idea that children at such an early age knew such a large amount about inflection, clause structure, and related syntactic processes, such as verb movement.

---

[3]Pierce provides systematic analysis of developmental trends and differences between children. None of those results alters our conclusions here.

[4]It is possible that there is a systematic difference between two small entries; the one representing a nonfinite verb appearing to the left of *pas* might be smaller than the entry representing a finite verb appearing to the right of *pas*. This possibility, though suggestive, has not yet been resolved in the literature.

The Optional Infinitive Stage

The method of correlating verbal morphology and word order in children's utterances was so obviously powerful that it was immediately extended to the study of many other languages and (to a lesser extent) other syntactic processes. For example, Poeppel and Wexler (1993) showed that a young German child (2;1) almost always placed finite verbs in second position in the sentence, as German main clauses require, and almost always placed nonfinite verbs in final position in the sentence, as German clauses also require. This is the same kind of fact as in French, but dependent on the particular parameter-settings of another language. German is a V2 language, one that requires finite verbs to be raised to second position in the sentence. Thus, the correlation of verbal morphology and word order was used in this study to show that German children knew the consequences of finiteness and nonfiniteness in German, including the fact that finite verbs and only finite verbs raise to second position. Thus, young German children know verb movement, why it is obligatory in certain conditions in German, and why it is forbidden in other conditions, essentially on the same kind of feature checking and economy grounds as in French, though with some particular differences.

One very important fact about the children's utterances must be immediately pointed out. As I have shown, the children produce many sentences with nonfinite verbs, almost always in the correct syntactic position for nonfinite verbs. But it should be puzzling that children this young produce nonfinite verbs. After all, nonfinite verbs are generally produced in embedded (subordinate) clauses of various kinds; simple main declarative clauses require finite verbs. It is known that children this young (for example, children younger than 2;0) produce hardly any embedded clauses. Why are nonfinite verbs in these children being observed?

In fact, the nonfinite verbs in these young children's utterances are *not* in embedded clauses. They are in simple clauses. Here are some examples in French (Pierce, 1992) and German (Poeppel & Wexler, 1993).

(9) a. pas tomber bebe
    not fall   baby
  b. pas attraper une fleur
    not catch   a   flower
(10) a. ich der Fos hab'n
    I   the frog have
    'I have the frog'
  b. Thorstn das haben
    T.   that have
    'Thorstn have that'

These nonfinite clauses exist as simple (main) clauses produced by children, although they may not appear as such in the adult language. Thus, it is not the case that children make no mistakes at all with respect to verbal inflection. A major difference between the children's utterances and adult utterances is that children at a certain stage seem to be willing to produce nonfinite main clauses.

This fact was first noted in a systematic fashion by Wexler (1990a, 1992, 1994a), who called these main clauses *(root) infinitives* or *optional infinitives*, to stress that for the most part the root infinitives coexist in children with root finite verbs. That is, most children appear to produce *both* finite and nonfinite main clauses.[5] Wexler (1994b) argued for the existence of an *Optional Infinitive (OI) Stage* in many languages.

(11) The optional infinitive stage has the following properties:
   a. There are main clauses with finite verbs.
   b. There are main clauses with nonfinite verbs.
   c. Nevertheless, the children know the difference between the finite and nonfinite verbs. That is, they know verb movement and all the morphosyntactic conditions associated with verb movement, both the UG conditions and the language-particular (parametric) conditions.

This description of the OI stage captures the examples previously discussed. For example, it allows for both finite (11a) and nonfinite (11b) main verbs in French at the OI stage, but requires that finite verbs precede *pas* and nonfinite verbs follow *pas* (11c). Similarly, in early German there are both finite and nonfinite verbs (11a, b), but finite verbs occur in second position and nonfinite verbs in final position (11c).

Wexler (1990a, 1992, 1994a) demonstrated that Danish, Dutch, English (to which I will return), French, German, Norwegian, and Swedish all showed the optional infinitive stage in young children of about the same age (e.g., in 2-year-olds). Since then, good evidence has appeared that Faroese (a Germanic lan-

---

[5]As Wexler (1994a) pointed out, it is quite possible, though not conclusively demonstrated, that there may be an early stage in children in which only nonfinite forms exist. Certainly, as children get older they produce a higher proportion of finite forms. Behrens (1993) analyzes data from seven German children and suggests that at the earliest observed recordings they only have nonfinite forms. However, this is not literally true. Rather, the youngest files from these children show a small number of finite utterances which Behrens analyzes as *formulaic* (i.e., memorized forms of some kind). Whether children's utterances are formulaic is a notoriously difficult judgment to make; Behrens follows some arbitrary criteria, but nobody knows if these criteria actually capture anything real about the children's grammars. On the other hand, the proportion of finite utterances in the grammars of extremely young children (1;6 or younger) is small enough in some files to make it plausible that if we could measure the children's grammars before they start speaking the proportion of nonfinite utterances would go to 100%. Barring further methods for investigating the grammars of children before they start talking, the question will probably be very difficult to answer. At any rate, it is extremely interesting, as I hope the text discussion makes clear, that very young children give so many nonfinite verbs, because they almost never hear declarative clauses without a finite verb.

guage; Jonas, 1995), Irish (a Celtic language; Wexler, 1994b), and Hebrew (Rhee & Wexler, 1995) all show the OI stage.[6]

Very Early Parameter Setting

What is the explanation of the OI Stage? Why does it exist? I return to (11b), the existence of nonfinite main verbs. (11a) is understandable because the adult language has finite main verbs; thus, we would expect children's grammars to have them. That children know universal morphosyntactic conditions (11c) is understandable if it is assumed, as the field basically does, that children know UG because they are born with it. (11c) also states that young children, in the OI stage, also know the relevant language-particular morphosyntactic conditions that are relevant to the previously studied issues. That is, children have set the parameters of basic clause structure/inflection correctly. This is a remarkable finding, remarkable enough to demand its own name, *Very Early Parameter-Setting (VEPS)*.

(12) *Very Early Parameter-Setting*: From the earliest observable stage, that is, from the time that children produce multi-word utterances, they have correctly set the basic inflectional/clause structure parameters. These include the V to TNS (verb-raising) parameter, the V2 parameter, and parameters of basic word order, including the relative order of verb and object.

Children are usually assumed to start producing utterances of more than one word at about 18 months of age (Brown, 1973). The hypothesis of VEPS (12) says that by this age the children already have set their basic clause structure/inflectional system parameters correctly, that is, to the value of the adult language that they are hearing.

What is the evidence for VEPS? Here I briefly mention some of the evidence discussed in considerable detail in Wexler (1990a, 1992, 1994a). Consider the V to TNS parameter. French is marked *yes* on this parameter; the finite verb raises on the surface to TNS, as was illustrated previously. English, on the other hand, is marked *no* on this parameter; the finite verb does not raise to the left of negation; (12′) is ill-formed.

---

[6]Wexler (1990a, 1992, 1994a), on the basis of data in Schaeffer (1990) also suggested that Italian did not show the OI Stage and gave a possible explanation in terms of agreement differences. Guasti (1993) has confirmed this finding for Italian, and it is now known that Spanish and Catalan (Grinstead, 1994; Torrens, 1995) do not show the OI Stage, nor does Tamil (Sarma, 1994). Rizzi (1994a), in connection with his Truncation theory of the OI stage, gives an explanation of why Italian does not show it; Wexler (1995) gave a generalization about which languages show the OI stage and which do not, and an explanation in terms of the minimalist theory of the OI stage which he develops there. Why a particular language shows the OI stage or not is a major fact about the OI stage that must be accounted for in any theory. It turns out that facts fit neatly into the theory that Wexler (1995) developed, and provide nice support for that theory.

(12′) a. *She owns not a book.

 b. *She goes not to the store.

I have shown that at a very early age French-speaking children know that the verb raises to TNS. Nobody has ever found evidence that English-speaking children mistakenly raise the finite verb to TNS, producing sentences like (13). Thus at very early ages, French- and English-speaking children appear to have set the V to TNS parameter correctly and to opposite values depending on the language.

Similarly, German is V2; English and French are not. Again, Poeppel and Wexler (1993) demonstrated that a 25-month-old child had set the V2 parameter correctly to the *yes* value, and there is a considerable literature now on the early correct setting of the V2 parameter in a number of languages. Rohrbacher and Vainikka (1994), in fact, analyzed the data from a 17-month-old German-speaking child and provided evidence that the child follows the Poeppel and Wexler pattern. That is, this child knows she is speaking a V2 language. On the other hand, there is no indication at all that English- or French-speaking children believe that their language is V2. They have correctly set the V2 parameter to *no*. Thus, German (and Dutch, etc.) children have set the V2 parameter correctly to *yes* at a very early age and French- and English-speaking children have set the V2 parameter correctly to its opposite value, *no*, at an early age.

Consider the value of the parameter that indicates that in underlying structure the direct object is to the left (Dutch, German) or right (English, French, Swedish) of the verb. Wexler (1990a, 1992, 1994) demonstrated that children have set this value correctly from a very early age also.

Much more empirical research needs to be done over a variety of languages and a variety of parameters. But it seems clear that at least a substantial number of core parameters are set correctly at a very early age. In fact, it is difficult to find evidence for the incorrect setting of a parameter in any language.!7·

To the extent VEPS is true, it is a remarkable result. Note that VEPS is not only about UG; rather, it is about language-particular parameter settings. Suppose it is true, as VEPS suggests, that basic clause/inflection parameters are set

---

[7]The classic case of a mis-set parameter involves the null subject parameter, which determines whether subjects may be phonetically empty, as in Italian. Hyams (1986) argued that missing subjects in English in young children was an instance in which the children had mistakenly mis-set the null-subject parameter as *yes* (null subjects are allowed). However, null subjects, although grammatical under certain conditions in OI-stage English (and other OI-stage languages) are not the result of a mis-set parameter, but rather the result of the OI stage. (They are grammatical for the children, as Hyams and Wexler (1993) have shown).

The best evidence I know at the current moment for a mis-set parameter involving clause structure in child language comes from Schoenenberger (1995), who has found two Swiss-German-speaking children who seem to quite consistently treat embedded clauses as V2, despite the fact that in the adult language the verb is at the end in these clauses. The generality and cause of this phenomenon is not yet clear.

correctly by the time children start putting words together to form primitive sentences. This means that the children have learned a good deal about abstract structural parameters before they start speaking. Thus, any kind of learning theory that demands that children do some kind of motor behavior or produce language in order to learn it (i.e., Piagetian theories, or various reinforcement theories) fails on empirical grounds. If VEPS is right, children learn the parts of language that they have to learn perceptually, without producing overt responses as part of the learning.[8] This should not really be surprising; other aspects of knowledge, for example, perceptual learning, are achieved in the same way. But it has not always been assumed about language.

Tense Optionality and Maturation

Most of the results I have described fall naturally into a biological theory of language development under assumptions that fit in with what is known about biology. The fact that children know much about UG at a very early age is consistent with the existence of a biological program that underlies UG. Because most other biological development is guided by a genetic program, the assumption that such exists for the species-specific complex called language is quite natural (see Wexler, 1990b). On the other hand, the fact that learning takes place early and well (i.e., the correct values of parameters are learned early) and easily is consistent with other kinds of learning that biologists have studied (i.e., imprinting).

There is only one part, though a major one, of the empirical results that I have described about the Optional Infinitive Stage, that is not subsumed under an innate genetic program or under easy and quick learning (based on a strong genetic program delimiting structure). This is (12b), the existence of main clause infinitives in the OI stage. These are not allowed by UG, generally, so they cannot be looked at as part of UG nor as the result of a particular mis-setting of a parameter. No parameter value will allow main clause infinitives as normal structures. Thus, how should we think of them?

Wexler (1990a, 1992, 1994a) suggested that the impossibility of optional infinitives at a certain stage follows from a maturational process. Until maturation takes place, children will allow optional infinitives, and once maturation takes place they will not.

That there are maturational processes in grammar was suggested in detailed studies by Borer and Wexler (1987, 1992) and by Felix (1984). Borer and Wexler specifically argued that maturational processes were ones that did not directly depend on learning but were guided by a genetic program, as in any case of physical maturation in biology (i.e., the growth of a second set of teeth or of

---

[8]For a theory of parameter-setting with many of the requisite properties developed in a formal, computational way, see Gibson and Wexler (1993).

secondary sexual characteristics, both of which occur at a considerable delay after birth).

If maturation in fact is the explanation of the existence and loss of optional root infinitives in natural language, then this phenomenon too (11b) fits naturally into biology. Maturation (i.e., genetically guided programs that unfold over time) is the basis for most of biological development. If it is assumed that human beings are a part of the biological world, and not a special category that does not conform to biological laws, then maturation, genetically guided programs unfolding over time, is what is expected.

What is the nature of the OI phenomenon? What is its cause? Since the description of the OI phenomenon, this has been the subject of much discussion. In my original papers (Wexler, 1990, 1994) I offered three theories involving some different knowledge of TENSE by the OI child.[9,10]

What could account for the existence of the OI stage? In what part of the child's grammar is there something different from the adult? Take as a description of the OI stage the first theory given in Wexler (1990a, 1992, 1994a), which simply says that:

(13) The OI Stage is TENSE-Optionality: TENSE is optional (All other relevant UG and language-particular facts are known to the child).

In many ways, (13) can be thought of as saying that TENSE is *underspecified* during the OI stage. The child often uses a form (the nonfinite form) that does not show grammatical features that are taken to be necessary in the adult grammar. Underspecification should not be interpreted to mean only that the actual form chosen as phonetic output has certain properties. Underspecification of TENSE has important syntactic consequences; TENSE must be actually

---

[9]Rizzi (1994a) proposed a truncation theory of the phenomenon, which accepts Wexler's (1990a, 1992, 1994a) description of lack of TENSE for the root infinitive sentences, but predicts a much larger class of structures that are different for the child. The theory predicts a number of interesting correlations in children's productions in the OI stage. Rizzi's theory is also a maturational theory. See Wexler (1995), Phillips (1995), Bromberg and Wexler (1995), Rhee and Wexler (1995), and Levow (1995) for some evidence that a number of the correlations and error patterns predicted by Rizzi's theory do not hold.

[10]There have been some attempts to treat the OI phenomenon as a learning phenomenon, with children learning that higher functional categories must appear, for example, Clahsen (1991), Vainikka (1994), but no learning theories are given, and it seems impossible to create them. In addition, such theories have empirical problems. Wexler (1990a, 1992, 1994a), DePrez and Pierce (1993), and many other sources have argued that the functional categories are all available in the OI stage. Radford (1990) assumed that the functional categories do not exist in the early child, but mature in the child. Although the nonexistence of functional categories at the early stage can be shown to be empirically false, the assumption of maturation at least provides a coherent account of the growth of the categories. The theory is simply empirically false as an account of the early stages. In contrast, the learning theories provide no account of the growth of functional categories.

missing in a serious enough syntactic sense for various checking (movement) operations to not take place.

There are many properties that follow from (13). To illustrate, in the next section, I apply (13) to English. But first, let us ask why it should be that (13) in fact exists and why it goes away. If (13) represents a biological process, then ultimately the answer is biological; the genetic program works this way. But can (13) be related to anything else in language acquisition?

Note that infinitives (nonfinite forms) exist in adult language; generally, they are just in embedded (subordinate) clauses, as in (14a), with structure (14b):[11]

(14) a. Mary wanted to leave.
     b. Mary wanted [PRO to leave].

The embedded verb *to leave* has a TENSE, a nonfinite TENSE. This means, following Enc (1987) and many other writers on TENSE, that embedded TENSE is dependent for its values on the TENSE of the matrix (main) clause, in much the same way that a reflexive pronoun is dependent for its referent on its antecedent. Note that in (14a) the time of leaving is *after* the time of *wanting*. The value of the finite TENSE is determined by context; it has not been specified in the sentence except that it is before the speech act; that is a property of the past TENSE of the finite main verb in (14a). Whatever time that is, however, the time of the embedded nonfinite verb *leave* is restricted to being *after* that time. Enc showed that many properties of embedded tenses relate to their being dependent on main tenses. Thus, it can be concluded that, for the adult, nonfinite tenses do not appear in main clauses because in a main clause there is no higher TENSE to be dependent on.

What I assume is that the OI child treats nonfinite tense as if it can be fixed by context, rather than as being necessarily dependent on a higher TENSE. When a French child says "Marie parler" ("Mary speak"; nonfinite), she is treating *parler* as if the time is being filled in by the context. The nonfinite verb depends on context (including discourse and other cotextual situational facts), rather than on a higher TENSE. It may be that OI children simply have this one difference from adults: Nonfinite tenses have an extra capacity to be dependent on (fixed by) context rather than by a higher TENSE.[12]

Thus, it is assumed that OI children's special property is that they have an interpretive (actually *pragmatic*) rather than structural difference from adults. They accept too wide a set of antecedents for nonfinite TENSE. Many studies have shown that children accept too wide a set of antecedents for pronouns;

---

[11]In (14b) the brackets represent an embedded sentence. PRO represents the phonetically empty subject of the embedded sentence, which is coreferential with *Mary*; that is, Mary wants *Mary* to leave.

[12]For a similar theory see Hyams (in press).

they assume that a pronoun can refer to antecedents that would not be accepted by adults (see Avrutin, 1994; Avrutin & Wexler, 1992; Chien & Wexler, 1990; Karmiloff-Smith, 1981; Wexler & Chien, 1985). Thus, the OI stage may be an example of a wider class of pragmatic problems, problems in determining what references are possible. TENSE is a referential category that refers to times or intervals of times, just like noun phrases refer to entities. Thus, it is not surprising that the functional category that children have trouble with is TENSE; it is quite analogous to pronouns, which also refer in a dependent way.

I should point out that the pronoun mistakes that ever since Wexler and Chien (1985) have been taken to indicate the existence of a particular kind of pragmatic problem often continue until a child is 4 or 5 years old, a time at which the OI stage is clearly over in normal children. Thus, the times do not match up exactly, which should make us careful about attributing these two different phenomena to the same underlying cause. But it may be that the two kinds of phenomena (optional infinitives and particular kinds of problems with pronouns) are indications of the same class of developmental problems, pragmatic ones that develop with different time courses in different domains (e.g., noun phrase reference vs. TENSE reference).[13]

I have been assuming that the nonfinite forms can have ordinary declarative meaning. In Wexler (1990a, 1992, 1994a), I proposed and rejected the possibility that the root infinitives resulted from a process of modal drop, that is, that all the root infinitive sentences had implicit modals. One reason I rejected this possibility is that it is very difficult to understand why modals would consistently be omitted from the phonetic output when they exist in the underlying representation. Another reason why I rejected this possibility is because Poeppel and Wexler (1993) showed that many of the optional infinitive sentences used by the German child they studied appeared to have ongoing activity meaning (i.e., they were not modal). Behrens (1993) confirmed this finding in a study of seven German children. She concluded that finite sentences were always used correctly, as an adult would use them, but that nonfinite root sentences (i.e., optional infinitives) were used in a wide variety of ways, including present and past ongoing activity meanings as well as modal meanings. It is possible that a modal with no meaning, a pleonastic modal, similar to English *do* is dropped; this is what Wexler (1994a) and Poeppel and Wexler (1993) called the Empty Dummy Modal hypothesis (EDM). Such a hypothesis has been argued for by Whitman (1994).

---

[13]See Wexler (1995) for a quite different theory, one that still involves the (dispreferred) possibility of contextual determination of nonfinite TENSE, but in which the cause of the OI stage is a particular relation between a particular functional category (Determiner) and an interpretive feature. The age differential between the OI stage and the pronoun problem does not hold for such a theory.

As Wexler (1994a) pointed out, it would be odd for the child to assume a dummy modal; sentences (15) are ungrammatical in English, except in emphatic contexts:

(15) a. *She does go to the store on Tuesdays.
     b. *She does like me.

These sentences would not substitute (except if used emphatically) for sentences (16):

(16) a. She goes to the store on Tuesdays.
     b. She likes me.

Thus, a child in English who uses a nonfinite form like "she like me" would, under the EDM analysis, be actually representing this sentence ungrammatically. There is no motivation for such an analysis. The problem is compounded if the language in question does not even have a visible dummy modal. As Jonas (1995) pointed out, Faroese is such a language. Because Faroese has an OI stage, Jonas argued that it makes no sense to think of Faroese as having an empty dummy modal.[14]

Thus, the assumption that optional infinitive (OI) sentences result from the surface dropping of a modal, possibly of no semantic content, does not appear to work, conceptually or empirically. The fact that many of the OI sentences have a descriptive, non-modal reading is important in understanding the correct analysis of these sentences. Because descriptive readings are not associated with infinitival root sentences in adult grammar, an understanding why they are possible in children's grammars is needed.

THE COMPLEX OF OI PHENOMENA: THE ENGLISH EXAMPLE

One of the intriguing properties of the OI theory of early inflectional and clausal development is that it predicts the existence of a tightly interrelated pattern of phenomena in many languages. In particular, the OI theory makes precise predictions about many grammatical representations that have been detailed in the literature. To illustrate, I concentrate on the case of English, following the presentation of Wexler (1990, 1994).

---

[14]Poeppel and Wexler (1993) also gave a distributional argument against the EDM hypothesis, an argument that Whitman (1994) answered by complicating the syntactic theory.

Recall that it is assumed that the OI stage is characterized by TENSE-optionality (13): TENSE is optional, but all other relevant UG and language-particular facts are known.

First, Wexler (1990a, 1992, 1994) predicted that, if English is an OI language, then an early stage of English should show forms like (17):

(17) a. She like me.
　　 b. She go to the store.

instead of the corresponding forms in (18):

(18) a. She likes me.
　　 b. She goes to the store.

This prediction is made because the English nonfinite verb (the verbal part of the infinitive) is constructed from the verb stem by a process of zero-morphology (i.e., the nonfinite inflection is phonetically empty). This corresponds to a phonetically nonempty inflection that is added in other languages, like French or German:

(19) To form −finite (i.e., nonfinite) *speak*:
　　 a. parl + er = parler (French)
　　 b. sprech + en = sprechen (German)
　　 c. speak + 0 = speak (English)

As Wexler (1990a, 1992, 1994a) pointed out, English is the only one of the Romance or Germanic languages that forms the nonfinite verb with a process of zero morphology. All of the other languages have phonetically audible inflections for the nonfinite forms of verbs [as in the French and German examples in (19)]. Thus, it is not surprising that it had not previously been noted that English-speaking children used nonfinite verb forms; however, it is clear that they do. Ever since Brown (1973) and Cazden (1968), it has been known that young children produce forms like (17), with -*s* missing from the verb stem. This has been analyzed as a process of -*s* drop. Wexler (1994b) argued that this is the incorrect analysis; forms like (17) should be seen as the use of the nonfinite verb form. (Because there is no marking that distinguishes present tense forms from nonfinite forms of the verb in anything other than third singular, it is difficult to check this process other than with third singular subjects in English. What happens when past tense is considered is discussed later.)

So (13) predicts that nonfinite forms like (17) exist, and they do; this is well-known. However, (13) says that TENSE is *optional*; sometimes it exists in a representation. Thus, tensed forms like (18) are predicted to exist. But it is well-known that such forms exist in early English, alongside forms like (17). Brown's central concept was "percentage correct in an obligatory context"; this was rarely 0 or 100% at the early stages. In the case of third singular -s, this means that it sometimes appears and sometimes does not; this is what is predicted, and this is what is empirically true.

Agreement

How about the use of agreement generally? I have shown that it is predicted that nonfinite forms exist. But suppose the form is finite? Is agreement correct?

Tense-optionality (13) assumes that language-particular facts are generally known to the OI child. I have not discussed (subject-verb) agreement, but in fact Poeppel and Wexler (1993) showed that German children in the OI stage know subject-verb agreement (this can also be assumed more generally). In particular, the claim is that if an agreement inflection is used on a verb (i.e., if the verb is finite in a language that shows agreement marking), then the subject will match the agreement inflection in relevant features.

There is only one agreement inflection in regular verbs in English, third singular -s. The prediction is that -s will only be used if the subject is third singular. Children are predicted to not utter forms like (20):

(20) a. *I goes to the store.
    b. *You likes me.

The analysis can be extended to *be* and *do*. Suppose these occur in finite form. Then, on the assumption that agreement is known, we predict that the subjects will match the forms of the verbs. Children will not utter ungrammatical forms like (21):

(21) a. *I is going to the store.
    b. *She am here.
    c. *He are here.
    d. *I does not like ice cream.
    e. *You does not belong here.

Do

Now consider *do*, a semantically empty auxiliary (modal), which is added to sentences under special conditions. The general assumption in syntactic theory is that *do* is added to sentences when, for some syntactic reason, the main verb cannot raise to TENSE. In such cases, *do* is added to TENSE. Consider (22):

(22) a. She TNS not goes.
    b. *She not goes.
    c. She does not go.

The assumption is that the existence of *not* in the underlying representation (22a) prevents the verb and TNS from getting together, which is necessary to check the verbal feature on TNS (i.e., TNS needs to be attached to a verb). (In some languages, the verb can raise over *not*; in others it cannot. It is assumed in accordance with (13) that the child knows which language she is speaking.) Thus, the form (22b) is ungrammatical; TNS has not had its verbal feature checked. Thus, *does* is inserted in (22) to check the verbal feature of TNS; (22c) is a good sentence.

TENSE-optionality (13) assumes that TENSE is optionally present in the OI stage. If TENSE is present, the underlying form of the sentence is (22a) and, because the child knows other UG and language-particular principles, (22c) results.

Suppose, on the other hand, that the child omits TENSE from (22), as is allowed by TENSE-optionality. Then (23) results:

(23) She not go.

Thus, TENSE-optionality predicts that in the OI stage, both (22c) and (23) should be allowed. Now consider (24):

(24) a. She TNS not goes.
    b. *She not goes.
    c. She do not go.

The ungrammatical sentence in (24b) can only be derived from the underlying representation in (24a), because the tense-marking on the verb *goes* must be checked against the functional category TNS. But, I have already shown that *not* prevents *goes* from raising to TNS and checking off its features. Thus (24b) will also be ungrammatical for the OI-stage child.

Now consider (24c). How could it be derived? I have shown that TENSE-optionality predicts that the OI child knows agreement. Thus (24c) cannot be a finite form with incorrect agreement. But (24c) might appear to correspond to the nonfinite use of *do*. Is this possible? After all, main verbs appear as nonfinite forms, a central property of the OI stage.

But consider how (24c) might be derived. If the underlying representation is (24a), with TNS in the representation, then a finite form of a verb, a form with TNS, must be used to check TNS. Thus, (24a) cannot be the underlying representation for (24c). On the other hand, TENSE-optionality allows TNS to be omitted from a representation. Thus, the underlying representation of (24c) might not include TENSE. But if TENSE is not in the representation, there is no syntactic reason to insert *do* in any form; by assumption (in the adult syntax), *do* is inserted only when a TNS category exists to which a verb has not raised. The insertion of *do* may only occur if TNS is in the representation. Thus, in this case, with no TNS in the representation, *do* may not be inserted. Thus (24c) may not be derived.

To summarize, I have predicted from the assumed description [TENSE-optionality, (13)] of the OI stage, the following distribution of sentences, where *OI+*

means grammatical for the OI-stage child, and *OI** means ungrammatical for the OI-stage child:

(25) a. OI+ She does not got.
     b. OI+ She not go.
     c. OI* She do not go.
     d. OI* She not goes.

(25b) represents the medial-neg stage of Klima and Bellugi (1966). For preliminary evidence on the nonexistence of (25d), see Wexler (1994a). For much more detailed evidence, from both experimental and natural-production studies, see Harris and Wexler (in press), who also showed that (25a) also amply coexists with (25b). The patterns predicted by the OI theory seem correct. This is rather surprising to many views that would have suggested that the medial-neg sentences (25a) exist because of lack of knowledge of dummy *do*. That cannot be true because *do* is quite known, and as Harris and Wexler (see also Wexler and Rice, in prep.) show, there are even more sentences like (25a) (with a form of *do*) than one would expect, given the proportion of TENSE omission with main verbs. See the quoted papers for a suggested explanation. At any rate, it is clear that *do* exists. Thus, again, the OI analysis seems to fare better than other explanations. This is a very subtle pattern of errors that is predicted, yet to the extent that data are in, it seems supported. (I put off a discussion of forms like [25c].)

Be

Now consider *be*. I have shown (21) that finite forms that do not agree with their subjects will not be used by OI children. But, more can be said. Suppose that *be*, like *do*, is semantically empty and is added to sentences only to provide a verbal feature for TENSE to check.[15] This is plausible because *be* does not appear to contribute any meaning to a sentence. In a progressive sentence like (26a), the aspect is contributed by the piece of inflectional morphology *ing* on the verb, and in (26b), the copular *be* also does not seem to contribute any meaning to the sentence; in fact, many languages have phonetically empty copula *be*.

(26) a. She is eating.
     b. She is quiet.

Thus, suppose *be* is added when a verb is needed that checks TENSE; progressive main verbs cannot do this. Perhaps for some morphological reason in English, TENSE cannot be a feature on a progressive verb. And in predicative

---

[15]For detailed syntactic argumentation, see Wexler (1995), who showed that *be* is not compatible with tenseless (small) clauses. Hyams and Jaeggli (1988) and Scholten (1988) have also linked *be* to tense.

constructions (e.g., 26b), there is no verb at all except for *be*. Thus, it is assumed that *be* is added in certain constructions when TNS is not otherwise bound.

TENSE-optionality (13) assumes that TENSE is possible in the OI stage. Thus, if TENSE is chosen, *be* will be necessary, and the child will have forms like (26a, b). But TENSE-optionality also allows TENSE to be omitted. If this happens, an economy condition will prevent *be* from being inserted, just as *do* may not be inserted if TENSE is not chosen. Thus (27) will also be possible for the OI child.

(27) a. She eating.
     b. She quiet.

In other words, *be* may be deleted from declarative sentences. It is quite clear that this is true; any perusal of early child English transcripts will show that forms like (27a) are quite common, perhaps typical. Because progressives are the normal way to represent present actions in English, and because young English children will omit TENSE quite often, we would expect forms like (28a), that is, progressive forms with *be* omitted, to be common. And they are. We also see forms like (28b), although perhaps less often, because predicate adjective constructions are probably less frequent. Quantitative studies will be welcome, of course.

Suppose TENSE is not chosen in the representation of a progressive or predicate adjective or predicate nominal sentence. Then there is no syntactic motivation for adding *be* to the representation, because *be* is only added to check off a verbal feature on TENSE. Thus, by an economy condition, *be* will not be added to the sentence. Thus, there is no way to derive a sentence with nonfinite *be*; it will be the wrong agreement form if finite, and it will not be allowed to be inserted if nonfinite. In other words, we have predicted that (28) will be ungrammatical for an OI child.

(28) She be going.

Again, one rarely sees forms like (28). The few nonfinite *be* forms that seem to exist are forms where *be* acts like a main verb, with semantic content, as in (29):

(29) You be a cowboy.

(29) does not mean that you are a cowboy, but rather that you are acting (or should act) like a cowboy. Thus, nonfinite forms of *be* would be expected in this context just as for any other main verb.

For quantitative data confirming most of the points that I have made about English, see Wexler (1990a, 1992, 1994a), Harris and Wexler (in press), and Rice, Wexler, and Cleave (1995).

The Verbal Morphology/Syntactic Position Correlation in English

Because English is not a verb-raising language (like French) or a V2 language (like German), it is not so easy to find large sets of sentences illustrating the verbal morphology/syntactic position correlation. However, English is residual V2, meaning that auxiliaries and modals raise up (or invert, in traditional terminology), so the correlation in these cases can be tested.

I have shown that finite verbs will not follow *not* (25d). Main verbs in English do not raise; because (13) assumes that OI children know the language-particular properties of their language, they will know that main verbs do not raise, a fact I have already discussed. Thus, (30) will not be allowed by an OI child:

(30) She goes not.

On the same assumption, that the OI child knows language-particular facts as well as universal ones, she will know that *be* and *do* raise in English, so that she will produce (31a), but not (31b):

(31) a. She is not going.
     b. *She not is going.

Furthermore, it is a property of English that only finite verbs invert in questions. Of course, only auxiliaries and modals invert in English, as I have just pointed out. But on the assumption that children know that nonfinite verbs do not invert, we can rule out forms in which questions are asked with nonfinite verbs, as in (32):

(32) a. Do she have a cake?
     b. Be she here?
     c. Be I going?

Rice, Wexler, and Cleave (1995) showed that this prediction is confirmed in 3-year-old OI children.

One area that I have not discussed is the phenomenon of null subjects in children in non-null subject languages, such as English (and the Germanic languages generally and French). The following seems to be the story. With nonfinite verbs null subjects would be expected because nonfinite verbs license null subjects (PRO) in syntax generally [see, for example, (14)]. As first pointed out by Wexler (1990a, 1992, 1994a) it turns out in language after language (for non-null subject languages) that there are proportionately far more null subjects of nonfinite verbs (optional infinitives, i.e., root nonfinite verbs) than of finite verbs. (For evidence of this fact in English, see Sano & Hyams, 1993.) Because nonfinite verbs license PRO, null subjects with nonfinite verbs are expected. The null subjects of finite verbs (the small number of them) might be cases where null subjects are allowed in the adult grammar, or perhaps they are a small discourse/pragmatic-based extension of them (i.e., the child might slightly over extend the discourse conditions under which such null subjects are allowed in non-null subject languages). These might be thought of as topic-drop-type null subjects (Hyams & Wexler, 1993), or diary-drop null subject following Rizzi (1994b) who bases his idea on diarydrop studies of Haegeman (1990). However, most null subjects in the OI stage are not of the topic-drop or diarydrop type, rather they are licensed by nonfinite verbs. Thus, contrary to the prediction of Rizzi (1994b) that null subjects will not exist in *wh*-sentences, they do exist, but only with nonfinite verbs, which I suggest means that they are licensed by the nonfinite verb. (For the data on *wh*-questions see Roeper & Rohrbacher, 1994, and Bromberg & Wexler, 1995.) Thus, we see that null subjects in non-null subject languages exist in early children because of the OI stage; they exist as a result of licensing by a nonfinite subject. See Wexler (1995) for an extensive discussion of subject types in the OI stage.

There is a fine-tuned analysis of a large variety of data, which basically seems to be correct. In no way can we say that there is some kind of general grammatical deficit with OI children, or even a general morphological deficit.[16] There is a particular problem descriptively captured by TENSE-optionality.

LEARNABILITY AND MATURATION

It has been known, since the linguistic arguments of Chomsky (1965), the mathematical learnability arguments of Wexler and Hamburger (1973), Wexler and Culicover (1980), and references cited there, that theories that assumed only general-purpose learning mechanisms could not explain language learning. Since Borer and Wexler (1987), there has been every reason to assume that certain properties of language mature (i.e., grow) with no special attention to learning from the input. At the same time, because languages vary, some properties must be learned.

I have shown that children know most of the central grammatical properties of clause structure at the earliest observed age and that they have even learned language-particular properties at the earliest observed age. One piece of competence stands out as distinct from the others, and it is not a language-particular one, but a universal one. This property is the obligatoriness of TENSE in declarative, assertive, sentences. This is the one piece of competence that children

---

[16]For example, it is well known (since Brown, 1973) that there is no trouble with progressive *ing*. TENSE is omitted, but *ing* is used from the start, probably in all or almost all necessary contexts. Also, there is not a problem with agreement; if the form is finite, correct agreement is used.

seem to lack, and this is true over many languages. Given children's abilities to learn all sorts of abstract language-particular properties, plus their (presumably genetically programmed) knowledge of the universal principles of grammar, the missing piece of competence stands out. The most natural explanation is that it matures somewhat later than the other pieces of grammar.

Suppose then that there is one piece of competence, the obligatoriness of TENSE, that matures somewhat late. It seems quite plausible that this late-maturing piece of grammatical competence could be subject to a genetically determined malfunctioning. Just as in some other genetic diseases, which only show up when a piece of biology that was supposed to grow at a certain time did not, it might happen that a certain piece of linguistic competence that matures at a certain time does not show up at that time in some children on genetic causes.

Rice and Wexler (this volume) and Rice, Wexler, and Cleave (1995) argue that this scenario might exactly be the case for Specific Language Impairment (SLI). Namely, they argue that SLI is to be characterized as the OI stage at a much later age, the extended optional infinitive (EOI) stage. The speculation is that a piece of maturation that takes place in normal children and that requires clauses to be finite, does not take place in SLI children, either until a much later age, or (possibly) never.

It is extremely difficult to think of any other account of all the facts that are known of in normal and nonnormal development. A learning-theory account seems hopeless, for the reasons given previously. Performance accounts also seem not to have the capacity to explain the phenomena. (See Hyams & Wexler, 1993, on lots of empirical arguments against performance accounts of null subjects in children, and Wexler, 1994a, on arguments against performance accounts of the OI stage.)

Thus, we seem to be left with maturation as an explanation, that is, biological growth guided by a genetic program. Such a result should be welcome, because it leads to scientific unification. So much of what is known about biology is explained by a "biological growth guided by a genetic program" model, in so many areas of biological structure and function, over such a wide variety of organisms, that the solid placement of language within that biological framework should be welcomed.

REFERENCES

Avrutin, S. (1994). *Psycholinguistic Investigations in the Theory of Reference*. Unpublished doctoral dissertation, Massachusetts Institute of Technology, Cambridge.

Avrutin, S., & Wexler, K. (1992). Development of principle B in Russian: Coindexation at LF and coreference. *Language Acquisition, 2*(4), 259–306.

Behrens, H. (1993). *Temporal reference in German child language*. The Haag: Koninklijke Bliotheek.

Bromberg, H., & Wexler, K. (1995). Null subjects in child Wh-questions. *MIT Working Papers in Linguistics* (MITWPL) *26*, 221–247.

Brown, R. (1973). *A first language*. Cambridge: Harvard University Press.

Borer, H. (1983). The projection principle and rules of morphology. In *Proceedings of NELS* (Vol. 14). University of Massachusetts, Amherst.

Borer, H., & Wexler, K. (1987). The maturation of syntax. In T. Roeper & E. Williams (Eds.), *Parameter setting* (pp. 123–172). Dordrecht, The Netherlands: Reidel.

Borer, H., & Wexler, K. (1992). Bi-unique relations and the maturation of grammatical principles. *Natural Language and Linguistic Theory, 10*, 147–189.

Cazden, C. B. (1968). The acquisition of noun and verb inflection. *Child Development, 39*, 433–448.

Chien, Y-C., & Wexler, K. (1990). Children's knowledge of locality conditions in binding as evidence for the modularity of syntax and pragmatics. *Language Acquisition, 1*(3), 225–295.

Chomsky, N. (1965). *Aspects of the theory of syntax*. Cambridge: MIT Press.

Chomsky, N. (1995). *The minimalist program*. Cambridge: MIT Press.

Clashen, H. (1991). Constraints on parameter setting: A grammatical analysis of some acquisition states in German child language. *Language Acquisition, 1*(4), 361–391.

DePrez, V., & Pierce, A. (1994). Crosslinguistic evidence for functional projections in early child grammar. In T. Hoekstra & B. D. Schwartz (Eds.), *Language Acquisition Studies in Generative Grammar*. Philadelphia: John Benjamins.

Enc, M. (1987). Anchoring conditions for tense. *Linguistic Inquiry, 18*(4), 633–757.

Fakui, N. (1988). Deriving the differences between English and Japanese: A case study in parametric syntax. *English Linguistics, 5*, 249–270.

Felix, S. (1984). Maturational aspects of universal grammar. In C. Cripper, A. Davies, & A. P. R. Howatt (Eds.), *Interlanguage* (pp. 133–161). Edinburgh, Scotland: Edinburgh University Press.

Friedemann, M.-A. (1994). The underlying position of external arguments in French: A study in adult and child grammar. *Language Acquisition, 3*(3), 209–255.

Gibson, E., & Wexler, K. (1993). Triggers. *Linguistic Inquiry, 25*(3), 407–454.

Guasti, T. (1994). Verb syntax in Italian child grammar: Finite and nonfinite verbs. *Language Acquisition, 3*(1), 1–40. (Original work published 1993)

Grinstead, J. (1994). *Tense, agreement and nominative case in child Catalan and Spanish*. Unpublished masters thesis, University of California, Los Angeles.

Klima, E. S., & Bellugi, U. (1979). Syntactic regularities in the speech of children. In J. Lyons & R. J. Wales (Eds.), *Psycholinguistic Papers*. Edinburgh, Scotland: Edinburgh University Press.

Harris, A., & Wexler, K. (in press). The optional infinitive stage in child English: Evidence from negation. In H. Clahsen (Ed.), *The acquisition of inflection*.

Haegeman, L. (1990). Understood subjects in English diaries. *Multilingua, 9*, 157–199.

Hyams, N. (1986). *Language acquisition and the theory of parameters*. Dordrecht: Reidel.

Hyams, N. (in press). The underspecification of functional categories in early grammar. In H. Clashen (Ed.), *The acquisition of inflection*.

Hyams, N., & Wexler, K. (1993). On the grammatical basis of null subjects in child language. *Linguistic Inquiry, 24*(3), 421–459.

Hyams, N., & Jaeggli, O. (1988). Morphological uniformity and the setting of the null subject parameter. *Proceedings of NELS, 18*(1), 238–253.

Jonas, D. (1995a). On the acquisition of verb syntax in child Faroese. *MIT Working Papers in Linguistics (MITWPL), 26*, 265–280.

Jonas, D. (1995b). *Clause structure and verb syntax in Scandinavian and English*. Unpublished doctoral dissertation, Harvard University.

Karmiloff-Smith, A. (1981). The grammatical marking of thematic structure in the development of language production. In W. Deutsch (Ed.), *The child's construct of language*. London: Academic Press.

Klima, E. S., & Bellugi, U. (1966). Syntactic regularities in the speech of children. In Lyon & R. Wales (Eds.), *Psycholinguistic papers*. Edinburgh, Scotland: Edinburgh Press.

Levow, G.-A. (1995). Tense and subject position in interrogatives and negatives in child French: Evidence for and against truncated structure. *MIT Working Papers in Linguistics (MITWPL)*, *26*, 281–304.

Lenneberg, E. (1967). *Biological foundations of language*. New York: Wiley.

Lightbown, P. (1977). *Consistency and variation in the acquisition of French*. Unpublished doctoral dissertation, Columbia University.

Lightfoot, D. (1982). *The language lottery: Toward a biology of grammar*. Cambridge: MIT Press.

Manzini, R., & Wexler, K. (Summer, 1987). Parameters, binding theory, and learnability. *Linguistics Inquiry, 18*(3), 314–444.

Marcus, G. (1993). Negative evidence in language acquisition. *Cognition, 46*, 53–85.

Nadel, L., & Wexler, K. (1984). Neurobiology, representations, and memory. In G. Lynch, J. L. McGaugh, & N. M. Weinberger (Eds.), *Neurobiology of learning and memory* (pp. 125–134). New York: Guilford Press.

Phillips, C. (1995). Right association in parsing and grammar. In C. T. Schutze, J. B. Ganger, & K. Broihier (Eds.), *MIT Working Papers in Linguistics (MITWPL)*, *26*, 37–93.

Pinker, S. (1994). *The language instinct*. New York: William Morrow.

Poeppel, D., & Wexler, K. (1993). The full competence hypothesis of clausal structure in early German. *Language, 69*, 1–33.

Pierce, A. (1989). *On the emergence of syntax: A crosslinguistic study*. Unpublished doctoral dissertation, Massachusetts Institute of Technology, Cambridge.

Pierce, A. (1992). *Language acquisition and syntactic theory*. Dordrecht, The Netherlands: Kluwer.

Radford, A. (1994). *Syntactic theory and the acquisition of English syntax*. Oxford: Basil Blackwell.

Rhee, J., & Wexler, K. (1995). Optional infinitives in Hebrew. *MIT Working Papers in Linguistics (MITWPL), 26*, 383–402.

Rice, M., & Wexler, K. (1995). Extended optional infinitives (EOI) account of specific language impairment. In D. MacLaughlin & S. McEwen (Eds.), *Proceedings of BUCLD, 19*, 451–462.

Rice, M., Wexler, K., & Cleave, P. L. (1995, August). Specific language impairment as a period of extended optional infinitive. *Journal of Speech and Hearing Research, 38*, 850–863.

Rizzi, L. (1994a). Some notes on Linguistic theory and language development: The case of root infinitives. *Language Acquisition, 3*(4), 371–393.

Rizzi, L. (1994b). Early null subjects and root null subjects. In T. Hoekstra & B. D. Schwartz (Eds.), *Language acquisition studies in genrative grammar*. Philadelphia: John Benjamins.

Roeper, T., & Rohrbacher, B. (1994). *Null subjects in early child English and the theory of economy of projection*. Unpublished manuscript, University of Massachusetts, Amherst.

Rohrbacher, B., & Vainikka, A. (1994). Verbs and subjects before age 2: The earliest stages in Germanic L1 acquisition. In J. Beckman (Ed.), *Proceedings of NELS 25* (Vol. 2, pp. 55–69). University of Massachusetts, Amherst: Graduate Linguistics Student Association.

Rumelhart, D. E., McClelland, J. L., & PDP Research Group. (1986). *Parallel distributed process*. Cambridge: MIT Press.

Sarma, V. (1994, October). How many branches to a syntactic tree? Disagreements over agreement. In *Proceedings of NELS 25* (Vol. 2), University of Massachusetts, Amherst: Graduate Linguistics Student Association.

Sano, T., & Hyams, N. (1993, February). Agreement, finiteness and the development of null arguments. In M. Gonzalez (Ed.), *Proceedings of NELS 24* (Vol. 2, pp. 543–558). University of Massachusetts: Amherst.

Schaeffer, J. (1990). *The syntax of the subject in child language: Italian compared to Dutch*. Unpublished masters thesis, State University of Utrecht, The Netherlands.

Scholten, C. G. M. (1988). *Principles of universal grammar and auxiliary verb phenomenon*. Unpublished doctoral dissertation, University of Maryland.

Schonenberger, M. (1995). Embedded V-to-C in early Swiss German. *MIT Working Papers in Linguistics (MITWPL), 26*, 403–450.

Suppes, P., Smith, R., & Leville, M. (1973). The French syntax of a child's noun phrases. *Archives de Psychologie, 42*, 207–269.

Torrens, V. (1995a). *The acquisition of syntax in Catalan and Spanish: The functional category inflection.* Unpublished doctoral dissertation, University of Barcelona, Barcelona, Spain.

Torrens, V. (1995b). The acquisition of inflection in Spanish and Catalan. *MIT Working Papers in Linguistics (MITWPL), 26*, 451–472.

Vainikka, A. (1994). Case in the development of English syntax. *Language Acquisition, 3*(3), 257–325.

Wexler, K. (1990a). *Recent studies in the development of inflection.* Paper presented at the symposium on current research in language acquisition at the Annual Meeting of the Society for Cognitive Science, MIT.

Wexler, K. (1990b). Innateness and maturation in linguistic development. *Developmental Psychobiology, 23*(3), 645–660.

Wexler, K. (1991). On the argument from the poverty of the stimulus. In A. Kasher (Ed.), *The Chomskyan turn.* Cambridge, MA: Basil Blackwell.

Wexler, K. (1992). *Optional infinitives, head movement and the economy of derivations.* (Occasional Paper #45). Center for Cognitive Science, MIT.

Wexler, K. (1994a). Optional infinitives, head movement and the economy of derivations. In D. Lightfoot & N. Hornstein (Eds.), *Verb movement.* Cambridge, England: Cambridge University Press.

Wexler, K. (1994b, April). Tense, agreement, aspect and root infinitives in young children. Paper presented at the Workshop on Language Acquisition, GLOW, Vienna.

Wexler, K. (1995, September). Feature-interpretability and optionality in early child grammar. Paper presented at the Workshop on Optionality, Utrecht, Amsterdam.

Wexler, K., & Chien, Y.-C. (1985). The development of lexical anaphors and pronouns. *Papers and Reports on Child Language Development, 24*, 138–149.

Wexler, K., & Culicover, P. (1980). *Formal principles of Language acquisition.* Cambridge: MIT Press.

Wexler, K., & Hamburger, H. (1973). On the Insufficiency of Surface Data for the Learning of Transformational Language. In K. J. Hintikka, M. E. Maravcsik, & P. Suppes (Eds.), *Approach to natural language* (pp. 153–166). Dordrecht, The Netherlands: Reidel.

Wexler, K., & Rice, M. (in preparation). *Do-support in negative sentences in the optional infinitive stage in normal and SLI children.* Unpublished manuscript, MIT and University of Kansas.

Weissenborn, J. (1988). *The acquisition of clitic object pronouns and word order in French: Syntax or morphology.* Unpublished manuscript, Max-Planck-Institut, Nijmegen, The Netherlands.

Whitman, J. (1994). In defense of the strong continuity account of the acquisition of verb-second. In B. Lust, M. Suner, & J. Whitman (Eds.), *Heads, projections, and learnability* (Vol. 1, pp. 273–287). Hillsdale, NJ: Lawrence Erlbaum Associates.

6

DEFINING THE OPEN AND CLOSED PROGRAM FOR ACQUISITION: THE CASE OF *WH*-QUESTIONS

Jill de Villiers
Smith College

In this chapter, I address the following guiding questions: What is known about the course of normal language acquisition? How important is linguistic experience for determining its course? If experience is required, does the child proceed by induction or deduction? Those theorists who believe language learning is highly dependent on amassed evidence from particular experiences, guided only by very general categorizing and memory mechanisms, are among the inductivists like Aristotle, Bacon, and Hume before them: The general principles and rules of language emerge from the painstaking compilation of raw data. In contrast, are the rationalist heirs to Plato, Descartes, and Kant: The child comes equipped with the principles of universal grammar and uses data only to confirm or deny deductions about which alternative manifestation of universal grammar surrounds the child. In this way, the induction/deduction dimension is related in significant ways to the ideas of empiricism/rationalism and to input-dependent versus input-independent development.

However, it is very clear that one cannot simply talk about language acquisition in its entirety. At the broadest level of distinction, many linguists would argue for a significant innate contribution to the acquisition of syntax, although they would argue that the vocabulary of any language must necessarily involve a major component of induction. After all, the principles of syntax are held to be universal, but the vocabulary of a language is specific to that language. That is not to say, however, that vocabulary acquisition could not involve innate principles that guide this search. For example, the child may know subtle constraints on what can be a possible word and on what kinds of meanings are likely to be highest in the hierarchy of possibilities that could be entertained (see Jackendoff, 1983; Markman, 1989; Clark, 1988; Nelson, 1988; and Carey, 1982).

Stepping beyond the syntax/vocabulary distinction, there are also proposals within the domain of grammar that differentiate some categories of acquisition from others. It is argued that certain features are universal and unvarying and, thus, should appear in fully fledged form early in the child's language, relying very little on exposure to data. In contrast, there may also be aspects of grammar that vary across languages, for example, different parameter settings, and these may not appear early but may await the right kind of input experience.

Alternative conceptions exist about subdividing grammatical properties that do not necessarily rely on the notion of parameter settings, for example, that some aspects of language are resilient, that is, input-independent and others are fragile, or input-dependent. An example of an input-independent aspect of grammar might be recursion, a property of all known human languages and one that occurs even in the invented language of deaf children (Goldin-Meadow, 1982). In contrast, a feature such as the use of auxiliary verbs to mark tense or aspect seems to be a feature that is highly input-dependent. Several studies have shown that the frequency or variety of auxiliaries in the input that children receive is related to the rate of acquisition of auxiliaries in the child's own grammar (e.g., Newport, Gleitman, & Gleitman, 1977).

The goals for this chapter can be described as follows:

1. To suggest a partitioning of the components of language acquisition into those that make up the *closed* part of the genetic program for language, that is, those that experience does not influence (though maturation might), versus the *open* part of the program, that is, those that would be subject to environmental influence.
2. To apply that frame of analysis to English question formation.
3. To describe evidence concerning the child's acquisition of questions.
4. To interpret that evidence in the light of the partitioning described.
5. To extend that reasoning to predictions about language delay/disorders in children.

SOME BACKGROUND

In what follows, I will assume a version of X-bar theory (X') (Chomsky, 1986; Radford, 1988), in which all grammatical categories are proposed to have the same canonical form, with a SPECIFIER, a HEAD (of the same type as the phrase) and COMPLEMENT(s):

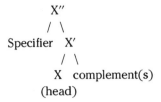

For instance, a noun phrase (NP) might be structured as follows:

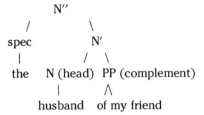

The head of the phrase selects its complements, so different complements can occur with different nouns or verbs. In some languages, the head appears on the left of the phrase as in English, but in other languages, like Japanese, it appears on the right. A fundamental difference between languages is, thus, captured by this head direction.

In addition to complements, phrases can be attached to other phrases as ADJUNCTS, in which case, they are not selected by the head. For example, adverbial phrases (e.g., *yesterday*) are typically not required or selected by particular verbs, hence attach as adjuncts to the verb phrase:

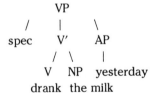

Under the assumptions of uniform phrase structure, the simple sentence (alias *S* in earlier grammars) is conceived as a similarly configured phrase with the subject as the specifier:

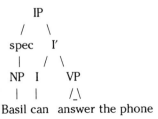

Under this analysis, the sentence becomes an IP, and the I (for inflection, sometimes called INFL) is the key constituent or head of that phrase. In a finite sentence, it contains the modal (e.g., *can*) or a form of tense; if nonfinite, it contains an infinitive particle, such as *to*. Because the head is also what selects the complements of the phrase, the I subcategorizes for a VP. In addition, the I constituent provides for the nominative case of the subject, and also tense markings. The category I is, thus, a central component of sentences in this new framework. I is called a functional category, in contrast to NP, VP, PP, and AP, which are lexical categories headed by lexical forms: noun, verb, preposition/postposition, adjective/adverb. On some analyses (Pollock, 1989), the I component is split into two separate functional categories independently responsible for tense (T) and agreement (A or AGR).

I is not the only functional category, however. IP is considered to be part of a larger constituent called CP:

In earlier grammars, this maximal projection was called an S-bar. In embedded sentences, the head of the CP, or C (comp or complementizer) is the constituent for complementizers, such as *that, whether, if,* and so on:

I know [*that* the earth is round].
She wondered [*if* the storm would arrive].

In main clauses, the C constituent is conceived as the landing site for the auxiliary in questions in languages like English. In yes/no questions, the C serves as the landing site for auxiliary movement (now called I-to-C movement):

Will he drink wine?

In languages like German, in which the verb occupies the second position in the sentence (V-2 languages), the C is considered to be the landing site for verb movement in declaratives also (Platzack & Homberg, 1989).

Notice the double movement involved in *wh*-questions in English:

What will he drink?

Because both auxiliaries and *wh*-questions move to the CP, it is proposed that the *wh*-word actually moves to the specifier position of CP and the Aux into the head of CP:

The CP is, therefore, another functional category with key roles: Question formation is dependent on the mastery of CP structure. So, also, are all embeddings of sentences either as complements, or as relative clauses in which a CP is embedded within an NP:

I saw the man [who you like].

Finally, the CP is important in long-distance movement of *wh*-questions (Chomsky, 1986). For example:

How did she think that she could open it?

Under one reading of the sentence, the answer could be:

With her key.

On this reading, the *how* is an adjunct (noncomplement) of the second verb, moved from the lower clause in the position marked by *t* for *trace* of the *wh*-question:

How did she think that she could open it *t*?

This is often referred to as *long-distance movement* because it takes place over one or even several clauses:

How did she think she could persuade Tim to let her open it *t*?

A short-distance *wh*-movement, in contrast, would occur within the main clause, for example, if the construal of the *wh*-question was that it was an adjunct to the verb *think* instead:

How did she think *t* that she could open it?

with an answer, such as *because she succeeded last time*. The movement of *wh*-words is argued to be *cyclic*, or one clause at a time, so that in long-distance movement the *wh*-word moves by cycling through the medial CP node before moving to the initial specifier of CP.

[$_{CP}$ How did she think ($_{CP}$ that she could open it *t*)].

Thus, the CP is an important component of the sentence that is responsible for both how complements attach to verbs and how questions move across these clause boundaries.

On this model, then, the phrase structure of sentences takes on universal hierarchical characteristics. The deep structure (D-structure) is generated before movement occurs, which results in the surface or S-structure. Under some accounts, there is another level of representation called logical form (LF), in which scope relations and meanings are computed. The variation that exists across languages is accommodated in a variety of ways to be discussed again later.

A CONTINUUM FOR LANGUAGE ACQUISITION

In the interests of promoting discussion, let me suggest that the differing proposals about language acquisition could be accommodated along a continuum. I will provide here only a brief description of the phenomena in a necessarily condensed preview, but in what follows I discuss many of them in greater depth.

(a) At the first level are universal properties of human language that exhibit no variation:

Recursion. The property that allows constituents to embed within each other, for example, NP within NP.

Hierarchical phrase structure. The idea that sentences are not linear strings, but structured trees of phrases; X′ theory discussed previously is one such refined proposal.

Structure dependence. All linguistic rules make reference to constituent structures (e.g., NP, VP, main verb) rather than nonstructures like "the third word" or "words beginning with *p*" (Chomsky, 1981).

Move alpha. All languages permit movement of structures within the syntax (Lasnik & Saito, 1992).

Empty category principle. This ensures that empty elements in the syntax (i.e., those that are left behind by movement) are appropriately licensed.

Logical Form (LF). The claim that there exists a hidden level of structure at which movement can also occur and which is responsible, for example, for scope relations and some aspects of meaning (Chomsky, 1981).

(b) Next come cases of parametric variations in universal principles:

Subjacency. A principle that dictates constraints on movement, namely, that constituents may not move across more than one bounding node. However, the particular bounding nodes, though drawn from a small set (CP, IP, NP), vary across languages (Cook, 1988).

Head direction. The fact that the structure of phrases (IP, CP, NP) in languages have a consistent direction of head placement. However, it can be either on the right (English) or the left (Japanese).

Binding principles. The ways that, for example, pronouns and anaphors (e.g., reflexives: *himself*) connect to their antecedent noun phrases. Although the principles are universal, there is some variation in the definition of the domain (roughly the level of the phrase) within which they must be free or constrained (e.g., Hyams & Sigurjónsdóttir, 1990; Wexler & Chien, 1985).

(c) Following those, there are parametric variations in properties/categories:

Pro-drop. This describes whether languages can omit a subject pronoun or not; Italian and Spanish can omit them, English and French cannot (Hyams, 1986).

The inventory of empty categories. This refers to the hidden or empty elements in the deep structure of sentences. Some languages (those with pro-drop) have *pro*, in reference to the empty subject pronoun; others do not. Other languages may lack *wh-trace*, the empty category left by *wh*-movement (Cook, 1988).

The properties and ordering of functional categories. The functional categories refer to a set of maximal projections or phrases that seem to have primarily a grammatical rather than a content function. Phrases include complementizer phrase (CP), inflection phrase (IP), and determiner phrase (DP) (Fukui & Speas, 1985). However, some languages are argued to have multiple phrases under IP, for example, tense and agreement, and their ordering appears to vary (Ouhalla, 1993). CP in some languages appears to be recursive, but not in others (Kraskow, 1994; Rudin, 1988).

(d) Finally, there are language-specific properties, though it is presently unclear whether this is an accurate characterization. Superficially, languages appear to differ in the elaborateness of marking, and the nature of the distinctions made in these areas. However, it is likely that deeper principles will be uncovered that will make many of these differences fall into a parametric framework, in which case the properties listed here could be absorbed into category (c):

Surface marking of case. Languages, such as English, mark case now only on pronouns (e.g., he, his, him), whereas other languages have rich case markings on NPs (e.g., German; Mills, 1985).

Elaborateness of tense markings. English has a relatively simple set of tense markings, but some languages have even fewer tense markings, and others have a richer system of markings. Furthermore, languages differ in the properties of events encoded in tense, for example, whether the event was witnessed or not witnessed (Slobin & Aksu, 1982).

Aspect. The durational quality of events. English makes fewer surface distinctions than say, Polish (Weist, Wysocka, & Lyytinen, 1991).

Number agreement. Found only on third person present singular verbs in English (e.g., he goes, I go), whereas other languages, such as Hebrew, have extensive agreement paradigms not just for number but for person and gender (Berman, 1985).

Specificity and definiteness. Relating to determiners. In English, distinctions are made based on definiteness (*a* vs. *the*) but many languages make further morphological distinctions based on specificity of the reference (Enç, 1991).

It must be said that in the current state of knowledge it is difficult to assign properties to one or another of these categories because as theories shift certain aspects of grammars take on or lose significance. For example, 10 years ago it would have been hard to foresee that the category of I, for inflection, would be assigned such a central role as the head of the sentence; the aspects such as tense and number agreement that it controls were seen as minor variants of the verb in older theories. Therefore, it must be noted that the theoretical status of the exemplars under these categories is bound to change. The proposal is that it could serve as a useful organizing framework for considering the relative contributions of nature and nurture, induction and deduction.

Consider now the implications of such a continuum for language acquisition. The first category, (a), should be absolute and innate and unlearnable: It sets the limits on what a human language can be. So the child's starting point is already situated within the hypothesis space that includes only these grammars, and induction plays no role in establishing this beginning point.

The second category, (b), is also largely innately given in the sense that the child's grammar must have these properties, but because there is language variation in how they are actualized, some input-dependence is possible. That is, though the principle of subjacency may be innately given (Chomsky, 1981), the child cannot know in advance what the bounding nodes are for subjacency in English until he has some exposure to the language (Cook, 1988). Neither can the child know in advance whether *wh*-movement occurs in the syntax (Huang, 1982), or whether the language permits the binding of anaphors across several clauses (Hyams & Sigurjónsdóttir, 1990). In each case, the relative contribution of induction and deduction to the learning process can be debated. The possibility arises that subtle triggers in one domain (i.e., crucial pieces of evidence that decide among parameter settings) allow the child to make certain choices that then impact the decisions in a quite different domain. That is, even where it is acknowledged that data from the language might be needed to set the switches, that data might be very subtle, deductive input rather than amassed inductive evidence of the sort, say, that contemporary connectionist models might require (Rumelhart & McClelland, 1987).

The third level, (c), parametric variation in properties, is even more input-dependent than the second, in the sense that more evidence might be needed to decide between competing alternatives. Take, for instance, the case of the empty categories. It seems to be a real possibility that a child might begin with a generic empty category and make finer distinctions as the need arises to distinguish, for example, *pro* from *trace*, or PRO from NP-trace. It has been proposed that the initial empty category might not be a variable (Perez-Leroux, 1993; Roeper & de Villiers, 1992). This is discussed further in a later section. Much work lies ahead to test this claim, but to the extent that there is linguistic variation in the inventory of empty categories, some learning must be involved. Similarly, different languages have been argued to have different ordering of the functional heads, tense and agreement, for instance. Exposure to the language would seem to be the only way to settle such an ordering, and it is less clear at present that a deductive trigger would do the trick. For example, a trigger (expletives such as *it* or *there* in sentences like *it is raining*) was proposed for setting the parameter for pro-drop (Hyams, 1986) though its adequacy and character is still disputed. Others argue that induction plays the larger role, with the child maintaining multiple hypotheses until sufficient evidence is amassed (e.g., Valian, 1990).

At the final level, the idiosyncratic properties of particular languages come into play. For example, take the fact that English exhibits no case marking or gender marking on its nouns or has the particular manifestations of specificity/definiteness in its determiners. These seem to be features beyond absolute principle, and beyond parametric variations in either principle or properties, accidental features of the history of the language. There are, however, arguments that would situate such variations within a larger picture of the language, claiming that any given language must end up with some average amount of complexity, or redundancy, or learning pitfalls. These are speculations about the evolution of a language, not about the child's preparation for learning it (Bates & MacWhinney, 1987).

Consider now a number of complications with this picture. The first is that the acquisition process might be affected by an additional variable, namely maturation. Maturation is invoked to explain why young children seem to have grammars that violate the principle of continuity with adult universal grammar. The argument is that, perhaps, certain features of the language faculty are on a maturational time course like other biological faculties and, thus, one would not expect the full principles or properties to show up before a certain point (e.g., Borer & Wexler, 1987). For this to be true maturation the development must not depend on particular input experiences but on biological growth within normal environmental limits. It might mean, however, that deductive triggers or evidence might not be processed before the maturation has taken place.

The second complication has to do with the possibility that the initial state of the parameter is some kind of default setting before data set it to correspond to that of the adult language, rather than being in an undecided state (Lebeaux, 1988, 1990). For example, the argument has been advanced that children begin with the default setting of the pro-drop parameter in favor of pro-drop, so that English-speaking children initially drop subject pronouns as if they were speaking Italian (Hyams, 1986). Others have argued that the child must begin with no such setting but must keep all hypotheses active (Valian, 1990). Many arguments have been made on both sides of this issue, but the possibility must be kept in mind that a child could appear to have adult knowledge if a default setting for some parameter coincided with the adult grammar of that language.

In the next section a major component of the mature grammar, the C-system, is discussed in more detail. Then, empirical data from children that address the question of how normal language acquisition of the C-system proceeds will be addressed. After discussing this developmental course, I return to the question of what might be input-dependent about this acquisition and what might be disrupted in the C-system in a child with specific language impairment.

COLLECTION OF FEATURES NECESSARY FOR A MATURE C-SYSTEM

In the contemporary syntactic approaches in which the sentence is headed by the maximal projection CP (see Fig. 6.1), that projection is responsible for a wide variety of different grammatical mechanisms involved primarily in the formation of questions in main clauses and in the connection of subordinate clauses to the main clause. In other languages like German, verb-second (V-2) phenomena also involve the CP because the verb moves into C. Here, I list with examples some of the phenomena involving the CP in English question formation.

Operator-Variable Relations

The *wh*-question in a *wh*-movement language, such as English, German, or Swedish, is held to move into the location in the specifier of the CP at the front of the sentence, leaving behind an empty category, a trace of itself. The *wh*-word in spec-CP is said to be an operator that binds the trace as a variable. This *wh*-operator behaves like a quantifier in that the question ranges over a set of individuals x_i, not a single individual. So, for instance, a question such as:

Who came to the party?

calls for a list answer such as *Jim, Mary, Belinda, Joe* and not for the name of a single individual who fits the description. This quantifier-like property of *wh*-

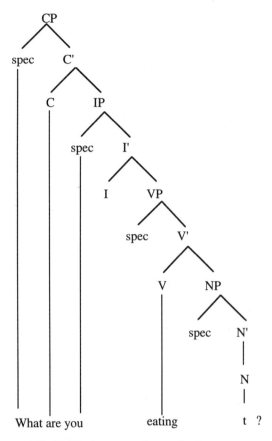

FIG. 6.1. The structure of a simple wh-question.

questions does not seem to apply when the *wh*-word is not in spec-CP but remains in situ as in an echo question:

You bought *what*?

in which case a legitimate answer may be to name the item in question (or dispute) and not list the entire contents of one's grocery bag.

In languages that do not have *wh*-movement in the syntax such as Chinese, Japanese, and Turkish, it has been convincingly argued that questions undergo movement to the CP at another hidden level, namely at logical form (Huang, 1982). In that way, the logical properties of questions in all languages can be described using the same machinery even though there is variation in whether there is overt syntactic movement. Consider this in terms of the decisions a child must make: Because the existence of the CP as a site for movement in syntax is subject to parametric variation, there must be some input-dependence to this grammatical development, but it would seem to be a relatively straightforward parameter to set. Some consideration of the complications is given in Roeper and de Villiers (1993), who set out a possible deductive learning course and discuss its potential hazards. Very recent arguments have disputed CP as the site of *wh*-movement for Romance languages, such as Italian, but the arguments are still in progress (Guasti, 1994).

Absorption and Superiority

There are conditions in which a *wh*-word does remain in situ in the syntax but with a quantificational reading, in the circumstances with double *wh*-questions in English. English requires one of the *wh*-words to remain in situ:

Who bought what?

Nevertheless, the question requires a pair-list answer in which each of the x_is pertaining to one question are paired with the y_is pertaining to the second:

Fran bought a book, Bill bought a watch, Latoya bought a CD.

Since Higginbotham and May (1981), this property is usually described as a consequence of a process called absorption occurring at the level of logical form in which both *wh*-words move into spec-CP, and their joint meaning is there computed.

However, Rizzi (1982) claimed that Italian does not allow multiple *wh*-constructions at all. In languages like Polish, Romanian, and Bulgarian, the two *wh*-words may move to the front of the sentence rather than one remaining in situ. There are good arguments that there may be differences even within this set of languages. In some, like Polish, one *wh*-question moves to spec-CP and the other to an adjunct position in IP, whereas in Romanian the CP node seems to be recursive and allows multiple *wh*-words to enter it (Kraskow, 1994; Rudin, 1988).

In English, a constraint known as superiority restricts the order of the two *wh*-questions so that certain orders are ungrammatical in ordinary circumstances in which a paired list reading is intended (Chomsky, 1973):

*What did who buy?

Explanations vary for this effect, but most invoke an appeal to the government relations in the sentence (Cheng & Demirdash, 1990). Recently, Chomsky proposed that an account could be developed based solely on the length of the movement involved, with an appeal to the concept of the least costly derivation (Chomsky, 1992). There are some doubts, however, about the universality of the superiority phenomena, which are reported to be weaker or even nonexistent in a language such as German (Bayer, 1992). The ordering of movements within those languages that allow double *wh*-movement may be discourse-governed rather than operating under a grammatical constraint, but not enough is presently known to be sure. In this realm, it is clear that there are also parametric variations that must be set by experience, possibly by yet-undefined deductive triggers.

Auxiliary Inversion

In English and some other languages, the movement of the *wh*-word to the front of the sentence is accompanied by movement of the auxiliary verb to a position in front of the subject, a rule known traditionally as *auxiliary inversion*. In the contemporary treatments, that movement is seen as a reflexive movement necessitated by spec-head agreement in which the presence of a *wh*-operator in spec-CP requires the C position to be filled also. The auxiliary undergoes head movement from the head of IP to the head of CP to fulfill this requirement, and a dummy form of the verb *do* is introduced if only a tense marker and not an auxiliary occupies I:

> Why did he come?
> *Why he did come?

Notice that in an embedded clause, spec-head agreement is not triggered:

> I wondered why he came.
> *I wondered why did he come.

Various explanations have been put forth to account for this fact. Most prominently, the fact that the verb selects the clause as its complement is proposed as the explanation for why the head of C is already marked by some feature that blocks inversion. Furthermore, there is at least one *wh*-question form in English that does not require inversion:

> How come she left?
> How come I didn't hear about it?

At least two possibilities exist for the analysis of *how come* questions: maybe *come* occupies the head of CP, blocking inversion, or maybe *how come* is itself a truncated matrix clause rather like *how did it happen that*, in which case the clause beneath it is embedded. Whatever the analysis, it is clear that the child learning auxiliary movement in English must attend to lexical aspects of the *wh*-word.

In addition, how inversion is played out has some language-specific features. In French main clause questions, inversion is optional (Hulk, 1993). In German and other V2 languages, the verb would normally occupy the C position even in nonquestions (Platzack & Holmberg, 1989). Given the amount of linguistic variation, the inversion of auxiliaries is predicted to be substantially input-dependent and the learning process more inductive than deductive.

Nevertheless, auxiliary inversion can be used to illustrate one well-established linguistic universal, structure dependence. Note that auxiliary inversion must involve the main clause auxiliary and not, for example, the first auxiliary in the string. That is, a child who is insensitive to the requirement that rules be structure dependent might generalize from cases such as these in which the first auxiliary is fronted in the question:

> The boy is swimming. . . . Is the boy swimming?
> The girl can draw. . . . Can the girl draw?
> The big fat turtle is burying himself. . . . Is the big fat turtle burying himself?

to a case where the first auxiliary is, in fact, not the main one:

> The man who can swim is saving the dog.
> *Can the man who swim is saving the dog?

where the correct, structure-dependent, question is:

> Is the man who can swim saving the dog?

Crain and Nakayama (1987) tested young children in an experimental game with a puppet to see if they could be induced to produce violations of structure dependence. The children were required to convert the carefully designed prompts by the puppet into yes/no questions, but they did not violate the requirements of structure dependence.

Cyclicity of *WH*-Movement

English along with many other languages has the further property that movement of a question can occur long distance, for example, across more than a single clause. Consider a question, such as:

> When did he say he hurt himself?

One can imagine circumstances in which the question would be ambiguous, for example, are you asking *when did he say it* or *when did he hurt himself?* The *wh*-question can thus originate in an embedded site and move to the front of the sentence in a number of steps, cycling through the intermediate spec-CP positions (see Fig. 6.2):

> What did she say she ate?

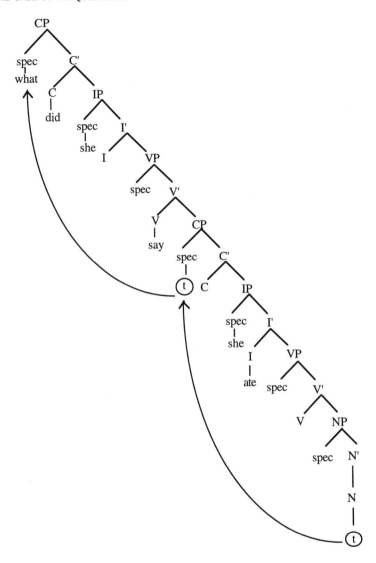

FIG. 6.2. Long-distance cyclic movement of a *wh*-question.

In languages with *wh*-movement, this cyclicity appears to be a requirement, though there are several different analyses of how many steps are involved (Chomsky, 1986; Cinque, 1990; Rizzi, 1990). The cyclicity of *wh*-movement might be predicted to be an automatic consequence of syntactic *wh*-movement, although there do not appear to be languages that only permit movement across a single clause. However, there is the following complexity to the picture. Some languages like German (McDaniel, 1989) simultaneously permit long-distance movement and partial movement, in which the following kind of structure occurs:

Was glaubst du mit wem Daniel spricht?
What think you with whom Daniel talks? or
Who do you think Daniel is talking to?
(McDaniel, Chiu, & Maxfield, in press)

In this type of sentence, the initial *wh*-word (*was* or *what* in German, but it can be other forms in other languages [Wayles Browne, personal communication, 1992]) serves as a kind of scope-marker, that is, an indicator that the sentence is a *wh*-question but otherwise contributing little meaning. The main question, the one to be answered, is the medial form *with whom*. The structures appear only with verbs that do not specifically subcategorize for a question (e.g., They couldn't occur with a verb like *wonder*), and coexist in the grammar of German with long-distance movement.[1] Hence, there are parametric choices to be made here too: Does the child's language permit partial movement? Yet a third possibility attested is *copying* in which the medial question is a copy of the initial question:

Wer glaubst du wer schwimmt?
Who do you think who swims?
Who do you think swims?

McDaniel (1989) and McDaniel, Chiu, and Maxfield (1995) provided some very interesting analyses of the relations among these forms in the languages that contain them. For our purposes, it is enough to note that they represent options in Universal Grammar that the child must rule out or in, and it is at least conceivable that the options could be deductively triggered.

Barriers to *Wh*-Movement

If one of the intermediate positions for cyclic movement is occupied by a *wh*-complementizer, the *wh*-chain (the *wh*-question and its linked traces) will fail, and the sentence is, thus, ungrammatical:

When$_i$ did he say how he hurt himself t_i?

The *wh*-question failed in its path because the landing site in the intermediate spec-CP was already filled, and so the reading was unavailable. Notice, however, that another reading is still possible, namely the one in which the *wh*-word originates next to the verb *say* (i.e., the short-distance reading).

---

[1]Subcategorization is used here as a very general term that refers to the selection of complements by their heads, but also to the particular requirement placed on such complements, that is, that a verb like *wonder* requires a *wh*-complement (*He wondered that she came. He wondered how she came.) whereas other verbs require the complementizer *that* (He reflected that he deserved it. *He reflected he deserved it.).

Phenomena such as this have been treated in various ways in the last 20 years or so since their discovery and have been variously referred to as constraints, islands, or, more recently, barriers to movement. However, argument questions (*who, what*) may circumvent barriers more easily than adjuncts (*how, when, where, why*) because of the different licensing conditions on arguments versus adjuncts. Various formulations of these licensing conditions have been tried, but consider one provided by Rizzi (1990):

> A nonpronominal empty category must be properly governed. Proper government can be satisifed either by lexical or antecedent government.[2]

The empty category principle (ECP) is designed to capture a striking difference between adjunct questions and argument questions in regard to the barrier effects. Argument questions refer to arguments of the verb: subjects and objects (and, possibly, indirect objects for verbs, such as *give*). On the other hand, adjunct questions occupy oblique roles of the verb: *Where, how, why*, and *when* are not required to fulfil the verb's thematic roles. When one compares the long-distance movement possibilities of different questions, this difference becomes significant. Compare the adjunct question in (a) with the argument question in (b):

(a) How did Mary ask who to help?
(b) Who did Mary ask how to help?

The answer to (a) should be unambiguous: The question is about how Mary asked the question, not how she wanted to help. But the answer to (b), with some thought, is ambiguous: It could be asking who Mary asked or it could be asking who she wanted to help. However, notice that the long-distance reading of *who* violates the barrier: It has apparently connected to a trace in the lower clause despite the intervening *how* question. The ECP captures this difference by allowing lexical licensing (head licensing) of arguments but requiring antecedent licensing for adjuncts. What this means is that object *wh*-traces (arguments) are always properly governed because they are lexically governed by the verb and, hence, can be found in the lower clause even with an intervening adjunct *wh*-word in the medial comp position:

Who$_i$ did Mary ask how to help t_i?

---

[2]*Government* means a particular configurational relationship between a governor and another element. So, all head of phrases can govern elements; a preposition can govern a NP or a verb can govern its object: This is lexical government. Antecedent government refers to a particular configuration of elements in the tree structure (see Cook, 1988, for a useful simplified discussion).

Because adjunct *wh*-traces (e.g., those of *how* or *when*), however, are not lexically governed, they require a process called *antecedent government* by being connected to an intermediate trace in spec-CP. However, if another *wh*-word intervenes in the medial comp, it acts as the nearest potential antecedent governor and the chain is disrupted:

   *How_i did Mary ask who to help t_i?

At present, it appears that these barriers to movement play out in similar ways across languages that permit long-distance movement, including the adjunct/argument distinction, though some features of the contrast may still be controversial for German (see Weissenborn, Roeper, & de Villiers, 1991). It is, thus, expected that the barriers should show up in children's grammar simultaneously with the possibility of long-distance movement, as should the argument/adjunct distinction.

Embeddedness and Complementation

Wh-complements constitute barriers to movement, but so also do a class of other grammatical phenomena roughly identified as adjuncts, rather than complements to a verb. Consider an example of why the complement status of a clause makes a difference for the movement of questions through it. Declarative complements are different from purpose clauses in the way they behave:

   Purpose clause: Last night he *called* to organize the party on Saturday.
   Complement: Last night he *wanted* to organize the party on Saturday.

The two clauses look the same, but their syntactic properties are distinct. If a *wh*-question is asked:

   When did he *call* to organize the party?

the answer must be when he called, for example, *last night*, not *on Saturday*. But, if the following is asked:

   When did he *want* to organize the party?

the answer could be *last night*, when he wanted it, or *Saturday*, when the party occurs. The difference between the clauses in this regard lies in their type of embeddedness: The clause under *called* is really an adjunct, not selected in any way by the verb. However, the clause under *wanted* is a complement, selected for and embedded under the verb. Complements can be considered obligatory with certain verbs, whereas adjuncts are not. Notice the difference if the clauses are omitted with the verbs previously mentioned.

John called.
*John wanted.

In short, in the adult language, *wh*-questions can extract from complements, but not from adjoined clauses. Compare the following in which the trace left by the *wh*-question is marked by *t*, so the required reading is the one with the trace in the lower clause:

Complement: Where did you decide you should go *t*?
Adjunct: *Where did you kiss your mother before she drove *t*?

The latter sentence does have a reading, but only if the question is construed to arise adjacent to the first verb, with an answer, such as *on the cheek*, that is, with the trace arising adjacent to *kiss your mother.*

Where did you kiss your mother *t* before she drove?

Recent work in syntactic theory has begun to refine the distinctions between complements and adjuncts (Hegarty, 1993). Several recent accounts suggest the simple two-fold distinction is inadequate to capture the range of possibilities in the syntax of natural languages. Though the barrier to extraction from adjuncts is presumed to be universal, how easy is it for a child learning a language to detect the status of a clause? Surely, the contrast is a subtle one for the purpose clause example, depending, as it does, on the child's initial determination of which verbs take complements. Gleitman (1990) argued that children are sensitive quite early to the syntactic frames of verbs and use them, in fact, to grasp meanings (the so-called *syntactic bootstrapping* hypothesis). Nevertheless, ambiguities could arise because many complement-taking verbs also permit simple NPs in that position:

John asked *to leave.*
John asked *Mary.*
Bill wanted *to write it.*
Bill wanted *the book.*

Consider then the ambiguity of sentences, such as:

Why did he want the butterfly to complete his collection?

Is *to complete his collection* a complement under *want* or not? On one reading it is and on the other, it is a purpose clause and, therefore, an adjunct. Thus, its status is ambiguous without a larger context. It is possible, therefore, that there might be a period in acquisition when children are conservative in long-distance movement because they do not yet know the complement/adjunct status of attached clauses. The lexical learning involved here could potentially take place over a protracted period, and be input dependent. More of this is considered next.

Lexical and Scope Effects

Consider the difference between the factive verb *forget*, whose complement remains true even when the verb *forget* is negated:

> He forgot that she was coming.
> He didn't forget that she was coming.

and a verb like *think*, nonfactive, where the truth of the complement is not known:

> He thought she was coming.
> He didn't think she was coming.

There are documented interactions between the semantics of factivity and the syntax of question-asking. For instance:

> (a) Why did he think they were driving?
> (b) Why did he forget they were driving?

In (a), one can answer both *why did he think it?* and *why were they driving?* In (b), it is not possible for adults to get a reading for the question, such as *why were they driving?* For example, *he forgot they were driving to go to the theater.* Instead, the question can only mean *why did he forget that fact?*, for example, *because his mind was on the math problem.* Thus, the factive verb operates as a barrier to the movement or interpretation of the *wh*-question from the lower clause (Szabolsci & Zwarts, 1990). The barrier effect is presumed to be universal (though no work to date has considered the interpretation of questions embedded under factive verbs in nonmovement languages). Nevertheless, because it depends on very subtle partitioning of the meanings of verbs and their scope, it might be guessed that acquisition in this area would be delayed and highly input dependent.

EVIDENCE ON THE ACQUISITION OF THE CP

In what follows, I consider what is known about acquisition of these features of the CP, focusing mostly on English. First to be discussed is the thorny issue of when the CP emerges at all as a functional category in children's grammar. After that, I have divided the acquisition evidence into parts corresponding to the phenomena discussed above.

The Emergence of the CP as a Functional Category

Controversy exists over whether the functional category CP is present from the start in children's grammars. In the last several years, a number of proposals have emerged that suggest children's grammars are limited at the start to lexical categories (e.g., NP,VP, AP), with the functional categories (e.g., CP, IP, DP) emerging only later. Some researchers attribute this lack to a maturational feature of early grammar but retain the view that all languages contain the same inventory of adult functional categories (e.g., Radford, 1988, 1990); others make the assumption that the default assumption of the child might be only that lexical categories are posited until positive evidence is provided about the nature of the additional functional nodes that must be built for that language (e.g., Clahsen, 1990; Lebeaux, 1988; Penner, 1992). This latter position gains credibility when the variation in functional categories is assessed. For example, some languages allow recursion in the CP so that multiple *wh*-movement is possible, but the English child must not adopt that option. Some languages move questions at logical form but not in the syntax. There are even questions about the existence of CP in the syntax of languages such as Japanese (Fukui & Speas, 1985).

There is, on the other hand, a growing body of opinion that the CP is, in fact, fully present in children's grammars from the start and that the observations that have led to the above proposals are insufficiently subtle. Here I can provide only a sense of why the debate cannot easily be settled by empirical data alone. Each view of the matter incorporates some assumptions, such as continuity of child and adult grammar (Boser, Lust, Santelmann, & Whitman, 1991; Hyams, 1992; Poeppel & Wexler, 1993; Santelmann, 1994; Valian, 1992; Verrips & Weissenborn, 1992), that are held to be as important as the facts in making the case. In addition, in the case of normally developing children, the controversy is over the very first months of the child's grammar, when production data are the only available, but ambiguous, sources of evidence (de Villiers, 1992). Recent work by Rizzi (1992, 1994) also raised good reasons from data on early subject-dropping for arguing that children's sentences may begin with an optional CP.

One argument in favor of the CP being present early in the child's grammar comes from the fact of *wh*-questions occurring with a *wh*-word that seems to have moved. Notice that subject *wh*-questions provide no evidence one way or the other, but object or adjunct *wh*-words are displaced from their usual position, and some have taken that as evidence that a CP must be present to receive them. Nevertheless, other possibilities exist and have been argued. For example, one viable option is that the *wh*-word might be moving into a topic position rather than a CP (e.g., Meisel & Muller, 1992). A second possibility is that it may be adjoined to the IP (proposed for early English by de Villiers, 1991, and for early French by Hulk, 1993). Santelmann (1994) explicitly tested this hypothesis for early child Swedish and found it lacking. In Swedish, apparently, null *wh*-questions are common in the early period (see also Penner, 1992, on Swiss

German). Radford (1994) provided a thorough analysis of the consequences, both developmentally and linguistically, of choosing different movement/attachment sites for early *wh*-questions. If any of the alternatives to spec-CP are valid, for example, topicalization or adjunction to IP, one might then expect the semantic properties of early questions not to be equivalent to the operator-variable relations of a true question. Hence, it is this issue that is discussed first in the next section.

Do Children Treat the Empty Category in *WH*-Questions as a Variable?

The idea that the *wh*-trace behaves as a variable has several facets. As mentioned, the *wh*-operator in spec-CP is considered to be a quantificational operator, and the trace it binds is not taken to refer to a fixed individual, but to a set. Takahashi (1991) and Maxfield (1991) both provided evidence of a sensitivity at age 3 or so to the differences between the discourse requirements of a real and an echo question. In an experimental study, Takahashi showed that 3- to 4-year-olds would answer a real question with a fully quantified (variable) answer, but would answer an echo question with a single name, or just the information supplied by the preceding discourse. For instance, in the context of a story children were told that:

The children had fruit for dessert.

and were shown a picture of children eating various fruits. When asked:

What did the children eat?

they answered, for example, *a strawberry, and a banana, and a cherry*, but when asked:

The children ate what?

they replied *fruit*. In short, the moved *wh*-question, but not the in situ question, was treated as requiring a variable answer by the child. Evidence such as this suggests that by age 3, *wh*-questions are derived by movement and involve a trace bound to an operator, which behaves like a quantifier.

What about the possibility that the CP is absent at the start and, thus, the earliest answers are not variable? Roeper and de Villiers (1992) and Penner (personal communication) both found support for this contention. The very earliest answers to questions that would normally require a set in reply offer instead only a single name. Unfortunately, there is less agreement over when this stage ends: Penner found evidence of nonvariable answers in Swiss German up to age 4, which is quite late in comparison to the age at which other arguments propose children have a missing CP. Although much more exploration of this phenomenon is needed to be sure it is a grammatical as opposed to a pragmatic failure, it is an important possibility to keep in mind in the controversy over the early existence of CP.

Acquisition of Absorption and Superiority

Experimental evidence (Roeper & de Villiers, 1992) revealed that children also provide exhaustive paired readings of double *wh*-questions such as:

Who brought what?

from about age 3 years, answering the following (for example): *The daddy brought the chips and the baby brought the apples* rather than *The daddy*. These results show that by 3, children are successfully combining the two *wh*-questions and relating the sets to one another in answering, achieving an outcome in keeping with absorption. Interestingly, though, data from a judgment task by McDaniel, Chiu, and Maxfield (1995) suggested that, given the choice, young children call such sentences ungrammatical! Thirty-nine percent of their responses consisted of a pattern in which children rejected all multiple *wh*-questions (the default, thus, being like adult Italian). The next most prevalent response, and one among older children, was that of accepting the multiple *wh*-questions that included one in situ, and rejecting double *wh*-movement cases such as:

Who thinks what Bert drank?

in which the lower *wh*-question has moved into the medial spec-CP. Unfortunately for our purposes, they did not test children on sentences that would be permissible in, for example, Polish:

Who what ate?

de Villiers and Plunkett (1992) explored children's obedience to the constraint known as superiority (Chomsky, 1981), designed to account for why certain orders of *wh*-questions are disallowed in sentences with more than one question. That is, certain questions seem to take priority over others in claiming the initial spec-CP slot:

Who slept where?
*Where did who sleep?

How will she make what?
*What will she make how?

de Villiers and Plunkett set up an elicited production task in which the child was presented with a situation containing multiple unknowns, and after hearing a couple of neutral models, the children were encouraged to ask a puppet what they needed to know to act out a requested scene with toys. For instance, the scene might contain three dolls and three different beds, and the child would be told that the dolls were all sleepy, but only Big Bird knew the sleeping arrangements. On half of the trials, children were prompted with a sentence beginning with the wrong *wh*-word, namely, one that would result in a superiority violation if completed in that order. For instance, if the target were something like:

*Where did who sleep?

The children might be prompted with:

Where . . . ?

In this way, 45 instances of double questions were elicited from seventeen 3- to 5-year-olds, and only one (dubious) case was a superiority violation. The children strongly resisted being misled into a wrongly ordered pair of questions and instead turned the misprompted sentences around, or rephrased them:

Where . . . ?
Where did this one and that one sleep?

One child even stopped the experimenter in midprompt and said *It's better if I start*!

Acquisition of Inversion

Auxiliary inversion evidence has also been used in the debates over the absence or presence of CP in early grammars. In the contemporary account, the auxiliary moves from the head of I to the head of C (see Fig. 6.1). Hence, evidence of early yes/no questions marked by auxiliary inversion could be taken as evidence for CP. As is well known, however, yes/no questions do not immediately get marked by inversion (Erreich, 1980; Ingram & Tyack, 1979). Those early ones that do appear have been interpreted, for example by Pierce (1989), as cases in which the subject is still within the VP, though others disagree with that interpretation. In early French, inversion of the auxiliary is also absent though that may also be the case in much of the colloquial French children hear (Hulk, 1993). In Swedish, the rate of inverted auxiliaries seems quite high (Santelmann, 1994).

Most strikingly to some observers, the auxiliary does not appear in inverted position in *wh*-questions (e.g., Weinberg, 1990) as early or reliably as in yes/no questions. Whether it is absent or in an uninverted position in *wh*-questions is more controversial, with the weight of data now in favor of the auxiliary being mostly absent from *wh*-questions before appearing in inverted position (Stromswold, 1994; Valian, 1992). In other words, if the auxiliary is present at all in a question, it is likely to be inverted.

In an earlier paper (de Villiers, 1991), however, I argued that there is a developmental coincidence concerning the appearance of inverted auxiliaries in *wh*-questions that may be of some significance for the debates concerning the development of the CP. In that paper, I reported doing some searches of the English CHILDES corpora (MacWhinney & Snow, 1985) for cases of *why* questions, because I was interested in the issue of whether auxiliary inversion was, as it seemed, especially late for *why* questions (see also Stromswold, 1994). I had culled all the sentences including the word *why* from the transcripts of Adam, and I was going through the printouts by hand to identify the first point of appearance of inverted auxiliaries. At a certain point I found several sentences appearing that I could not code because the *why* questions were in embedded contexts such as *I wonder why he did that*. In the next few transcripts, I found the first evidence of inverted auxiliaries for main questions, and the two developments, embedded questions with *why* and inversion of auxiliaries with *why*, proceeded then in tandem, with increasing numbers of both types of sentence. I turned to the next subject's corpus and the same thing was immediately apparent: The inversions occurred right at the same time as the first embeddings. The coincidence was always most apparent for *why* questions, but followup analyses looking at *what, when, where,* and *how* revealed the same general trend. Furthermore, the coincidences were lexically specific: For Adam, inversions appear at quite different points for *what, how,* and *why,* and the order of those developments was mirrored in the order of the development of embeddings of those questions. More refined analyses are probably needed on these data, because proving coincidence of linked developments is no easy statistical task given the small size of the samples.

One could use these data to argue that the child might be acquiring a CP at this point. That is, the first embedded questions provide the first real evidence of a complementizer (*that* and *if* are much rarer and later; only *to* appears early; see also Bloom, Rispoli, Gartner, & Hafitz, 1989), so they may represent the point of emergence of a real CP. Parallel claims exist for using this point for the CP for French by Hulk, (1993), and for German, by Meisel and Müller (1992), and Gawlitzek-Maiwald, Tracy, and Fritzenschaft (1992). Having a real CP allows the child to move main *wh*-questions into it and also provides a place for inverted auxiliaries for the first time. Of course, the lexical specificity is an intriguing puzzle. Does this really occur piecemeal for each *wh*-question?

These views go against very good arguments (based mostly on V-2 in German) in favor of an early and complete CP that preserve continuity of grammar between child and adult (e.g., Boser et al., 1991; Poeppel & Wexler, 1993). However, it would be unreasonable not to keep this alternative alive particularly until linguists resolve whether there is variation in the existence/structure of the CP across languages. To the extent that there is such variation, the possibility of an open program for acquiring it must be kept in mind.

Acquisition of Cyclicity and Barriers

Consider the evidence provided by *wh*-question interpretation on the issue of whether children successfully partition complements and adjuncts. In two studies (de Villiers, Roeper, & Vainikka, 1990; Maxfield & Plunkett, 1991), I showed that children by age 3 or so do allow long-distance movement of *wh*-questions, which is possible only from obligatory complements. I presented children with stories followed by ambiguous questions that permitted the children a choice between two interpretations. For example, take the following short story (accompanied by pictures for the children):

Story:
A little girl went shopping one afternoon, but she was late getting home. She decided to take a short way home across a wire fence, but she ripped her dress. That night when she was in bed, she told her mom, "I ripped my dress this afternoon."

Question:
When did she say she ripped her dress?

Given the story, there are two possible interpretations of the question, depending on where the listener interprets the trace to be for the *wh*-question *when*: Is it connected as an adjunct to *say*, or to *rip*?

(a) When$_i$ did she say t_i she ripped her dress?
(b) When$_i$ did she say she ripped her dress t_i?

That is, you might answer *at night*, if (a), or *that afternoon*, if (b), depending on your interpretation. Remember that the answer corresponding to (a) is referred to as a short-distance movement, because the *wh*-word moves within the first clause. The answer to (b), on the other hand, involves long-distance movement because the *wh*-word moves from the lower, or embedded, clause.

I showed that 3- to 6-year-olds are quite happy to provide either answer, suggesting that they do readily permit long-distance movement. However, it is also necessary to prove that the interpretations children provided were not just based on inference from the story, but were grammatically derived. To do this, I exploited the barrier phenomena discussed earlier. If the medial position in a question is already filled, a *wh*-question cannot cycle through that position and leave a trace, so the resulting sentence is ungrammatical:

(c) *When$_i$ did she say how she ripped her dress t_i?

The long-distance interpretation *in the afternoon* was all right in (b), but it is not permissible as an answer for (c). The question in (c) sounds grammatical, but only under the interpretation of *when did she say it* not *when did she rip the dress*: The long-distance reading is, therefore, excluded. Hence, if children's interpretations of questions are grammatically derived, rather than pragmatic inferences from the story, they should be sensitive to these grammatical requirements; they should block long-distance interpretation if the medial position is filled by another *wh*-complementizer. I controlled the stimuli so that for any given story, half of the children received a story and sentence without a medial complementizer; the other half received the same story followed by the sentence with a medial complementizer.

The findings showed that 3-year-old children only rarely give long-distance answers to questions such as (c), which contains a medial *wh*-word. This means they do respect the barrier to movement of a medial *wh*-question. Together with colleagues, I have explored children's knowledge of this barrier in other *wh*-movement languages: German, French, (Weissenborn et al., 1991), Caribbean Spanish (Perez-Leroux, 1991), and Greek (Leftheri, 1991). Young children exposed to the translations of these sentences have provided very similar data showing early evidence of long-distance movement, barrier effects, and the argument/adjunct distinction (Maxfield & Plunkett, 1991; Roeper & de Villiers, 1994). Kudra, Goodluck, and Progovac (1994) reported appropriate obedience in child speakers of Serbo-Croatian also.

Other work has shown that children in this same age range do not permit *wh*-extraction out of noncomplements, such as temporal adjuncts (Goodluck, Sedivy, & Foley, 1989) or relative clauses (de Villiers & Roeper, 1995). Furthermore, in some unpublished work on adjunct rationale clauses, I have never obtained a long-distance reading from 4-year-olds. These are cases such as:

Where did the boy find a pen to write?

in which an answer such as *on his book* (long distance) is inappropriate, but answering *in his bag* is fine. By age 4 or so, children apparently have the entire grammar of *wh*-movement and traces established. It would appear that they make a ready distinction between adjuncts and complements from an early age. In the section following, I offer some qualifications of this position that point to some late acquisitions.

The Acquisition of Complementation

There are other facts that make it premature to reach the conclusion that the child has the entire problem of cyclicity and barriers in place by age 4 years. The facts are well-documented yet may initially strike the reader (as they did me) as artifactual, that is, not reflecting the grammar so much as some performance variable. Children in all of the previously mentioned languages show a characteristic error at 3 and 4 years of age: they treat the medial question as if it were a real question, rather than a complementizer. So, for (c):

(c) When did she say how she ripped her dress?

children age 4 or so will say *with the wire on the fence,* answering the medial *how.* Simple explanations spring to mind for this phenomenon: Perhaps the children just answer the last question word they hear? We have vigorously pursued several control experiments that rule out various nongrammatical explanations (e.g., de Villiers & Roeper, 1995). Furthermore, children at the same stage produce the forms in elicited production tasks (Thornton, 1991) and judge them to be grammatical in judgement tasks (McDaniel, Chiu, & Maxfield, 1995), so the mistake occurs across performances.

The important feature of this mistake is that it is similar to a real option for interpretation that appears in some dialects of German and other languages (McDaniel, 1989). Hence, it appears that this option in universal grammar is making its appearance in the grammars of children speaking languages where it does not appear in the adult syntax. Why would this option be chosen by the children? It cannot be coming from the input, so does it represent that category of difference called a default setting?

Critical to our argument is the fact that partial movement occurs in the languages that allow it only where there is no subcategorization. In those languages, the medial question is the real question, not a complementizer, and the front question word is just a marker that the utterance is a question, but contributes no meaning.

We claim that children at this stage do not appropriately subcategorize the lower clause and, hence, allow partial movement (see also Roeper & de Villiers, 1994). In this way, the medial question is considered a real one, rather than a complementizer. We are not alone in arguing that children do not immediately represent subcategorization distinctions for verbs (Bowerman, 1974, 1982). Consider these examples of children's utterances, from our files, that indicate failures with the subcategorizations of complement-taking verbs:

"Did I seem that I was under the covers."
"It's clear to hear."
"Listening what you think."
"Listening what I whistled."
"I'm thinking why it's broken."
(Roeper & de Villiers, 1994)

In some ways, these observations are at variance with theories that argue that the child proceeeds to acquire meanings via syntactic bootstrapping, that is, that children can establish the syntax very early and use this to guess at the meaning of new verbs (Fisher, Gleitman, & Gleitman, 1991; Gleitman, 1990). However, much of that work has been concerned with relatively simple distinctions in argument structure and not with the subtle variants in clause connectives and structures that individual verbs exhibit, so I see no inherent contradiction between these possibilities.

In technical terms, we proposed recently that the child's complement clause is attached at a higher level than for the adult: It is not syntactically subcategorized by the verb though it is thematically governed by it (Roeper & de Villiers, 1994). The children's grammar permits partial movement structures primarily because they do not yet have the embedded question subcategorized under the main verb, though they recognize that it is thematically connected to it. By thematic connection we mean that the verb in question requires the satisfaction of a particular thematic role such as theme or goal (e.g., that *say* does not stand alone as an intransitive verb). But the embedded question cannot be a true adjunct or long distance movement would not be possible out of it. Hence, a child may make a basic differentiation between adjuncts and complements and yet still lack the full syntactic subcategorization of a mature grammar. One possible analysis would be to locate the difference hierarchically by arguing, for instance, that the clause is not attached as a sister to the V but to the V' (see Fig. 6.3).

Other analyses of the medial question phenomena have been provided by Thornton (1991) and by McDaniel, Chiu, and Maxfield (1995). Thornton's analysis of equivalent mistakes in the child's production assumes the child has full knowledge of complementation, except for a parametric setting involving agreement between the spec of CP and its head. It is an attractive analysis but does not cover the whole range of facts (see McDaniel et al., 1995) or the fact of the

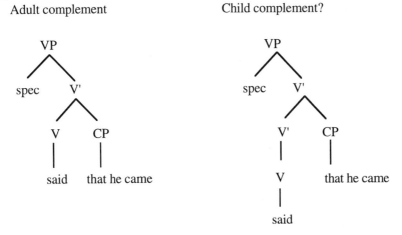

FIG. 6.3. Possible complement attachment for the adult versus the child's grammar.

child's answering the medial question. McDaniel et al., using primarily judgment data from young children, traced the development of grammars permitting partial movement, and presented convincing evidence that it is a stage through which children pass as they acquire full competence in English. They attributed it to the lack of a feature in the spec-CP, arguing that the complement is then structurally on a par with a relative clause. However, they argued that this does not mean that the clause will be interpreted as a relative clause: It refers only to a restriction on which elements might occur in spec-CP and CP. The feature to be acquired is one proposed by Rizzi (1990) called [pred] for *predicate*, and acquiring that feature allows children to make a structural distinction between *wh*-complements and relative clauses. The analysis has some parallels with my own and may indeed constitute the mechanism I have been seeking for my intuition that the structural connection of the *wh*-complement is higher on the tree than a true complement.

Many details remain to be worked out, but there is some convergence of opinion that partial movement represents a genuine default assumption for children. McDaniel, Chiu, and Maxfield (1995) even propose a particular deductive trigger that could result in the establishment of the [pred] feature in children's grammar. In my treatment, because moving away from partial movement depends on getting the particular subcategorizations right for particular verbs, the process could be drawn out and lexically specific. Clearly, this area is a ripe one for further research that could decide among different learning accounts.

Children's Knowledge of Lexical and Scope Effects on *WH*-Movement

The full early mastery of *wh*-movement and barriers is also called into question by work on children's understanding of the barriers introduced by factive verbs, negation, and quantificational adverbs. In contrast to children's early obedience to the barrierhood of medial-*wh*-questions, Philip and de Villiers (1992) and de Villiers and Philip (1992) collected substantial evidence that 4- to 6-year-olds do not respect these so-called weak islands to *wh*-movement. For adults, negation, adverbs, and factive verbs all block long-distance interpretation of the *wh*-question. Imagine a scenario where Jim's aunt is coming to town to go to the ballet and ask yourself if you can ever give that reason in response to the following *why* questions:

Factive: *Why$_i$ did Jim forget his aunt was coming t_i?

[Note: The reading "because he lost his diary" is fine, but "to go to the ballet" is disallowed.]

Quantificational adverb: *Why$_i$ did Jim always say his aunt was coming t_i?

Negative: *Why$_i$ did Jim not say his aunt was coming t_i?

Children allow long-distance readings quite readily with such questions, in striking contrast to their strict obedience to the barrier for *wh*-medial:

Wh-medial: *Why$_i$ did Jim say how his aunt was coming t_i?

Philip and de Villiers (1992) explored the possibility that the delay in acquisition of these weak barriers may have to do with delayed acquisition of the inferential/semantic properties of the terms in question, which is an explanation of the weak barriers put forth originally by Szabolsci and Zwarts (1990, 1992). Notice that the elements in question, factives, adverbs, and negatives, do have an effect on the inferences one can draw from propositions. Consider the following:

Jim forgot that his aunt was arriving by train, so he went to the bus station to pick her up.

When asked:

Did Jim forget that his aunt was coming?

adults say, "*No, he just forgot she was coming by train.*" In contrast, 4-year-olds answer confidently, "*yes!*" The particular feature that such a sentence manifests is the feature of not being monotonic increasing, that is, with most ordinary sentences, the inference from one statement to a more inclusive statement works out fine. Compare:

(a) Jim said that his aunt was coming by train.
(b) Did Jim say his aunt was coming? The answer is yes.

The sentence with *said* has the property of being monotonic increasing, that is, one can make an inference from the subset of events described by (a) to the larger set described by (b). However, certain elements create environments in which the inference is not allowed:

Negation: (a) Jim didn't say his aunt was coming by train.
           (b) Did Jim say his aunt was coming? Not necessarily.
Adverbs: (a) Jim always ate spaghetti with chopsticks.
           (b) Did Jim always eat spaghetti? Not necessarily.
Factives: (a) Jim forgot his aunt was coming by train.
           (b) Did Jim forget his aunt was coming? Not necessarily.

Philip and de Villiers (1992) and de Villiers and Philip (1993) tested the strong hypothesis that it is only once the children succeed at these inferences that long distance movement is blocked for them. At present, the best we have achieved is to show that children age 4 to 6 have corresponding difficulties in the two domains, at least hinting that the two are connected. The age of mastery on either task is not yet known, but the possibility of a deductive connection of this sort is being pursued.

DEFINING THE OPEN AND CLOSED PROGRAM FOR QUESTIONS

Turning back to the discussion from Section one of this chapter, what can be concluded from this survey of the acquisition of questions? How are the claims about relative degrees of input dependence borne out in the empirical data? I have encapsulated the following provisional discussion in Fig. 6.4.

At the most universal level, there seem to be a few very general properties that a child will not have to learn from the input. One is a feature such as structure dependence; evidence from Crain and Nakayama (1987) supported the idea that children will demonstrate it at a young age and make no mistakes. A second is the basic distinction between adjuncts and complements, with the qualification that a child must learn something of the thematic requirements of the verb before he or she could know if it takes a complement. It seems clear that children also use and understand questions involving movement from the earliest ages. What is still uncertain is whether they necessarily move the question initially to spec-CP. Most significantly, it seems from the evidence on barrier effects that children know some variant of the empty category principle as early as they show evidence of movement rules.

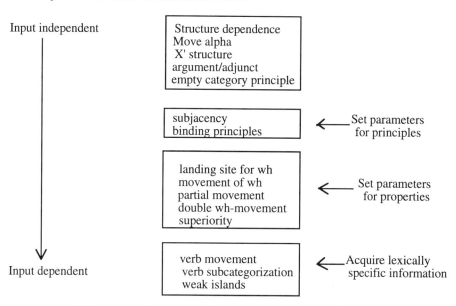

FIG. 6.4. Input-independent and input-dependent features of grammar.

With respect to the second category, namely, principles that have parametric variation, there seems to be rather little work on the CP that is relevant here. No one to date has studied children's sensitivity to subjacency in languages that exhibit different bounding nodes: The only work has been on English, where subjacency seems to be early respected (de Villiers & Roeper, 1995; Otsu, 1981). It would be very interesting to test children on languages, such as Italian, where the bounding nodes differ from English: Do children assume all the possible bounding nodes initially, then restrict them? What triggers the restriction? Given that this is one of the domains where negative evidence, or corrective feedback, is very unlikely, more work is clearly needed on the issues.

Let me offer the guess that movement to spec-CP and the cascade of consequences such as the variable properties of traces and absorption properly belong in the third category, namely, that they require some exposure to the language before that option can be chosen. Future work will be needed to test how much exposure is necessary and whether maturation plays any role.

In the third category may fall further properties that require exposure to the language to establish the right parametric choice. The existence of partial movement is clearly dependent on evidence from the language being learned, though that evidence may be in the form of discovering how complements of particular verbs are properly attached—that itself might be a slow process of learning. Further examples may be the potential for double *wh*-movement, and superiority effects. Also, input dependent should be the placement and movement of the verb, the distinction between V-2 languages and languages like English that move the auxiliary into C. Notice that even though this is input dependent, there is no assumption here that it will be late acquired, in fact, it will probably be early established given the centrality of that placement to other decisions about the grammar, and its pervasiveness in the sentences that the child will hear.

The category of weak islands consisting of negation, adverbs, and factive verbs may be more difficult to categorize. One would expect negation to be relatively early and universal, but the category of quantificational adverbs and factive verbs is less clearly determined, depending, as it does, on nuances of meaning difference in particular lexical items that surely await experience. But the possibility for all these weak islands is that they await some cognitive development, perhaps concerning inference and set relationships that are not established until late in the preschool years. At present, it would be premature to decide whether they belong together in this third category, or the fourth, namely, relatively idiosyncratic features of particular languages. Cross-linguistic evidence on the factive/nonfactive verb distinction and its reliability with respect to other meaning properties would be very helpful here.

At the most language-specific level, what aspects of *wh*-questions are left? Clearly, some aspects of auxiliary inversion are language-specific and even lexically specific. For example, it is not quite true that auxiliaries, and only auxiliaries, invert in English following *wh*-questions. The well-known exception to inversion is with *how come*:

How come he can go?

Further, American English allows one main verb, *be*, to invert:

Why is he here?

and British English also allows the main verb *have*:

Where has he a house?[3]

Given that variation, it may not be surprising if children acquire the auxiliary system in a piecemeal fashion and invert at different times with different *wh*-forms.

Finally, the lexical variation in complements needs further attention before it can be properly characterized. Many theorists argue for principles of argument structure that may be universal (Pinker, 1989), but the problem for the child is that the lexical items must still be differentiated and learned, and this process will surely be highly input-dependent. As a result, the barriers associated with various kinds of verbs and the complements they take should be late acquired even if there is some cross-linguistic consistency in the principles that they manifest.

IMPLICATIONS FOR LANGUAGE DELAY AND LANGUAGE DISORDERS

Having outlined the possibilities for normal acquisition, consider now a child with specific language impairment (SLI). What are the implications of the continuum discussed in the introduction for language learning in such a case? Being that there is no theoretical agreement yet about what is disturbed in SLI, I am free to speculate about possible disorders. I assume that no theory of SLI considers it a disorder so encompassing as to rule out the selection of a grammar with the properties in (a), namely the universal features of human grammars. It is theoretically possible that there could exist a developmental disorder so profound that it would result in a brain with no special capacity for language, just retaining the general cognitive capacities that are shared with other higher primates. All existing accounts of the language of individuals affected with SLI seem to consider it a human language, possessing the properties of hierarchical phrases, movement rules, structure dependence, and the empty category principle.

---

[3]I find myself hesitating about this one, although British English clearly allows main verb *have* inversion with yes/no questions like *have you a pen?* Without more informants, I can't tell if it is the decay of my dialect after 25 years or whether there is a genuine difference here!

In fact, some initial progress on this question has been made with respect to obedience to the empty category principle and the barrier effects on long-distance movement previously addressed. Seymour, Bland, Champion, de Villiers, and Roeper (1992) presented evidence that older children (aged 6 to 12) identified as having significant language delays nevertheless showed long-distance *wh*-movement, barrier effects, and the argument/adjunct distinction. As argued previously, given that these properties seem to apply for all languages that have syntactic *wh*-movement, their input-independence, and their early appearance in young normally developing children, the result is a natural one. Whether the same results would obtain if the sample of children chosen were limited to a more precise subgroup of specific language impairment is not yet known.

What about the second level, principles that require parameters to be set? There may be more uncertainty about this level because very little work to date has been guided by linguistic theory of sufficient sophistication (Seymour, Roeper, de Villiers, & Connell, 1992). It is at least plausible that individuals with a specific language impairment might exhibit the wrong parameter settings for aspects of universal principles, such as binding or barriers, because they may lack the deductive mechanism that normally would control the setting of these parameters or lack the means of processing the relevant input data. However, they should exhibit the principles themselves to the extent that these are part of universal grammar. That is, all of the deviations from the adult language should represent possible parametric settings of other languages. Whether or not this is true does not seem presently to be known.

At the third level, where parametric properties of language must be set by experience (deductive or inductive), there should be evidence for departures from adult grammar in SLI-affected individuals. An individual might exhibit a grammar whose combination of (unset? mis-set?) properties resembles no extant human language. For example, it could have prodrop, like Italian, and no *wh*-movement, like Korean. Does this ever occur and would it be recognized if it did? Clearly, work in this field will take considerable linguistic sophistication on the part of the diagnosticians who assess the child if it is to succeed.

Finally, at the level of the idiosyncratic properties of English, it might be expected that individuals with a language impairment would show considerable disturbance. It is at this point that any general deficits in rule acquisition, or in processing and analyzing data, should show their greatest effects. However, if the properties of the language at this level are unprincipled, they might somehow be compensated for by non-linguistic cognitive processes, such as a good long-term memory for idiosyncratic forms or phrases.

There is one illustration from the acquisition of the CP that makes this point, namely, the acquisition of inversion. To the extent that is highly input-dependent and lexically specific, children with difficulty in processing linguistic evidence might show prolonged difficulty in auxiliary inversion. If we are correct in our analysis of the lexical specificity of much complement learning, it is probable that this area would also show up as causing particular difficulty, and medial *wh*-answers should be seen to persist in affected individuals for many years.

Furthermore, for those input-dependent aspects, notice that the kind of information that will trigger the building of extra maximal projections—such things as aux inversion, or multiple determiners, or tense markings—might, in fact, depend on cues from unstressed functors, which are usually held to be especially vulnerable in the case of specific language impairment.

To add to these possibilities, what if the child with SLI does not, in fact, lack certain fundamental parts of the language faculty or processing mechanisms, but is just developmentally on a slower maturational timetable than normal, so all of the aspects that depend on maturation are deferred for several years past the normal case? Then, language acquisition might be seen occurring in an individual without access to those guiding principles that are hypothesized to mature. Is such development self-correcting? Or, do compensatory mechanisms take over so that the resulting grammar is a kind of Rube Goldberg invention?

Laying out even these limited possibilities has been an exhausting process, and the reader probably feels quite faint at the complexity of it all. However, the complexity must be faced if serious work is to proceed in this domain.

REFERENCES

Bates, E., & MacWhinney, B. (1987). Competition, variation and language learning. In B. MacWhinney (Ed.), *Mechanisms of language acquisition* (pp. 157–193). Hillsdale, NJ: Lawrence Erlbaum Associates.

Bayer, J. (1992). *On the origin of sentential arguments in German and Bengali.* Unpublished manuscript, University of Dusseldorf.

Berman, R. (1985). The acquisition of Hebrew. In D. Slobin (Ed.), *The crosslinguistic study of language acquisition: Vol. 1: The data.* Hillsdale, NJ: Lawrence Erlbaum Associates.

Bloom, L., Rispoli, M., Gartner, B., & Hafitz, J. (1989). Acquisition of complementation. *Journal of Child Language, 16*, 101–120.

Borer, H., & Wexler, K. (1987). The maturation of syntax. In E. Williams & T. Roeper (Eds.), *Parameter setting.* Dordrecht: Reidel.

Boser, K., Lust, B., Santelmann, L., & Whitman, J. (1991). The syntax of CP and V-2 in early German child grammar. *Proceedings of NELS North East Linguistic Society 22.*

Bowerman, M. (1974). Learning the structure of causative verbs: a study in the relationship of cognitive, semantic and syntactic development. Papers and reports on child language development (Stanford University).

Bowerman, M. (1982). Reorganizational processes in lexical and syntactic development. In E. Wanner & L. Gleitman (Eds.), *Language acquisition: The state of the art* (pp. 320–346). New York: Cambridge University Press.

Carey, S. (1982). Semantic development: the state of the art. In E. Wanner & L. Gleitman (Eds.), *Language acquisition: The state of the art* (pp. 347–389). New York: Cambridge University Press.

Cheng, L., & Demirdash, H. (1990). Superiority violations. *MIT working papers in linguistics.*

Chomsky, N. (1973). Conditions on transformations. In S. Anderson & P. Kiparsky (Eds.), *A festschrift for Morris Halle* (pp. 232–286). New York: Holt, Rinehart & Winston.

Chomsky, N. (1981). *Lectures on government and binding.* Dordrecht: Foris.

Chomsky, N. (1986). *Barriers.* Cambridge, MA: MIT Press.

Chomsky, N. (1992). *A minimalist program for linguistic theory* (MIT working papers in linguistics).

Cinque, G. (1990). *Types of A'-dependencies.* Cambridge, MA: MIT press.

Clahsen, H. (1990). Constraints on parameter setting: A grammatical analysis of some stages in German child language. *Language Acquisition, 1,* 361–391.

Clark, E. (1988). On the logic of contrast. *Journal of Child Language, 15,* 317–335.

Cook, V. (1988). *Chomsky's universal grammar.* London: Blackwells.

Crain, S., & Nakayama, M. (1987). Structure dependence in grammar formation. *Language, 63,* 522–543.

de Villiers, J. G. (1991). Why questions? In T. Maxfield & B. Plunkett (Eds.), *The acquisition of wh.* University of Massachusetts Occasional papers in Linguistics.

de Villiers, J. (1992). General commentary. In J. Meisel (Ed.), *The acquisition of verb placement: functional categories and V-2 phenomena* (pp. 172–173). Dordrecht: Kluwer.

de Villiers, J. G., & Philip, W. (1993, July). *The relation between logical and syntactic development.* Paper presented in Symposium on cognitive and semantic factors in quantification and wh-movement. International Child Language Congress, Trieste, Italy.

de Villiers, J. G., & Plunkett, B. (1992). *Children show superiority in asking wh-questions.* Unpublished manuscript, Smith College.

de Villiers, J. G., & Roeper, T. (1995). Relative clauses are barriers to wh-movement for young children. *Journal of Child Language, 22,* 389–404.

de Villiers, J. G., & Roeper, T. (1991). Introduction. In T. Maxfield & B. Plunkett (Eds.), *The Acquisition of wh.* University of Massachusetts Occasional papers in Linguistics.

de Villiers, J. G., & Roeper, T. (1993). The emergence of bound variable structures. In E. Reuland & W. Abraham (Eds.), *Knowledge and language: Orwell's problem and Plato's problem.* Dordrecht: Kluwer.

de Villiers, J. G., Roeper, T., & Vainikka, A. (1990). The acquisition of long distance rules. In L. Frazier & J. G. de Villiers (Eds.), *Language processing and acquisition.* Dordrecht: Kluwer.

Enç, M. (1991). The semantics of specificity. *Linguistic Inquiry, 22,* 1–26.

Erreich, A. (1980). *The acquisition of inversion in wh-questions: what evidence the child uses?* Unpublished doctoral dissertation, City University of New York.

Fisher, C., Gleitman, H., & Gleitman, L. (1991). On the semantic content of subcategorization frames. *Cognitive Psychology, 23,* 331–392.

Fukui, N., & Speas, M. (1985). Specifier and projection. *MIT working papers in linguistics,* 8.

Gawlitzek-Maiwald, I., Tracy, R., & Fritzenschaft, A. (1992). Language acquisition and competing linguistic representations: the child as arbiter. In J. Meisel (Ed.), *The acquisition of verb placement: functional categories and V-2 phenomena* (pp. 139–179). Dordrecht: Kluwer.

Gleitman, L. (1990). The structural sources of verb meanings. *Language Acquisition, 1,* 3–55.

Goldin-Meadow, S. (1982). The resilience of recursion. In E. Wanner & L. Gleitman (Eds.), *Language acquisition: The state of the art* (pp. 51–77). New York: Cambridge University Press.

Goodluck, H., Sedivy, J., & Foley, M. (1989). Wh-questions and extraction from temporal adjuncts: a case for movement. *Papers and Reports on Child Language Development, 28,* 123–130.

Guasti, T. (1994, April). *The acquisition of interrogatives in Italian child language.* Paper presented at the Generative Linguists of the Old World (GLOW) workshop on Theoretical Linguistics and Language Acquisition, Vienna.

Hegarty, M. (1993). *The derivational composition of Phrase structure.* Unpublished manuscript, University of Pennsylvania, Department of Linguistics.

Higginbotham, J., & May, R. (1981). Questions, quantifiers and crossing. *Linguistic Review 1,* 41–80.

Huang, C-T-J. (1982). *Logical relations in Chinese and the theory of grammar.* Unpublished doctoral dissertation, Massachusetts Institute of Technology.

Hulk, A. (1993, September). *Some questions about questions.* Paper presented at GALA conference, Durham, England.

Hyams, N. (1986). *Language acquisition and theory of parameters.* Dordrecht: D. Reidel.

Hyams, N. (1992). The genesis of clausal structure. In J. Meisel (Ed.), *The acquisition of verb placement: functional categories and V-2 phenomena* (pp. 371–400). Dordrecht: Kluwer.

Hyams, N., & Sigurjónsdóttir, S. (1990). The development of "long distance anaphora": a cross-linguistic comparison with special reference to Icelandic. *Language Acquisition, 1,* 57–94.

Ingram, D., & Tyack, D. (1979). Inversion of subject NP and auxiliary in children's questions. *Journal of Psycholinguistic Research, 8,* 333–341.

Jackendoff, R. (1983). *Semantics and cognition.* Cambridge: MIT Press.

Kraskow, T. (1994, January). Slavic multiple questions: evidence for wh-movement. Paper presented at the LSA, Boston, MA.

Kudra, D., Goodluck, H., & Progovac, L. (1994, January). *The acquisition of long distance binding in Serbo-Croatian.* Paper presented at the Linguistics Society of America (LSA), Boston, MA.

Lasnik, H., & Saito, M. (1992). *Move alpha.* Cambridge, MA: MIT Press.

Lebeaux, D. (1988). *Language acquisition and the form of the grammar.* Unpublished doctoral dissertation, University of Massachusetts, Amherst.

Lebeaux, D. (1990). The grammatical nature of the acquisition sequence: adjoin-a and the formation of relative clauses. In L. Frazier & J. G. de Villiers (Eds.), *Language processing and acquisition* (pp. 13–82). Dordrecht: Kluwer.

Leftheri, K. (1991). *Learning to interpret wh-questions in Greek.* Unpublished honors thesis, Smith College.

MacWhinney, B., & Snow, C. (1985). The child language data exchange system. *Journal of Child Language, 12,* 271–296.

Markman, E. (1989). *Categorization and naming in children.* Cambridge, MA: MIT Press.

Maxfield, T. (1991). Children answer echo questions *how?* In B. Plunkett & T. Maxfield (Eds.), *The acquisition of wh* (pp. 203–211). University of Massachusetts Occasional Papers in Linguistics. G.S.L.A., Amherst.

Maxfield, T., & Plunkett, B. (1991). *The acquisition of Wh.* University of Massachusetts Occasional papers in Linguistics, Graduate Student Linguistic Association (GSLA), Amherst.

McDaniel, D. (1989). Partial and multiple wh-movement. *Natural Language and Linguistic Theory, 7,* 565–605.

McDaniel, D., Chiu, B., & Maxfield, T. (1995). Parameters for wh-movement types: evidence from child English. *Natural Language and Linguistic Theory, 13,* 709–753.

Meisel, J., & Muller, N. (1992). On the position of finiteness in early child grammar: evidence from simultaneous acquisition of French and German bilinguals. In J. Meisel (Ed.), *The acquisition of verb placement: functional categories and V-2 phenomena* (pp. 109–138). Dordrecht: Kluwer.

Mills, A. (1985). The acquisition of German. In D. Slobin (Ed.), *The crosslinguistic study of language acquisition: Vol. 1: The data.* Hillsdale, NJ: Lawrence Erlbaum Associates.

Nelson, K. (1988). Constraints on word learning? *Cognitive Development, 3,* 221–246.

Newport, E., Gleitman, L., & Gleitman, H. (1977). Mother, I'd rather do it myself: Some effects and noneffects of maternal speech style. In C. Snow & C. Ferguson (Eds.), *Talking to children: language input and acquisition.* New York: Cambridge University Press.

Otsu, Y. (1981). *Universal grammar and syntactic development of young children.* Unpublished doctoral dissertation, Massachusetts Institute of Technology.

Ouhalla, J. (1993). *Functional categories and parametric variation.* London: Routledge & Kegan Paul.

Penner, Z. (1992). The ban on parameter setting, default mechanisms, and the acquisition of V-2 in Swiss German. In J. Meisel (Ed.), *The acquisition of verb placement: functional categories and V-2 phenomena* (pp. 245–281). Dordrecht: Kluwer.

Perez-Leroux, A. (1991). The acquisition of long distance movement in Caribbean Spanish. In T. Maxfield & B. Plunkett (Eds.), *The acquisition of wh* (pp. 79–99). University of Massachusetts Occasional papers in Linguistics.

Perez-Leroux, A. (1993). *Default systems in the acquisition of Wh-movement in Spanish*. Unpublished doctoral dissertation, University of Massachusetts, Amherst.

Philip, W., & de Villiers, J. (1992). Monotonicity and the acquisition of weak wh-islands. In *The Proceedings of the Twenty-Fourth Annual Child Language Forum* (pp. 99–111). Stanford: Center for the Study of Language and Information.

Pierce, A. (1989). *On the emergence of syntax: a cross-linguistic study*. Unpublished doctoral dissertation, Massachusetts Institute of Technology.

Pinker, S. (1989). *Learnability and cognition: the acquisition of argument structure*. Cambridge, MA: MIT Press.

Platzack, C., & Holmberg, A. (1989). The role of AGR and finiteness in Germanic VO languages. *Working papers in Scandinavian syntax, 43*, 51–76.

Poeppel, D., & Wexler, K. (1993). The full competence hypothesis of clausal structure in early German. *Language, 69*, 1–33.

Pollock, J. Y. (1989). Verb movement, universal grammar, and the structure of IP. *Linguistic Inquiry, 20*, 365–424.

Radford, A. (1988). Small children's small clauses. *Transactions of the Philological Society, 86*, 1–46.

Radford, A. (1990). *Syntactic theory and the acquisition of English syntax: the nature of early child grammars of English*. Oxford: Blackwell.

Radford, A. (1994). The syntax of questions in child English. *Journal of Child Language, 21*, 211–236.

Rizzi, L. (1982). *Issues in Italian syntax*. Foris: Dordrecht.

Rizzi, L. (1990). *Relativized minimality*. Cambridge, MA: MIT Press.

Rizzi, L. (1992). Early null subjects and root null subjects. *Geneva Generative Papers*.

Rizzi, L. (1994, January). *Root infinitives as truncated structures in early grammar*. Paper presented at Boston University conference on language development, Boston, MA.

Roeper, T., & de Villiers, J. G. (1991). Ordered decisions in the acquisition of wh-questions. In H. Goodluck, J. Weissenborn, & T. Roeper (Eds.), *Theoretical issues in language development*. Hillsdale, NJ: Lawrence Erlbaum Associates.

Roeper, T., & de Villiers, J. G. (1992). *The one feature hypothesis for acquisition*. Unpublished manuscript, University of Massachusetts.

Roeper, T., & de Villiers, J. G. (1994). Lexical links in the Wh-chain. In B. Lust, G. Hermon, & J. Kornfilt (Eds.), *Syntactic theory and first language acquisition: Cross linguistic perspectives Vol. II: "Binding, dependencies and learnability*. Hillsdale, NJ: Lawrence Erlbaum Associates.

Rudin, C. (1988). On multiple questions and multiple wh-fronting. *Natural Language and Linguistic Theory, 6*, 445–501.

Rumelhart, D., & McClelland, J. (1987). Learning the past tense of English verbs: implicit rules or parallel distributed processing? In B. MacWhinney (Ed.), *Mechanisms of language acquisition*. Hillsdale, NJ: Lawrence Erlbaum Associates.

Santelmann, L. (1994, January). *Early wh-questions: evidence for CP from Child Swedish*. Paper presented at Boston University conference on Language development, Boston, MA.

Seymour, H., Bland, L., Champion, T., de Villiers, J., & Roeper, T. (1992, November). *Long distance wh-movement in children with divergent language backgrounds*. Poster presented at the Convention of the American Speech and Hearing Association, San Antonio.

Seymour, H., Roeper, T., de Villiers, J., & Connell, P. (1992, November). *Theoretical work in language acquisition and applications*. Miniseminar at the Convention of the American Speech and Hearing Association, San Antonio, TX.

Slobin, D., & Aksu, A. A. (1982). Tense, aspect and modality in the use of the Turkish evidential. In P. J. Hopper (Ed.), *Tense-aspect: between semantics and pragmatics*. Amsterdam: John Benjamins.

Stromswold, K. (1994, January). The nature of children's early grammar: evidence from inversion errors. Paper presented at the LSA, Boston, MA.

Szabolsci, A., & Zwarts, F. (1990). *Islands, monotonicity, composition and heads.* Paper presented at GLOW conference, University of Leiden, The Netherlands.

Szabolsci, A., & Zwarts, F. (1991, August). Unbounded dependencies and the algebraic semantics. *Lecture notes of the Third European Summer School in Logic, Language and Information*, Saarbrucken.

Takahashi, M. (1991). The acquisition of echo questions. In B. Plunkett & T. Maxfield (Eds.), *The Acquisition of wh.* University of Massachusetts Occasional Papers in Linguistics. G.S.L.A. Amherst.

Thornton, R. (1991). *Adventures in long distance moving: the acquisition of complex wh-questions.* Unpublished doctoral dissertation, University of Connecticut.

Valian, V. (1990). Logical and psychological constraints on the acquisition of syntax. In L. Frazier & J. de Villiers (Eds.), *Language processing and language acquisition* (pp. 119–145). Dordrecht: Kluwer.

Valian, V. (1992). Categories of first syntax: be, being and nothingness. In J. Meisel (Ed.), *The acquisition of verb placement: functional categories and V-2 phenomena.* Dordrecht: Kluwer.

Verrips, M., & Weissenborn, J. (1992). Routes to verb placement in early German: the independence of finiteness and agreement. In J. Meisel (Ed.), *The acquisition of verb placement: functional categories and V-2 phenomena* (pp. 283–331). Dordrecht: Kluwer.

Weinberg, A. (1990). Markedness versus maturation: the case of subject-auxiliary inversion. *Language Acquisition, 1*, 165–194.

Weissenborn, J., Roeper, T., & de Villiers, J. G. (1991). The acquisition of wh-movement in French and German. In B. Plunkett & T. Maxfield (Eds.), *The Acquisition of wh* (pp. 43–74). University of Massachusetts Occasional Papers in Linguistics, Graduate Student Linguistic Association (GSLA), Amherst.

Weist, R., Wysocka, H., & Lyytinen, P. (1991). A cross-linguistic perspective on the acquisition of temporal systems. *Journal of Child Language, 18,* 67–92.

Wexler, K., & Chien, Y-C. (1985). The development of lexical anaphors and pronouns. *Papers and reports on child language development* (Vol. 24). Stanford University.

COMMENTARY ON
CHAPTER 6

Catherine E. Snow
Harvard Graduate School of Education

de Villiers' chapter presents a marvelously clear and thoroughly systematic discussion of her answer to the question, "What might the environment be able to affect the acquisition of?" What is particularly impressive about her chapter is her willingness to acknowledge the counterevidence to every claim—thus, to present the case in a way that represents a search for truth, rather than simply an attempt to win an argument.

de Villiers starts, though, with a slight oversimplification of the possibilities for the role of environment, by contrasting the strong inductivist position ("language learning is highly dependent on amassed evidence from particular experiences . . .") to the deductivist stance ("the child . . . uses data only to confirm or deny deductions about which . . . grammar," p. 145). I would argue that she has, in fact, ignored the constructivist position, which would emphasize the child's dependence on input to provide information about language, but also the child's capacity for structuring and analyzing language data in ways that are neither immediately dependent on input nor directly given by innate principles.

de Villiers also reminds the reader that *language* is not synonymous with *syntax*, and that as developmentalists we should be seeking explanations for the acquisition of vocabulary as well as syntax; I would add communicative skills, conversational skills, and skills at extended discourse to the list of phenomena that require explanation. Nonetheless, her basic approach of looking for splits within the language system that might define components with different developmental trajectories and subject to different developmental influences is one that I, too, have pursued and do endorse (e.g., Pan, Snow, & Willett, 1994).

The work done by de Villiers is firmly embedded in a particular analytic framework, a discussion of which would go far beyond the scope of this brief commentary. The framework, X-bar theory, proceeds from a starting assumption that seems to be motivated more by aesthetic elegance than by either processing constraints or data about languages, namely, that all constituents have an identical two-part structure, in which one part (consistently the first or the second, depending on the language) is defined as the *head*. Thus, what used to be called a noun phrase is seen as being composed of two units, the specifier and the noun; in English, the order is specifier-noun, and English is a right-branching (or head-on-the-left) kind of language, so the specifier has to be the head. These rules, which were designed to assure uniformity of structure across a variety of structural types, then force analysts into minor contortions, for example, that *aux* ends up as the head of the complementizer in questions like "what will he drink?" because, in effect, there is not any other place at the beginning of the sentence for it to land. As de Villiers points out, a theory like this is almost sure to change, so perhaps it makes little difference that details of it are implausible, but it is important to note the criteria on which this theory was proposed and accepted—criteria of consistency, rather than conformity to data. The degree to which theory drives analysis is made clear in the invocation of the invisible logical form as a level at which languages that clearly operate differently at the audible level are said, nonetheless, to be identical, and at which problems of syntactic description which are intractable elsewhere can be solved.

There is no denying the elegance of the system proposed, and of de Villiers' presentation and discussion of it. However, not all the distinctions she makes are as clear-cut as they might seem. In fact, it seems that pragmatic and semantic factors (factors that it is typically assumed children become good at because they learn about them, not because they have innate access) massively interact with the syntactic explanations offered by de Villiers. Thus, for example, the distinction she offers between complement clauses, in which the verb selects the complement ("he planned to swing from the tree") and adjunct clauses ("he climbed to swing from the tree") can be conceived of as a structural difference, but certainly also as a semantic difference (the sense of close connection between matrix verb and complement) and a pragmatic difference (the new information in focus in the first case is "planned to swing" and in the second, "climbed"). Similarly, the distinction between

(1) Why did he think they were driving?

in which *why* can refer to either clause and

(2) why did he forget they were driving?

in which only the local interpretation is possible must relate to factivity differences between think and forget. But note that with a different *wh*-word the potential for movement shifts:

(3) Where did he think they were driving?
(4) where did he forget they were driving?

When the question word is *where*, forget supports a long-distance interpretation ("he forgot they were driving to gramma's"), though, admittedly, the pragmatics of the answer (and thus the question) are slightly peculiar.

de Villiers identifies a class of items in the main clause that block the long-distance interpretation—but, in fact, this class is not particularly well-defined or limited. The blocking of the long-distance interpretation from *not* in

(5) Why didn't he say she was driving?

strikes me as no stronger than the blocking effect of any adverb, quantificational, negating, or whatever. Thus, for example, a sentence like

(6) Why did he suddenly say she was driving?

is certainly more likely to receive the local interpretation, as is

(7) Why did he shout she was driving?

or any other form that puts the first clause in focus. Both adverbs and verbs marked for manner bring the matrix verb into focus; in contrast, I can get the long-distance reading for sentences like (8), (9), and (10)

(8) Where did he say she was driving?
(9) Where didn't he say she was driving?
(10) Where did he suddenly say she was driving?

in which the presence of *where* creates an expectation that a verb of motion or location will be in focus. Obviously, like de Villiers' young subjects, I cannot get a long-distance reading for

(11) Where did he say how she was driving?

but interpreting *where* as moved out of the subordinate clause there would quite obviously be perverse on pragmatic, semantic, syntactic, and simple processing grounds.

The data de Villiers presents about children's relatively early understanding of the possibility of long-distance interpretations of *wh*-words in two-clause questions are convincing about children's language capacities, but only consistent with, not proof of, her particular description of the nature of those capacities. She states "We showed that children by age three or so do allow long distance movement of *wh*-questions, *which is possible only from obligatory complements* (italics added)" (p. 170). In other words, she seems to be assuming there are no other conceivable explanations for this capacity, which, in fact, could well have been discovered precisely through hearing and understanding sentences where the long-distance interpretation was greatly favored by pragmatic or semantic factors, for example:

(12) When do you think we'll get there?
(13) How did mummy say you should sit?
(14) What time did daddy think you should go to bed?

The easy accessibility of the long-distance interpetation here suggests that children might well have had experiences supporting their discovery of long-distance possibilities.

de Villiers' basic point, though, is a crucial one worth repeating—that the language system children acquire can be seen as composed of various components, some of which will be heavily dependent on input for their normal development, and others less so. For those aspects of language learning that are relatively robust and, thus, less sensitive to the effects of input, we cannot, however, be satisfied with an explanation of the form "it is innate" or "it is genetically determined." A plausible mechanism is needed to link the biological preparedness to the behavior or knowledge displayed by the child. de Villiers and I disagree about what constitutes a plausible explanatory system for the data presented here, but it is clear that the data themselves are made accessible, the structuring of the whole chapter as a discussion of more- or less-environmentally influenced phenomena is very clear, consistent, and nicely adhered to, and I enjoyed it a lot!

REFERENCE

Pan, B. A., Snow, C. E., & Willett, J. B. (1994). *Modeling growth in lexical, syntactic, morphological, and conversational skills.* Harvard Graduate School of Education. Manuscript in preparation.

III

LANGUAGE IMPAIRMENTS

7

GENETIC AND ENVIRONMENTAL CONTRIBUTIONS TO THE RISK FOR SPECIFIC LANGUAGE IMPAIRMENT

J. Bruce Tomblin
University of Iowa

The acquisition of spoken language is a universal property of humans. For most children, language is successfully acquired with relative ease in the span of a few years without formal training by adult caregivers. There are some children, however, who are challenged by the demands of language acquisition and, as a result, are unable to meet the communicative, educational, and social expectations of their society. These children with developmental language disorders fall into several groups. Some children with language impairment may have concomitant problems such as pervasive cognitive, social/emotional, or motor problems, that interfere with language learning and use. However, a number of children exhibit unexpected and, to date, unexplained difficulties in the acquisition of spoken language that are not associated with other developmental disorders. These have been referred to as children with specific language impairment (SLI).

DIAGNOSTIC DEFINITION OF SLI

Although there is no universally accepted diagnostic standard for SLI, a representative definition of SLI is that it is a limitation in the acquisition of the language of the child's community, occurring in the context of:

- Normal hearing sensitivity.
- Normal nonverbal intellect.

- Normal social and affective status.
- Normal motor skills.

Further, although not a part of the diagnostic criteria, research with these children has shown that they have been provided with largely ordinary linguistic experience (Leonard, 1987), and the differences that exist in the linguistic behavior of mothers of children with SLI are likely to be a product of the child's language behavior (Conti-Ramsden, 1985). Thus, the current consensus is that the language-learning problems they demonstrate are not likely to be a product of insufficient opportunity to acquire the language spoken in their homes.

Status of SLI as a Diagnostic Entity

Although the term *specific language impairment* is relatively new, the construct often termed developmental or congenital aphasia has been in the literature for a number of years (Benton, 1964; Ewing, 1930). Despite this relative long history, the status of SLI as a distinct diagnostic entity is certainly suspect. The cause(s) of this language learning problem are still unknown and, therefore, there are no means to define SLI in terms of an underlying explanatory theory. Due to the lack of an etiology on which to base a diagnosis of SLI, the definition is based on the presence of certain behavioral characteristics. These characteristics, however, are very general. For instance, speech–language pathologists do not require certain features of language to be impaired, rather, any deficits that are sufficient to place the child somewhere between 1 and 2 standard deviations (*SD*s) below age expectations are likely to lead to a diagnosis of language impairment (Tomblin, Freese, & Records, 1992). Thus, the diagnostic standard described earlier does not necessarily yield a homogeneous group of language learners that is distinctively different from borderline normal language learners. Further, although Kamhi and Johnston (1982) have reported qualitative differences between the grammatical achievements of children with SLI- and mental age (MA)-matched children with mental retardation, it is still unclear how distinctive the language characteristics of children with SLI are from other clinical conditions.

The status of SLI as a diagnostic entity is not unlike many of those in psychiatry, such as attention deficit disorder, manic depression, schizophrenia, and dyslexia. What is different is that under a medical model, psychiatrists have taken their diagnostic entities seriously and, thus, a considerable effort has been made to test the validity of these entities. In this work, it is assumed that each of these conditions has a unique etiology and course. Speech-language pathology rejected the medical model for a number of years and adopted instead a behavioral/educational model. As a result, little work has been invested in testing and refining the diagnostic standards for many developmental communication disorders, including SLI.

Currently, the clinical management of language impairment is not conditioned on the diagnostic entity and, therefore, we are not pressed to be concerned with the etiology. In research concerned with developmental language impairment, however, it becomes more difficult to avoid this issue. As was noted by Tomblin (1991) and Johnston (1991), an essential part of a theory of SLI, if it is to be a scientific theory, must have to do with statements of cause and, therefore, etiology is an essential topic.

Most of my following remarks will be concerned with research conducted to uncover the etiology of SLI as it is currently defined. Because this diagnosis is based for the most part on the absence of explanations, it should be assumed that SLI is etiologically heterogeneous. Thus, my null hypothesis in this work will assume an extreme position that SLI occurs in a sporadic, random fashion among individuals and that the etiology for each affected individual is unique. The alternate hypothesis is that the diagnostic category of SLI comprises subgroups or, in the most unlikely case, consists of a single etiologically homogeneous group. As I examine this null hypothesis as well as its alternatives, I will be searching for ways to modify my diagnostic criteria to yield more etiologically homogeneous groups of language learners.

DESCRIPTION OF THREE RESEARCH POPULATIONS

My effort to identify risk factors for SLI will come from three ongoing studies. I emphasize that in each case more data will be gathered before the studies are complete and, therefore, everything stated here is preliminary.

Family Study

One data set I use is the result of studying 45 families ascertained through an SLI proband.[1] In each family, I directly examined all first degree relatives of the SLI proband for SLI as well as any first degree relatives of affected parents of the proband. Thus, in all cases, I determined the status of the siblings and parents of the SLI proband; if one of the parents was found to be affected, I then attempted to examine the parents (proband's grandparents) and siblings (aunts and uncles) of this affected parent. In addition to directly testing these relatives for SLI, I also obtained a historical report of speech, language, and reading problems in the first, second, and third degree relatives of the proband even when these were not directly examined. The data from this study have been gathered primarily for the study of familial risk factors for SLI.

---

[1]A *proband* is an individual through whom the family is sampled. Usually, the proband has a particular trait or phenotype that is the focus of the study, and the presence of this affection status is what leads the proband and, thus, the family to be sampled.

The diagnosis of SLI among family members has presented a challenge because the same language measures cannot be used for all family members, regardless of their age. The language skills of those with normal and impaired language change considerably with age and, therefore, measures that will reveal SLI in a preschool or young school-age child will not be sensitive to individual differences found in the adolescent and adult population. Therefore, I constructed a protocol that employs different language measures at different ages. Across the age range, however, I have attempted to use a similar diagnostic standard. Specifically, in order for family members to be diagnosed as SLI, they needed to meet the exclusionary requirements concerned with hearing and nonverbal IQ and also have language scores that are associated with high probabilities (greater than .30) of clinical diagnosis of language impairment (Records & Tomblin, 1994; Tomblin, Freese, & Records, 1992).

Epidemiologic Study

A second study currently underway is an epidemiologic study of SLI, using a large population sample of kindergarten children screened for language problems. I expect to have sampled over 6,000 children by the end of the study; however, the data to be used in this chapter will represent the first 849 children sampled.

Our objective in this particular analysis was to identify all SLI children who are members of the population sampled. The sampling method used was a stratified cluster sample. The stratification variable was concerned with the child's residential setting, specifically, whether the child was living in an urban, suburban, or rural environment. Within each of these strata, all elementary schools with kindergarten classrooms represented potential clusters that could be randomly sampled. The exception to this was the exclusion of schools that chose not to participate. The remaining schools were then randomly sampled to produce an equal number of children in each strata.

Once a school was sampled, all kindergarten children were administered a 10 minute language screening test. Subsequent to the screening, all children who failed the screen and an equal number of randomly sampled children who passed were administered a diagnostic exam that included a hearing test, two nonverbal subtests (picture completion and block design) of the WISC-R (Wechsler, 1989), a language battery consisting of the TOLD-2:P (Newcomer & Hammill, 1988), and the narrative test developed by Cullatta, Page, and Ellis (1983). Using these measures, I constructed an overall composite score as well as five subcomposite scores based on the TOLD-2:P and the narrative test. The subcomposite areas were vocabulary, grammar, narrative, receptive language, and expressive language, as shown in Fig. 7.1.

The five subcomposite areas represented in this scheme encompassed three domains of language, vocabulary, grammar, and narrative, as well as the two modalities of language use, reception and expression. Each subcomposite score

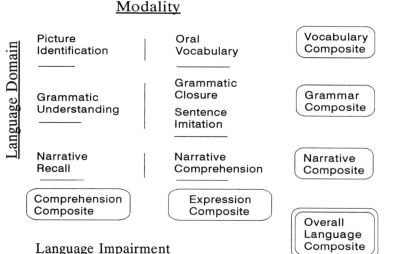

Modality

Language Domain

Picture Identification	Oral Vocabulary	Vocabulary Composite
Grammatic Understanding	Grammatic Closure / Sentence Imitation	Grammar Composite
Narrative Recall	Narrative Comprehension	Narrative Composite
Comprehension Composite	Expression Composite	Overall Language Composite

Language Impairment

2 or more domain or modality composite scores below -1.25 s.d. or an overall composite score below -1.2 s.d.

FIG. 7.1. A schematic of the diagnostic system employed within the epidemiologic study. The matrix is formed by three domains of language (vocabulary, grammar, narration) and two levels of modality (comprehension and production). Marginal composite scores and an overall composite score can be computed from which a diagnosis of language impairment can be made using the criterion provided.

was based on the combination of the subtests in that area and, thus, from 2 to 4 subtests contributed to these subcomposite scores. The diagnostic standard for SLI required that the child have normal hearing, a performance IQ of 85 or greater, and an impairment of language as defined in Fig. 7.1. This standard was selected by comparing language test scores with clinical judgments made by the examiner after testing a group of more than 100 children, several of whom had been identified as language impaired. Thus, this standard was intended to be reflective of the diagnostic standards used by practicing speech-language clinicians and is very consistent with the diagnostic standards for developmental language impairment reported by Records and Tomblin (1994).

A telephone questionnaire was administered to the parents of all those children who were diagnosed as SLI and those selected as control subjects. This questionnaire was concerned with the health of the parents and study child, the pregnancy and birth characteristics of the study child, and general characteristics of the rearing environment of the study child.

The data from this epidemiologic study allowed me to address questions concerned with the prevalence of SLI, the nature of children with and without SLI, and risk factors that are associated with SLI.

Twin Study

The third study being conducted is a twin study of SLI. In this study, I recruited twinships in which one or both are thought to present SLI. These twins were sampled by several different methods. Many of the twins were obtained by means of letters sent to speech-language clinicians practicing in the upper midwest. These clinicians were asked if they had any members of a twinship in their caseload that presented SLI. To reduce the bias of clinicians referring only concordant twinships, the letter emphasized that it was not necessary for both twins to be in the caseload. The clinicians then were provided a recruitment letter to be forwarded to the parents of each twinship, describing the study and asking the parents to contact me if they were willing to have their children participate. I also recruited twins by advertising in magazines for parents of twins and by sending letters to twin clubs in the Midwest.

Once a twinship was recruited, an examiner visited them in their home and administered a diagnostic battery for SLI and some additional language measures designed to probe more extensively into grammatical and lexical skills. At the time of the visit, the twins' parents also completed a questionnaire designed to estimate the zygosity of the twinship (Cohen, Dibble, Grawe, & Pollin, 1973). At this time, I have studied 43 twin pairs in which at least one was SLI. Thirty of these twins were monozygotic, and 13 of them were dizygotic.

The purpose of this twin study is to determine the extent to which SLI is heritable and, further, to determine if certain linguistic disabilities are more heritable than others.

WHAT IS THE PREVALENCE OF SLI?

I begin my inquiry into the epidemiology of SLI by first considering the prevalence rate of SLI in the sample of 849 kindergarten children. I found 25 children that met the diagnostic criteria for SLI by screening 2.8% of the population; this value, however, underestimated the true prevalence rate in this population. There were several children who were selected to be given the diagnostic test who were not tested either because their parents refused to give consent or because their parents did not respond to repeated efforts to obtain consent. In all, 55% of the parents of children eligible to be diagnosed consented; therefore, I had to adjust the total number of children screened by that figure before computing the prevalence. This adjustment resulted in an effective sample size of 464, and when this value was used as the denominator for computing the prevalence rate, the prevalence estimate was 5.4% ($\pm$ 2% @ 1.95). It is important to recognize that this prevalence rate is directly influenced by the diagnostic standards I employed, which are described later in Table 7.2. If more stringent standards, such as the requirement of language scores below 2 standard deviations, were used, the rate would be lower.

There are very few studies that provide clear evidence of the prevalence of SLI, and those that provide estimates vary considerably with respect to the diagnostic standard used and, hence, the definition of SLI. Randall, Reynell, and Curwen (1974) found that 1.1% of the 176 3-year-old children they studied had language skills below 2 standard deviations without concomitant mental retardation. This value is considerably lower than that obtained in the current study; however, Randall et al. (1974) also required that a child have a more severe language delay in order to be considered SLI. Stevenson and Richman (1976) have reported that severe specific expressive language impairment occurred at the rate of .57% in a sample of 705. Again, this value is much lower than that found in the current study, but Stevenson and Richman's low rate may be due, in part, to the restriction of their study to expressive language impairment. Finally, Fundudis, Kolvin, and Garside (1979) estimated that among the 7-year-old children in their study, 2.5% had specific speech and language impairment. The prevalence rate found in the current work, therefore, is greater than that reported previously; this suggests that our clinically based diagnostic standard is more inclusive that those employed previously. Because there is no existing "gold standard" for the diagnosis of language impairment in general or SLI, in particular, it is not possible to argue who is correct.

RISK FACTORS

The prevalence rate of 5.4% can serve as a reference point against which I begin searching for factors that alter this prevalence rate. These factors associated with elevation of prevalence rates can be viewed as risk factors. It is important to emphasize that risk factors are not necessarily the same as causal agents. When risk factors are present in a person's background, that person has a greater probability of presenting the condition or disease in question than the person who does not have risk factors in his or her background. Thus, risk factors are statements of association. Once a risk factor is identified, it is necessary to learn the basis for the association.

Sex/Gender

Sex/Gender as a Risk Factor. There are many developmental speech, language, and learning impairments believed to have higher rates of occurrence in males than females. The sex/gender property of children is then an important potential risk factor for SLI. My confidence in this variable as a risk factor has been shaken recently by research on dyslexia, which has suggested that males and females are equally likely to present with dyslexia (Shaywitz, Shaywitz, Fletcher, & Escobar, 1990; Wadsworth, DeFries, Stevenson, Gilger, & Pennington, 1992). These recent reports have employed sampling methods in which the dyslexics were systematically sampled rather than sampled from a clinically referred population. Data from clinically referred populations always have the potential of being biased by factors influencing the clinical referral. In situations in which the subjects are self-referred, the bias may stem from variables related to persons seeking help. In those situations in which clients are referred by professionals, the bias may be due to factors that draw the child to the attention of a professional. Shaywitz, Shaywitz, Fletcher, and Escobar (1990) believed that the prior evidence, showing more males than females in the dyslexic population, may be due to teachers referring boys more often than girls for reading problems due to elevated rates of associated behavior problems in boys.

Because dyslexia and SLI are often found in the same individuals and may be variants of the same underlying etiology, it is necessary to ask whether, in samples that are not clinically referred, evidence of an elevated rate of SLI in males compared with females is found. Two of the studies provide such samples. Those SLI children identified in the epidemiology study were systematically obtained from the population of kindergartners in general. Also, although the probands in the family study were clinically referred, the relatives of the probands were not referred and, therefore, should represent an unbiased sample. Within the epidemiology study, 25 SLI children were identified. Of these, 16 were males and 9 were female. Of the first degree family members of SLI probands, there were 24 who were identified as SLI, 17 were male and 7 were female. When these two samples were combined, I found that the ratio of males to females was 2.06:1 (33 males and 16 females), which was significantly different from a 1:1 ratio ($p = .01$). These data then support the belief that males are more liable for SLI than are females.

How Specific Is Sex/Gender Influence on Language Performance? A simple explanation for this heightened liability in males is that males may be less adept language learners in general. That is, the distribution of language skills in males is lower than that of females and, therefore, more males will fall below some standard threshold than will females. This explanation rests on the common belief that females are more skilled in language learning than boys. The literature on this topic, however, does not provide strong evidence for this view. Hyde and Linn (1988) recently performed a meta-analysis on the literature concerned with gender differences in verbal ability and concluded that "the gender difference in verbal ability is currently so small that it can effectively be considered to be zero" (Hyde & Linn, 1988). The epidemiology study provides an opportunity to examine the influence of sex and gender on the specific language skills used to diagnose SLI and within the same population from which these children were identified. As a part of the study, I administered a short form of the TOLD-2:P to 432 kindergarten males and 417 females. On this test, the males averaged a score of 43.03 ($SD = 14.81$), and the females averaged 43.32 ($SD = 13.21$), yielding a nonsignificant difference ($t = .68$, $df = 847$, $p < .49$). These data are consistent with Hyde and Linn's conclusions and suggest that the oral language skills of kindergartners are not different according to sex or gender factors. Thus, the sex difference found for SLI is not simply due to the fact that there are two different distributions for language skills.

What has been shown so far is that more boys than girls present with SLI, but that within the total population of kindergartners, boys have very similar language skills as girls. Is it possible, however, that the elevated rate of SLI boys is not unique to SLI but, rather, is found in all forms of language impairment? I address this question as well from the epidemiology study because there was a group of children who presented with language impairment, but who also were found to have nonverbal IQs below 85. Out of the 22 children with low language and nonverbal cognitive skills, 9 (41%), were boys. Thus, it appears that the sex modification found in SLI is somewhat specific for SLI and does not extend to other children with depressed language skills. These data provide some support for the contention that the diagnostic entity of SLI is unique, at least with respect to the influence of gender factors.

Familiality of SLI

Family History as a Risk Factor. There are several studies now reporting a familial aggregation for specific language-learning impairments (Beitchman, Hood, & Inglis, 1992; Neils & Aram, 1986; Tallal, Ross, & Curtiss, 1989; Tomblin, 1989) Most of these used a mixture of direct measures and historical report. Further, the phenotype often included both speech, language, and reading problems. The family study provided data for measuring the familiality of SLI when the phenotype is restricted to SLI and is based on direct diagnosis.

Figure 7.2 presents the rate of SLI in the parents and siblings of the SLI probands. If a prevalence rate of 5.4% in the general population is assumed, SLI is much higher in the families of SLI probands—on the order of about 4 times higher. Furthermore, the data show that the rate of SLI in the parents is the same as that in the siblings and, also, the sex effect is the same for both generations.

The overall rate of affected family members in these 44 families was 21%; however, this rate of SLI was not true for all families. In fact, 53% of the families had no other affected first degree relative; therefore, the SLI proband in these cases may be viewed as a sporadic case, and their families can be termed simplex families. Also, in 47% of the families there were at least two affected nuclear family members. In these multiplex families, the rate of SLI was 42%.

This difference in familial concentration could mean that SLI is an etiologically heterogeneous condition. One form of SLI could be strongly influenced by shared genes and/or environment and this variant may be found in the probands of the multiplex families. The other variant demonstrated by the isolates from the simplex families may be caused by nonfamilial factors. If the simplex and multiplex probands have different etiologies, finding differences in the language characteristics of these two groups might be expected.

Tallal, Townsend, Curtiss, and Wulfeck (1991) reported that 65% of their SLI probands had parents who had histories of speech, language, or reading problems. The remaining 35% were isolates. These investigators examined the lan-

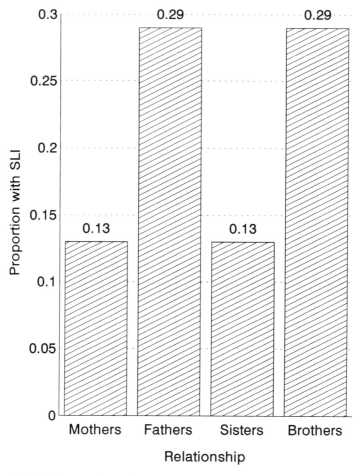

FIG. 7.2. The rate of specific language impairment (SLI) in first degree family members of SLI probands.

guage characteristics of those probands from the multiplex families (those with positive family histories) against those from simplex families (those with nega- tive family histories) and found no significant differences between the Leiter IQ scores, receptive language scores, or expressive language scores. I have also examined the language characteristics of the simplex and multiplex probands. In Table 7.1, I compare the language performance of the probands from multiplex and simplex families in certain areas of language. The data mirror those of Tallal et al. (1991), in that no differences were found between these groups on any of these measures. Thus, in both studies, there is no evidence of a difference in language or nonverbal cognitive characteristics between SLI isolates (simplex families) and SLI children from multiplex families. It still may be possible that the form of SLI found in the multiplex families is etiologically different from that

TABLE 7.1
Comparison of Simplex and Multiplex Probands With Respect to
Nonverbal IQ and Receptive and Expressive Language Measures

Measure	Simplex Probands	Multiplex Probands
Performance IQ	100.66	101.38
	(11.88)	(9.01)
PPVT-R	85.75	91.05
	(13.05)	(8.84)
TACL-R	11.9	15.58
	(9.81)	(10.31)
ITPA grammatic closure	27.20	26.33
	(7.19)	(6.95)

in the simplex families, but if this is true, I have yet to find a distinctive phenotype characteristic that reveals this.

Explanation for the Familialty of SLI? Earlier, it was hypothesized that SLI was a random event and, therefore, all children faced the same risk for SLI. I have now found evidence supporting a rejection of this hypothesis. First, males are more likely to be identified as SLI than females and, second, relatives of SLI children are more likely to present SLI than relatives of children without SLI. Before exploring other risk factors, it seems reasonable to explore the basis of the familiality of SLI further. Family members typically share their environment and their genes. To what extent do each of these factors contribute to the elevated rate of SLI in certain families?

There are several ways to address this question. The ideal situation would be to find family members in which the sharing of genes and sharing of environment are not confounded. Siblings who have been adopted out to different homes provide one approach to this by reducing the amount of shared environment among siblings. There are numerous challenges facing the researcher who attempts this type of design. For example, for a trait such as SLI, it will be necessary to diagnose the biological parents, the adoptive parents, and the adopted children. Therefore, all these individuals would need to be located and examined which requires access to legal records that are often restricted and the consent of the biological parents to participate.

Another approach is to use a design in which the degree of gene sharing varies across sibships and the amount of environmental sharing remains similar. Studies of twins provide this type of design because monozygotic twins (MZ) share all their genes, whereas dizygotic twins (DZ) share on average 50% of their genes. Thus, if a trait such as SLI, is genetically influenced, there should be a greater similarity between MZ twins than DZ twins.

The literature contains a handful of case studies reporting concordance for SLI in single MZ twinships, but these studies cannot be used as support for a genetic basis of SLI. Recently, Lewis and Thompson (1992) reported probandwise concordance rates for a history of speech and/or language therapy of .86 in MZ twins and .48 for DZ twins. It is likely that many of these twins presented SLI. Very recently, Bishop (1992a) published the preliminary results of a twin study of SLI she has been conducting in Britain. Bishop did not report concordance rates for her twins; however, when one twin was found to have SLI, Bishop reported that the other MZ cotwin also had either a history or current diagnosis of SLI 66% of the time, in contrast with 42% of the time for the DZ twins. In the twin study I have been conducting, I have obtained a pairwise concordance rate for MZ twins of .62 and .37 for DZ twins.

I conclude from the twin data that the pattern of twin concordance rate for SLI is consistent with a moderate genetic contribution to SLI and justifies the belief that the familiality of SLI is due at least in part to shared genes among family members. These twin data also, however, provide evidence that there must be other factors influencing the occurrence of SLI because I find that the concordance rate for MZ twins, who share the same genes, is not at 100%. Thus, a genetic explanation alone is not sufficient to explain all instances of SLI. These instances of nonconcordant MZ twins may be due to nongenetic phenocopies or, more likely, they reflect the influence of environmental factors that reduce the penetrance of the genetic expression of an SLI genotype.

If it is necessary to use the differential rate of SLI in MZ versus DZ cotwins as evidence of a genetic contribution, at least two important assumptions need to be made. First, it must be assumed that the environment of MZ twins and DZ twins is no different and, thus, differences in concordance rates for these two twin types are due to the differences in a genetic source rather than an environmental source. Second, if twin data is used to generalize to the population at large, it must be assumed that there is not something about twinning that influences the liability for SLI.

With regard to the first issue—the equal environment assumption—there are several studies that have provided evidence supporting the "same environment" hypothesis. Although I know of no studies concerned specifically with the nature of linguistic input to DZ and MZ twins, there have been several studies addressing this question with respect to general experiential characteristics such as dressing alike, playing together, etc. Plomin, DeFries, and McClearn (1990) provided a review of this literature and concluded that the "data ... strongly support the reasonableness of the equal environments assumption" (p. 319).

With respect to the other assumption—the effect of twinning on language development—the situation is less comforting. Twins have been found to be delayed in language development compared to singletons—particularly during the preschool years (Hay, Prior, Collett, & Williams, 1987; Tomasello, Mannle, & Barton, 1989). Wilson (1983) provided evidence that the influence of twinning on language development diminishes with age and is not present after the early school years. The influence of twinning on language development may be due to conditions in the womb, greater rates of prematurity in twins than singletons, or, most likely, limited access to the undivided attention of adult caregivers.

Some sense of the extent to which this special twinning influence contributes to the concordance rate in twins can be obtained by comparing the concordance rate in our DZ twins to the concordance rate in same-sex siblings from the family study. The same-sex siblings in the family study have the same genetic relationship as the DZ twin, but they do not have the special shared environment associated with twinning. Of the 45 probands in the family study, 32 had at least one same-sex sibling. In cases where there was more than one same-sex sibling, I selected the sibling closest in age to the proband. Nine of the 32 same-sex siblings, all of whom were males, were also SLI. Thus, the pairwise concordance rate for the singletons was .28, which is lower than our DZ twin concordance rate of .37. Thus, the concordance rate in twins may be affected by the special circumstances of twinning, but this effect does not appear to be great.

Bishop (1992a) arrived at a similar conclusion about the influence of twinning on the concordance rate of SLI in twins. The twin effect is a special twin specific environmental influence. Therefore, some twins may present language impairment due to the special influence of twinning, and this form of language impairment will not be of a familial form. In other words, the twin SLI population may be more etiologically heterogeneous than the singleton population, and this additional etiologic source should reduce the degree of familiality of SLI among twins. This is, in fact, what Bishop found. There was a lower familiality of SLI (22%) in the twin families than in the families of singleton SLI probands (32%); however, this was not statistically significant. The data with regard to the degree to which twinning influences language development are still very limited; however, my sense of these data is that some aspect of twinning may be elevating the rate of SLI somewhat among twins, and it must be assumed that the twin population contains more phenocopies than the singleton population. I also believe, however, that the twinning effect does not account for the difference in concordance rates between MZ and DZ twins, and it is this differential rate that provides the strongest evidence for a genetic contribution to SLI.

Risk for SLI as a Function of Relationship. It is possible to summarize the data discussed so far by viewing the rates of SLI in family members and twins as indices of risk for SLI. A change in risk for SLI, with respect to varying degrees of relationship, can then be observed. The kindergarten children, participating in the epidemiology study, are unrelated and, thus, I can use the rate of SLI in this population as a benchmark for the risk of SLI to the population-at-large. The risk of SLI to first-degree family members increases to .21. The risk to same-sex first-degree family members moves to .28 and, if the same-sex siblings are twins, the risk climbs to .33. Finally, if the siblings share all their genes, the risk of SLI for one twin, given that the cotwin has SLI, is .71. Across these studies, I have attempted to use a rather similar standard for the diagnosis of SLI. With different standards, the rates are likely to change; however, the absolute values of these rates are not as important as the relative values. The liability for SLI increases substantially as a function of family relationship and, in particular, genetic relationship. Collectively, these data suggest that a genetic etiology is an important aspect of an etiologic account of SLI and support further efforts to identify the particular nature of this genetic contribution to SLI.

Environmental Factors

The data just reviewed provide evidence of a genetic contribution to the cause of SLI; however, the data from the twin studies also suggest that a portion of the risk for SLI may also come from environmental sources. Those working in genetic epidemiology and behavior genetics have always emphasized that, with respect to complex behavioral phenotypes, the appropriate question is how genes and environment both contribute to the phenotype.

If we are to become serious about the inclusion of environmental factors in more complex models of gene-environment interactions, we need to have some idea of what to include. At this point in time, there is little evidence pointing to any environmental factor, biological or experiential, associated with SLI. One candidate factor examined has been parental language input. Although it is not possible to find evidence proving that language exposure is not a contributor, there remains little convincing evidence in support of the alternate claim that inadequate language experience contributes substantially to SLI (Bishop, 1992; Leonard, 1987).

The epidemiology study using kindergartners contains a component that obtains a wide range of exposure information. In this case, exposure refers to a wide range of biological and experiential events in the life of the child. I review only a few of these, focusing on those that are either of theoretical interest or those in which there is some evidence of a relationship between exposure and SLI. I have tried to place these in categories having to do with biological exposures or experiential exposures; however, in some cases, the variable of interest could affect the child through either or both paths.

Home Background. The first set of variables pertains to some basic characteristics of the home that may collectively reflect the social, educational, economic, and living climate of the homes of these children. Figure 7.3 presents the distribution of parents education for the two groups. These data show that the parents of the SLI children tended to stop school earlier than the parents of the control group. This trend, however, is not significant with the current sample size. Despite the trend toward lower educational levels in the parents of the SLI children, there was very little suggestion of a difference in the family incomes of these two groups, as is shown in Table 7.2. The parents' educational level and income levels are variables that provide indirect indices of the home environment. Another index of the home environment is the birth order and number of siblings in the home. There have been several studies (Beitchman, Peterson,

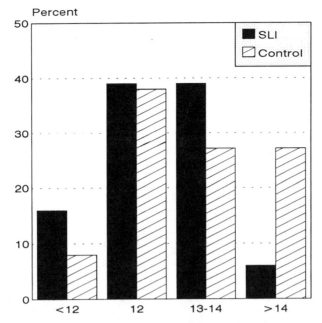

Years of Formal Education (Fathers')

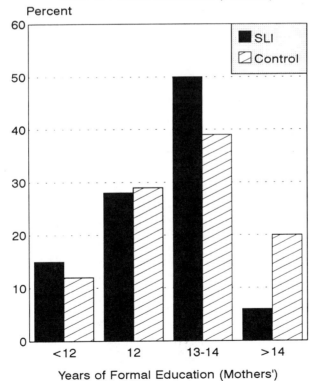

Years of Formal Education (Mothers')

FIG. 7.3. The highest grade attended by the mothers and fathers of control children and children with specific language impairment (SLI).

TABLE 7.2
Percentage of Families With Income in Brackets of
Ten Thousands of Dollars for SLI and Control Families

	Income Brackets				
Group	< 10	10–20	20–30	30–40	> 40
Control	19	20	21	17	23
SLI	19	22	16	22	22

& Clegg, 1988; Butler, Peckham, & Sheridan, 1973; Fundudis, Kolvin, & Garside, 1979) that have reported that children who are language impaired are more likely to be later born than those who have normal language development. In contrast, I have noted (Tomblin, 1990) that this birth order effect may be due to a confounding with parental education and socioeconomic status of the parents because family size and parental education are associated such that larger families are associated with lower parental education. When parental education was controlled by matching, the birth order was not found (Tomblin, 1990). Table 7.3 shows the birth order of the children in the current study. These data show that a greater proportion of the SLI children were later born than were the control children. However, this birth order trend is not significant, and it is likely that the trends in these data are associated with the trends toward the SLI parents being less well educated. I do not have a sufficient sample size to permit matching SLI and controls for parental education to determine if this trend would disappear with the effect of parental education removed.

The data obtained on the home characteristics of the children in the epidemiology study revealed that there were no robust factors that were associated with SLI. There are trends that suggest a difference between the SLI and control groups with respect to socioeconomic background. However these trends were not statistically significant.

Biological Exposures

The second set of variables to be considered pertain to the health or health practices of the study child's mother during the 12 months preceding the child's delivery. The results of our inquiry concerning these questions are contained

TABLE 7.3
Percent of Study Children at Each Birth Order

	Birth Order			
Group	First	Second	Third	Fourth (or Greater)
Control	42	36	14	8
SLI	38	28	19	16

TABLE 7.4
Percent of Maternal Exposures During Pregnancy With
Kindergarten SLI Cases and Language Normal Controls

Exposure Variable	SLI	Control
Infections	34%	38%
Medication	56%	73%
Alcohol	43%	57%
Smoking	56%	37%

in Table 7.4. The mothers participating in this study were asked a series of questions concerning any infections or illnesses such as toxemia, diabetes, hypertension, or thyroid problems, they may have experienced during the specified time interval. The mothers of the SLI children were not statistically different from the control mothers on the aggregate rate of these exposures, and in fact, the SLI mothers reported fewer exposures of these kind. These mothers were also asked about their use of alcohol and tobacco during their pregnancy with the study children. These two variables yielded interesting results because I found that the control mothers reported more alcohol consumption than the SLI mothers, whereas the opposite pattern was true for smoking tobacco.

As with the variables having to do with the children's home background, I found no statistically significant differences in the health of the mothers of these two groups of language learners. In fact, except for the greater rate of smoking, which again is likely to be associated with the trends in lower parental education, the health of the mothers of the SLI children was slightly better.

I also inquired about the child's birth. There is very little evidence to date supporting the belief that perinatal events are associated with SLI (Bishop, 1992a; Tomblin, 1992), and the current data provide additional evidence for such conclusions. The mothers of the study children were asked if the child was born through cesarean section, if forceps were used, if labor was induced, or if there were any other complications during delivery. No differences in the rate of any of these events were found between the SLI and control groups. Further, no difference ($t = -1.05$, $df = 91$, $p < .29$) was obtained between the birthweights of the SLI children ($M = 122$ grams, $SD = 16.54$) and the control children ($M = 117$, $SD = 21.65$).

These data are preliminary. Certainly, as data are obtained for more children, the power of the tests will improve and, no doubt, some of the data that show only nonsignificant trends will become statistically significant. These data were based on 32 SLI children and 114 controls, which is a very small number for epidemiologic research. I expect a substantial improvement in the power of our statistical tests when this study is complete and there is data from approximately 200 SLI children and 800 controls. However, this increase in power will not necessarily lead to an increase in the effect size between these variables and the diagnostic outcome. Therefore, these results at least suggest that there are few highly potent exposures in the background of the SLI children that place them at risk for SLI.

Before I leave this topic, however, I should consider whether these questionnaire data can reveal any known risk factors. I noted earlier that family history of speech, language, and learning problems has been shown to be a risk factor for SLI. I questioned the parents about a history of these problems in their background. Of the 32 mothers and fathers of the SLI children 13% and 19%, respectively, reported positive histories of these problems whereas 5% of the control mothers and 4% of the control fathers reported such histories. The difference between these rates was significant for the fathers ($\chi^2 = 7.4$, $df = 1$, $p < .007$), but was not significant for the mothers. The two known risk factors, maleness and familiality, are robust enough to show up within this risk questionnaire.

CONCLUSIONS

This chapter began by posing the null hypothesis that SLI is a condition that occurs randomly among children. The data addressing this hypothesis are only beginning to accrue, but it is becoming clear that this hypothesis can be rejected with considerable confidence. Children who are male and who have other family members with language and learning problems are more likely to have SLI than are those without such a family history, particularly females. The emerging twin data, furthermore, provide evidence that the familiality of SLI is likely to have strong genetic roots although this may be influenced by additional nongenetic factors. These conclusions are heartening in that they provide strong encouragement for those who wish to apply behavioral genetics, genetic epidemiology, and molecular genetics to the understanding of SLI.

REFERENCES

Beitchman, J. H., Hood, J., & Inglis, A. (1992). Familial transmission of speech and language impairment: A preliminary investigation. *Canadian Journal of Psychiatry—Revue Canadienne De Psychiatrie, 37,* 151–156.

Beitchman, J. H., Peterson, M., & Clegg, M. (1988). Speech and language impairment and psychiatric disorder: The relevance of family demographic variables. *Child Psychiatry and Human Development, 18,* 191–207.

Benton, A. (1964). Developmental aphasia and brain damage. *Cortex, 1,* 40–52.

Bishop, D. (1992a). Biological basis of developmental language disorders. In P. Fletcher & D. Hall (Eds.), *Specific speech and language disorders in children* (pp. 2–17). San Diego CA: Singular.

Bishop, D. (1992b). Biological bases of specific language impairment (developmental aphasia). In I. Kostovic, S. Knezevic, H. Wisniewski, & G. Spilich (Eds.), *Neurodevelopment, aging, and cognition* (pp. 253–271). Boston: Birkhauser.

Butler, N. R., Peckham, C., & Sheridan, M. (1973). Speech defects in children aged 7 years: A national study. *British Medical Journal, 3,* 253–257.

Cohen, D., Dibble, E., Grawe, J. M., & Pollin, W. (1973). *Archives of General Psychiatry, 29*, 465–469.

Conti-Ramsden, G. (1985). Mothers in dialogue with language-impaired children. *Topics in Language Disorders, 5*, 58–68.

Culatta, B., Page, J. L., & Ellis, J. (1983). Story retelling as a communicative performance screening tool. *Language Speech and Hearing Services in Schools, 14*, 66–74.

Ewing, A. W. G. (1930). *Aphasia in children.* London: Oxford University Press.

Fundudis, T., Kolvin, I., & Garside, R. (1979). *Speech retarded and deaf children: Their psychological development.* New York: Academic Press.

Hay, D. A., Prior, M., Collett, S., & Williams, M. (1987). Speech and language development in preschool twins. *Acta Geneticae Medicae et Gemellologiae Twin Research, 36*, 213–223.

Hyde J. S., & Linn, M. (1988). Gender differences in verbal ability: A meta-analysis. *Psychological Bulletin, 104*, 53–69.

Johnston, J. R. (1991). The continuing relevance of cause: A reply to Leonard's "Specific language impairment as a clinical category." *Language, Speech, and Hearing Services in Schools, 22*, 75–79.

Kamhi, A., & Johnston, J. (1982). Towards an understanding of retarded children's linguistic deficiencies. *Journal of Speech and Hearing Research, 25*, 435–445.

Leonard, L. B. (1987). Is specific language impairment a useful construct? In S. Rosenberg (Ed.), *Advances in applied psycholinguistics* (Vol. 1, pp. 1–39). New York: Cambridge University Press.

Lewis, B. A., & Thompson, L. A. (1992). A study of developmental speech and language disorders in twins. *Journal of Speech and Hearing Research, 35*, 1086–1094.

Neils, J., & Aram, D. M. (1986). Family history of children with developmental language disorders. *Perceptual and Motor Skills Part 1, 63*(2), 655–658.

Newcomer, P., & Hammill, D. (1988). *Test of language development-2 primary.* Austin: Pro-Ed.

Plomin, R., DeFries, J. C., & McClearn, G. E. (1990). *Behavioral genetics: A primer.* New York: W. H. Freeman.

Randall, D., Reynell, J., & Curwen, M. (1974). A study of language development in a sample of three-year-old children. *British Journal of Disorders of Communication, 9*, 3–16.

Records, N., & Tomblin, J. B. (1994). Clinical decision making: Describing the decision rules of practicing speech-language pathologists. *Journal of Speech and Hearing Research, 37*, 144–156.

Shaywitz, S. E., Shaywitz, B. A., Fletcher, J. M., & Escobar, M. D. (1990). Prevalence of reading disability in boys and girls. Results of the Connecticut Longitudinal Study [see comments]. *Journal of the American Medical Association, 264*, 998–1002.

Stevenson, J., & Richman, N. (1976). The prevalence of language delay in a population of three-year-old children and its association with general retardation. *Developmental Medicine and Child Neurology, 18*, 431–441.

Tallal, P., Ross, R., & Curtiss, S. (1989). Familial aggregation in specific language impairment. *Journal of Speech and Hearing Research, 54*, 167–173.

Tallal, P., Townsend, J., Curtiss, S., & Wulfeck, B. (1991). Phenotypic profiles of language-impaired children based on genetic/family history. *Brain and Language, 41*, 81–95.

Tomasello, M., Mannle, S., & Barton, M. (1989). The development of communicative competence in twins. *Revue Internationale de Psychologie Sociale, 2*, 49–59.

Tomblin, J. B. (1989). Familial concentration of developmental language impairment. *Journal of Speech and Hearing Disorders, 54*, 287–295.

Tomblin, J. B. (1990). The effect of birth order on the occurrence of developmental language impairment. *British Journal of Disorders of Communication, 25*, 77–84.

Tomblin, J. B. (1991). Examining the cause of specific language impairment. *Language, Speech, and Hearing Services in Schools, 22*, 69–74.

Tomblin, J. B. (1992). Risk factors associated with specific language disorder. In M. Wolraich & D. K. Routh (Eds.), *Developmental and behavioral pediatrics* (pp. 131–158). Philadelphia: Jessica Kingsley.

Tomblin, J. B., Freese, P. R., & Records, N. L. (1992). Diagnosing specific language impairment in adults for the purpose of pedigree analysis. *Journal of Speech and Hearing Research, 35,* 832–843.

Wadsworth, S. J., DeFries, J. C., Stevenson, J., Gilger, J. W., & Pennington, B. F. (1992). Gender ratios among reading-disabled children and their siblings as a function of parental impairment. *Journal of Child Psychology and Psychiatry and Allied Disciplines, 33,* 1229–1239.

Wechsler, D. (1989). *WPPSI-R manual: Wechsler preschool and primary scale of intelligence-revised.* New York: Psychological Corporation.

Wilson, R. S. (1983). The Louisville twin study: Developmental synchronies in behavior. *Child Development, 54,* 298–316.

COMMENTARY ON
CHAPTER 7

Shelley D. Smith
Boys Town National Research Hospital

Dr. Tomblin's chapter presents the preliminary findings of three interrelated projects. The overall goal is to define the genetic and environmental causes of specific language impairment, and each project contributes to that goal from a different vantage point.

To achieve this goal, additional questions of phenotype definition and frequency must be addressed. The primary problem is that of phenotype definition because that affects any subsequent analyses. Tomblin has taken a very careful approach to this, considering the separate domains of language, the level of functioning that should be considered to be impaired, and possible changes in phenotype with age. Most importantly, he also allows for the possibility of refining the phenotypic criteria based on later information about etiologically similar subgroups.

The three projects that are described are a family study, an epidemiological study, and a twin study. The family study is designed to examine the evidence for genetic etiology by looking at recurrence risks across different degrees of relationship and between families. This study is notable in that the first degree relatives of the proband are directly tested, and diagnosis is based on reliable criteria. The recurrence risks obtained from these assessments are consistent with those obtained by other studies (Felsenfeld, 1994; Gilger, 1992; Lewis, 1992). A particularly interesting result is that there was definite interfamilial heterogeneity in that 53% of the 45 families studied thus far did not have any affected first-degree relatives beyond the proband, but in the other 47% of families, 42% of the first-degree relatives were affected. This striking contrast would be consistent with a major genetic effect in the multiplex families. Some of the simplex families could also be the result of genetic influences with low recurrence risks (i.e., multifactorial, recessive, or dominant with reduced penetrance), but that group of families may be more likely to include nonfamilial causes. Comparison of risk factors in these two groups, then, could reveal important environmental influences. At the level of phenotypic assessment used so far, there was no difference between the two groups, but this could also be an important avenue to pursue. Finally, as Tomblin alludes to at the end of the paper, genetic epidemiological approaches, such as complex segregation analysis, could identify a model for the modes of inheritance in these families. This could determine whether there are single gene, and/or multifactorial influences, and verify whether there is a high proportion of sporadic, nongenetic cases. If there is evidence for the influence of only a few genes in the multiplex families, these would be ideal for linkage analysis, as well.

The family study apparently did not collect as much data on environmental risk factors and cannot give estimates of frequency of SLI. The epidemiological study fills in this information by systematically screening a kindergarten population and collecting environmental and family information on children who are found to be affected as well as matched control children. Again, the criteria for diagnosis are appropriately clear and conservative. The family history information, collected by questionnaire rather than direct testing, shows increased familiality in the relatives of SLI children, although the frequencies are less than was seen in the family study where relatives were tested. Several potential environmental risk factors are assessed, including history of pregnancy complications, prenatal trauma, and low birthweight. None of these were found to be significantly different from control children. Aspects of the home environment were also assessed, including parental education, socioeconomic status (SES), and sibship size. Again, no significant differences were found, although Tomblin points out that the small sample size in these preliminary analyses could mask subtle differences. Indeed, parental education tended to be lower, and sibship size tended to be greater in the SLI children's families. However, it is possible that these are related through SES. Moreover, parental language disability could result in less success in school, which would then affect degree of education and SES. Such an effect has been well documented in specific reading disability (Finucci, 1986), so it is possible that the parental genotype has an influence on language through the child's environment as well as through his or her genes. Interestingly, the only factor found to be significantly different between SLI children and controls in the epidemiological survey is the family history of speech, language, and learning problems.

Familiality of a condition does not prove that the cause is genetic, but this can be assessed by a twin study. First, it is important to demonstrate that twinning alone is not an environmental risk factor for SLI. Evidence from previous studies, including within family comparisons of twins and their singleton siblings, does not support a strong risk of twinning, although it should be pointed out that the important comparison is whether monozygotic (MZ) twins are at greater risk than dizygotic (DZ) twins, because this would inflate the differences in concordance. Monozygotic twinning does carry higher risk of morbidity and mortality, particularly in monoamniotic/monochorionic twins, because of factors such as intrauterine constraint and anastomoses of vessels in the shared placenta. However, it is not clear that these produce increased risk of SLI; in fact, the lack of effect of pre- and perinatal factors in the epidemiological study suggests that these factors rarely produce SLI. Other potential risk factors, such as prematurity or less attention by caregivers, should affect both MZ and DZ twins. The work of Wilson (1983) indicated that twins do have a transient language delay compared to their nontwin siblings, which could indicate that prenatal factors might have an initial influence on language; however, Wilson's paper did not report the extent of such delays and whether they would actually constitute SLI. The paper did show that concordances in mental ability between DZ twins became essentially the same as concordances between the twins and their siblings as they got older, particularly beyond 8 years of age. The ages of the twins in Tomblin's study are not given, although it is likely that they were somewhat older because they were receiving services from a speech pathologist and, thus, may have already experienced the "catch up" from any adverse effects of twinning alone. As Tomblin notes, the twin sample is still quite small, although the concordance rates certainly support a genetic influence on SLI. As a larger sample is collected, he may wish to address the issues of twin environment and history of early language problems. The proportion of MZ to DZ twins should also normalize because at present it appears heavily skewed toward ascertainment of MZ twins. Finally, with a larger sample, quantitative methods of analysis may be more informative (e.g., DeFries & Fulker, 1988); this would also get around any debate of whether pairwise or probandwise concordances should be figured.

An important finding in this and previous twin studies is that MZ concordance for SLI is not 100%. Thus, although genetic factors appear to operate, there are still nongenetic factors that need to be uncovered. These factors could be identifiable environmental risk factors or, in some cases, they could be simply random variations in central nervous system (CNS) development. As Kurnit, Layton, and Matthysse (1987) pointed out, genes do not set out an absolute blueprint for neuronal development and migration. Purely stochastic effects and chance may affect how connections are made, without input from specific genetic or environmental influences. Chance effects would result in discordance in MZ twins and in sporadic phenocopies in family studies.

Although the ideal study would be to gather all possible genetic and environmental information and test data on the extended families of singletons, twins, and matched controls, all ascertained through an unbiased epidemiological screening protocol, such a study would, unfortunately, be difficult for the researcher and the families involved. Tomblin's three projects provide the next best set of information, covering genetic and environmental risk factors in family, twin, and epidemiological approaches.

REFERENCES

DeFries, J. C., & Fulker, D. W. (1988). Multiple regression analysis of twin data: Etiology of deviant scores versus individual differences. *Acta Geneticae Medicae et Gemellolgiae (Roma)*, *37*, 205–216.

Felsenfeld, S. (1994). Developmental speech and language disorders. In J. C. DeFries, R. Plomin, & D. W. Fulker (Eds.), *Nature and nurture during middle childhood*. Cambridge, MA: Blackwell.

Finucci, J. M. (1986). Follow-up studies of developmental dyslexia and other learning disabilities. In S. D. Smith (Ed.), *Genetics and learning disabilities*. San Diego: College Hill Press.

Gilger, J. W. (1992). Genetics in disorders of language. *Clinics in Communication Disorders*, *2*, 35–47.

Kurnit, D. M., Layton, W. M., & Matthysse, S. (1987). Genetics, chance, and morphogenesis. *American Journal of Human Genetics*, *41*, 979–995.

Lewis, B. A. (1992). Genetics in speech disorders. *Clinics in Communication Disorders*, *2*, 48–58.

Wilson, R. S. (1983). The Louisville twin study: Developmental synchronies in behavior. *Child Development*, *54*, 298–316.

8

A Phenotype of Specific Language Impairment: Extended Optional Infinitives

Mabel L. Rice
University of Kansas

Kenneth Wexler
Massachusetts Institute of Technology

The search for genetic contributions to language acquisition has recently focused on children with specific language impairment (SLI). These are children whose language development is significantly delayed or incompletely developed, even though the putative prerequisite abilities, in the areas of hearing, cognition, psychosocial development, and neuromotor functioning, seem to be in place. What is interesting about these children is that they seem to demonstrate variation where none would be expected. By the age of 5 years, grammatical fundamentals are well in place for most children. Yet, children with SLI are well behind their age peers in certain key areas of grammatical development.

The reasons for this variation are unknown. Attempts to attribute the children's grammatical problems to variations in linguistic input or parental interactions have proven inconclusive. Attention has shifted from environmental factors to genetic influences with the recent discovery that the condition of SLI is quite likely to aggregate in families and to be highly concordant in monozygotic twins (see Bishop, North, & Donlan, 1995; Crago & Allen, Gilger, and Tomblin, this volume). These findings, among others, suggest the possibility that a condition characterized by variation in language aptitude is heritable. The clear implication is that it should be possible, in principle at least, to identify the genes that contribute to this inheritance.

The search for genetic contributions to higher cognitive processes, however, depends on a clear understanding of the trait that is being passed on, and the behavioral manifestations of that trait (i.e., the behavioral phenotype; see Brzustowiscz, Gilger, Pennington, and Smith, this volume; Lander & Schork, 1994). Procedurally, it must be possible to sort the population into those individuals with the trait versus those without it. Individuals with the trait who are over-looked by the procedure are cases that will contribute to error in analyses. Accurate identification of affected individuals is essential, and identification depends upon the definition of the trait. At another level, an understanding of the trait is crucial for interpretation of possible genetic contributions and related effects on the central nervous system (CNS) and cortical structures (cf. Plante & Poeppel, this volume). It is important to know which aspects of language acquisition are influenced by genetic encoding and the ways in which that influence plays out. The chain of influence will involve RNA, enzymes, structures and functions of the CNS, linguistic representations, and environmental influences. To sort out the genetic etiology and intervening processes, it is essential to have a clear model of the linguistic end state.

Central to the genetics issue, then, is the question of how to characterize the condition of SLI. In this chapter, we report on recent advances in our understanding of the linguistic qualities of SLI. We draw upon theoretical developments in normative language acquisition to formulate predictions for the grammars of children with SLI. The area of interest is that of morphosyntax, an area known to be problematic for these children. For example, it is well-known that young children with SLI typically demonstrate a protracted period of acquisition of verbal morphology. The work described here (a) contributes an account of this symptom in a way that places the problem in the linguistic domain (related psychological processes, such as auditory processing or memory, are not primary explanatory factors), (b) predicts related grammatical symptoms, (c) accounts for what the children *know* about grammar as well as *don't know*, and (d) shows the ways in which the grammar of affected individuals compares to those of unaffected individuals.

The phenomenon of interest is that of an extended optional infinitive (EOI) stage of morphosyntax in children with SLI. Theoretical underpinnings of this model can be found in Wexler (1994) and his chapter in this volume. Detailed empirical evidence is reported in Rice, Wexler, and Cleave (1994) and Rice and Wexler (in preparation). In this chapter we review the relevant linguistic developments, specify particular predictions, and summarize the evidence in support of an EOI stage of SLI. We then turn to interpretive issues. One is the way in which this stage enhances our understanding of what constitutes a delay of language. We explore possible reasons for an EOI and conclude that fully satisfactory explanations have yet to be worked out. Finally, we discuss the implications of an EOI stage for specification of a phenotype of SLI.

ACQUISITION OF MORPHOSYNTAX
IN NORMALLY DEVELOPING CHILDREN:
OPTIONAL INFINITIVE STAGE (OI)

Theories of the knowledge of verbal inflection in young children have been greatly revised in recent years. A crucial new insight is that inflection is intimately related to syntax, hence the terminology *morphosyntax*. In particular, inflection can be seen as the phonetic instantiation of inflectional categories that have their own reality in the *structural representation* (the phrase-marker) of a sentence (see Wexler, this volume). Thus, for example, there is a category of tense (TNS) and a category of agreement (AGR) which show up in the representation of a sentence. Categories like TNS and AGR are called *functional categories* because they are involved in the formal structure of a sentence, as opposed to *lexical* categories like nouns (N) or verbs (V).

Verbs can be marked for tense and agreement. If so, they are +finite. Nonfinite forms, or infinitival forms, are *–finite*. In (1) the verb *talk* is finite, as indicated by the third person singular affix -*s*. Similarly, in (2) *talk* is finite because of the past tense marking.

(1) The scientist talk*s*.
(2) The scientist talk*ed*.

In (3), *talk* is nonfinite, as in (4). In neither of these contexts can a tense marker be inserted (as indicated by the *, a conventional symbol for an ungrammatical structure).

(3) The scientist likes *to talk/*talks/*talked*.
(4) The scientist made the student *talk/*talks/*talked*.

Other languages, such as French, are more consistent in overtly marking finiteness on verbs; that is, the infinitival form shows up less frequently as a finite form. For example, the verb *parler* (talk) appears in an infinitival form, *parler*, and in finite forms marked for person and number and tense (e.g., *parle* for first and third person singular, present tense; *parles* for second person singular).

The crucial observation is that +finite verbs obey certain syntactic principles. This can be clearly demonstrated with regard to placement of the negative particle, *pas*, in French. Nonfinite verbs follow *pas* in what is thought to be the base position (5).

(5) Elle ne peut pas *parler*.
   She can not talk [–finite].

In contrast, finite verbs move or raise to the left of *pas*, as in (6).

(6) Elle ne *parle* pas.
   She does not talk [+finite].

Syntactic phenomena, such as these, attested to across multiple languages (cf. Wexler, 1994, this volume) show the existence of functional maximal projections in the phrase structures of clauses. Two such projections are the grammatical categories of TNS and AGR. Among their functions is that they serve as landing sites for the movement of verbs around the negative marker in French.

The notion of morphosyntax and functional maximal projections in the phrase structure contributes several important insights to our understanding of the grammar of children with SLI. First, it forms the basis for a new model of children's acquisition of morphosyntax, leading to the identification of the optional infinitive stage (OI) (cf. Wexler, 1994, this volume). Second, it provides a formulation of English verbal morphology that captures ways in which different surface morphemes carry out common grammatical functions, thereby allowing for evaluation of a cluster of morphemes and avoiding the analytic problems of a one-morpheme-at-a-time piecemeal approach. Third, it allows for an interpretation of the commonly attested bare forms of verbs in which grammatical morphemes seem to be omitted. This interpretation also explains why, when the forms do appear, they are very likely to be used accurately. Finally, it allows for a rather precise way of characterizing what the children do not know about morphosyntax.

Optional Infinitive Stage

Wexler (1994; this chapter) has shown that there is a stage in the development of young nonimpaired children in which they do not obligatorily mark tense in main clauses but in which they know, nevertheless, the grammatical properties of finiteness. This is known as the optional infinitive stage. In non-English languages young children sometimes use infinitival forms of verbs where they should use finite forms. For example, samples from French-speaking children yield declarative utterances in which the main verb is an infinitive such as:

(7) Voir l'auto papa.
    See [−finite] the car of daddy.

During the same stage of infinitival use, the children show that they know about the related linguistic processes that apply to finite verbs. For example, in French (see Pierce, 1992; Weissenborn, 1994) even very young children know that finite verbs precede the negative marker, *pas*, as in utterances such as (8):

(8) Il est pas mort.
    He is [+finite] not dead.

and that nonfinite verbs follow *pas*, as in utterances such as (9):

(9) Pas manger la poupee.
    Not eat [−finite] the doll.

The interpretation is that French-speaking children know that verbs raise, or move to precede the negative marker, if the verbs are marked +finite, that is, if they are marked for tense and agreement features. What children do not seem to know is that such features must be marked on the main verb of a clause. That is, they sometimes choose an infinitive, in which case the finite features are not registered and the related processes do not appear. At the same time, children do seem to know that finite verbs, when used, must show agreement and tense and that they must appear in certain positions in the clause.

What is interesting is that young normally developing children in an OI stage do not, at the initial emergence of their grammar, seem to honor an important principle of their morphosyntax. Some time elapses before they begin to consistently use finite forms in main clauses even though they, early on, show that they understand related underlying principles. Wexler concluded that they are relatively late in marking tense where tense marking should appear. This is somewhat surprising in that tense is a fundamental requirement of a fully formed clause in adult grammar.

Predictions for English Morphology

The relevance of the infinitival stage in French- and German-speaking children for our understanding of the acquisition of English was pointed out by Wexler (1994). He argued that in utterances such as (10) and (11) the bare stem form of lexical verbs, such as *talk*, can be interpreted as infinitival forms.

(10) *The scientist talk.
   [The scientist talk*s*.]
(11) *Yesterday, the scientist talk.
   [Yesterday, the scientist talk*ed*.]

Thus, in English, the OI stage could be revealed by children's optional use of -*s* and -*ed* as markers of present and past tense on lexical verbs.

In English, finiteness is also marked on nonlexical verbs (i.e., those that do not contribute semantic information). *Do* and *be* forms function in this way, as illustrated in (12) and (13):[1]

(12) Do/did you want something?
(13) Is/was the scientist happy?

These linguistic facts lead to the prediction that English-speaking children in an OI stage of development should show optional use of the set of morphemes that mark finiteness. More specifically, they should show omissions of -*s*, -*ed*, *be*

---

[1]BE and DO (all caps) are used here to denote the citation form of the set of BE and DO verbs of interest in this discussion. This would include the phonological variants, that is, *am, are, is,/do, does*. For BE it includes the grammatical classes of copula (e.g., "he is happy") and auxiliary verbs (e.g., "she is running"). For DO it includes the auxiliary form of the verb but excludes main verb contexts (e.g., "you do it.").

and *do* forms in contexts where these forms are obligated in the adult grammar. More precisely, in the linguistic framework adopted here, children in an OI stage have a grammar that allows optional use of a finite or nonfinite version of a given form, F, in linguistic contexts where the adult grammar specifies a finite version of the form. The children's grammar allows an option not allowed in the adult grammar. Use of a nonfinite form is an allowable option in the child grammar and, in that sense, is not an omission or error. To avoid excessive wordiness, we use terms like *omissions, errors,* and *correct use* here, to mean with regard to the adult grammar. The reader should keep in mind that these are not viewed as ungrammatical within an OI child grammar.

Note that the predicted set of affected morphemes includes different ways to phonetically and morphologically represent finiteness: *-s* and *-ed* are affixes; *be* and *do* are free-standing morphemes that can appear in different positions within a sentence and have different forms, depending on person, number, and tense (e.g., *am, is, are; do, does*). The point is that they are not defined by their surface properties. They are, instead, defined by whether or not they carry markings of tense and agreement and, if so, the word order rules that apply. Notice also that the notions of tense and agreement are formal grammatical notions. Tense is not the same as a child's notion of time. Tense can be marked as present or past. Children can know about presentness and pastness but not know about the need to mark grammatical tense in sentences. The distinction is especially evident in constructions with *do*, as in (12), where the marker of present/past tense appears on a grammatical form that does not contribute to the meaning of the sentence.

Further predictions of the OI stage are that children who omit *-s, -ed, be* and *do* will, at the same time, know that if they use the targeted forms, certain constraints apply. Thus, the way in which their usage will differ from that of the adult grammar will be in the omission of finiteness, not in overt errors of agreement. It is predicted that, if the forms appear, they will show correct agreement. This is most evident in the *be* forms of English, where person and number agreement show on different forms (e.g., *I am, you are, he is*).

MORPHOSYNTAX OF CHILDREN WITH SLI: EXTENDED OPTIONAL INFINITIVE

For some time, it has been reported in the literature that one way in which the grammar of English-speaking children with SLI differs from that of control children is in a lower likelihood that they will use certain grammatical forms in contexts where those forms are obligated in the adult grammars. This is well documented for *-s, -ed,* and, to a lesser extent, *be* forms (Bishop, 1992; Leonard, 1989; Rice, 1991). The OI account raises the possibility that these facts are part of the same underlying phenomenon, that of an EOI stage in the morphosyntax of children with SLI. If children with SLI are adhering to the linguistic constraints that guide normative acquisition of morphosyntax, then we would expect that they, too, would demonstrate an OI stage. What would distinguish them would be (a) a later-than-expected emergence (i.e., first uses) of the targeted grammatical forms; (b) once finiteness emerges, a lower-than-expected optional use of finite forms in contexts where the adult grammar requires finiteness; and (c) a longer-than-expected period of OI, perhaps into adulthood.

In this chapter, we discuss (b) and (c). Although it is generally presumed that (a) is an accurate characterization, there is surprisingly little direct evidence available about emergence of grammatical forms. Hadley and Rice (in press) studied the emergence of *be* and *do* and document a late onset for children with SLI. With regard to (b), predictions that apply to -*s*, -*ed*, *be*, and *do* are as follows:

1. For -*s* and -*ed* markings on lexical verbs, bare stems (i.e., the nonfinite forms of the verb) may optionally be used where inflected forms are required.
2. For -*s*, in contexts other that third person singular, there will be no overt marking. That is, *they walks* is predicted to not be a productive error.
3. -*ed* will be restricted to past tense contexts.
4. Auxiliary and main verb *be* may be omitted.
5. Auxiliary *do* may be omitted.
6. When *be* and *do* forms are used in contexts where the adult grammar requires a finite form, children will give correct agreeing forms.[2]

These predictions were tested in a recent study (Rice, Wexler, & Cleave, 1994).[3] The participants included 18 children with SLI with a mean age of 5 years

---

[2]The use of nonfinite forms of *be* in contexts where allowed in the adult grammar are not ruled out within an OI or EOI stage. These are contexts, such as infinitival complements (e.g., "I want to be happy"), verb phrases containing modals (e.g., "I will be happy"), and imperatives (e.g., "Be happy"). What should not appear are nonfinite forms of *be* in grammatical contexts which require finiteness; for example, we do not expect forms like "She be happy." In fact, as Wexler (1994) pointed out, it is a general property of the OI stage cross-linguistically that auxiliaries do not appear as nonfinite forms in contexts where adult grammar requires that they be finite. The prediction is that the OI and EOI stage of English follow this pattern.

At the same time, it is recognized that some dialectical variants of English allow main verb uses of *be*, and may also allow other surface features similar to those of the OI stage, such as omission of third person singular -*s* on lexical verbs. The specific details of how an OI stage may be manifest in dialectical variants of English have yet to be worked out.

[3]Since the writing of this chapter, the findings of Rice, Wexler, and Cleave (1995) have been replicated in Rice and Wexler (in preparation), who investigated a separate second sample of children with SLI and nonaffected children in two control groups, one matched for chronological age and the second for mean length of utterance. Because the findings across the two studies are very similar, the details reported in this chapter can be considered illustrative of the findings for Rice and Wexler (in preparation), as well. Rice and Wexler go on to present evidence of distributional nonoverlap of the children with SLI and their age peers on the targeted morphemes, and develop further arguments for tense as a clinical marker.

(range of 55–68 months). These children were all clinically identified as SLI, were enrolled in intervention programs, met inclusionary criteria for receptive and expressive delays in language acquisition, and met the exclusionary criteria of no clinically significant impairments of hearing, cognition, psychosocial development, or neuromotor functioning. There were two comparison groups of nonimpaired children, one a group of 22 children of the same chronological age (referred to as the 5N group) and the other a group of 20 children at an equivalent level of general language acquisition (as indexed by their mean length of utterance in spontaneous samples). These children's mean age was 2 years younger than that of the group of SLI children, with a mean of 36 months (range of 30–40 months); they will be referred to as the 3N group.

To evaluate the children's use of the targeted morphemes, data were collected from transcripts of their spontaneous utterances. In addition, to ensure sufficient numbers of occurrence of the targeted forms, the children also were administered a series of probe tasks, designed to elicit -s, -ed, be and do forms. Pictures were used to elicit -s and -ed contexts. Play with stuffed animals was used to elicit be and do contexts. Be was elicited in copula and auxiliary contexts for questions, as in (14) and (15), and in declarative contexts, as in (16).

(14) Is the bear cold?
(15) Is the bear crying?
(16) The bear is cold.

Do was elicited in questions, as in (17).

(17) Does the bear want a blanket?

Although the evidence was submitted to detailed analyses (see Rice, Wexler, & Cleave, 1995 for a full report), the general findings can be summarized as in Table 8.1. It is apparent that the children with SLI show a lower than expected optional use of finite forms of -s, -ed, be and do in their sentences, relative to the control groups of children. This is evident even when the comparison group is two years younger (i.e., the 3N group shows a higher proportion of finite forms than does the SLI group). As a group, the SLI children's proportion of finite forms for -s was 30%, 23% for -ed, 45% for be, and 46% for do. In comparison, their age peers were at nearly adult levels, 90% or more. The 3N group was closer to the adult levels than were the children in the SLI group, with mean levels of optional finite forms of 45% for -s, 53% for -ed, 65% for be, and 62% for do.

In further analyses to evaluate predictions 1 through 6 it was found that the ways in which the children were not accurate was that they omitted the targeted forms in contexts where the adult grammar required them. This generalization held for the SLI group as well as the control groups. Thus, the children seemed to regard the target morphemes as optional, even in the tested contexts where in the adult grammar they are obligatory. As predicted, the children did not

TABLE 8.1
Percent Correct in Obligatory Contexts, Collapsed Over
Elicited and Spontaneous Samples

	% SLI	% 3N	% 5N
-s	30	45	90
-ed	23	53	90
be	45	65	99
do	46	62	98

assume the morphemes could be applied in contexts not allowed in the adult grammar. They did not say things like (18) or (19).

(18) *You walks.
(19) *Yesterday he walks.

The EOI predicts that when children choose to use a *be* form, they will select the form that goes with the person and number of the subject. Thus, children will generate utterances such as those in (20) but not those in (21).

(20) She is happy/you are happy.
(21) *She are happy/ *you is happy.

This prediction was upheld, perhaps most surprisingly for the children in the SLI group. Even though they omitted *be* forms at an average rate of 55%, in the 45% of the utterances in which an overt form of *be* did occur, it was highly likely to be the right choice, with accuracy over 90%. This shows that these children, who demonstrate such striking limitations in their morphosyntax relative to children a full two years younger, nevertheless know some important properties of English grammar. The *be* form of English is an unambiguous context in which to observe agreement. What these findings suggest is that the SLI children, by the age of those children studied here, do show consistent marking of agreement, at the same time that they do not show consistent marking of tense.

EXTENDED DEVELOPMENT THEORY OF SLI: IN WHAT WAY IS GRAMMAR DELAYED?

The Logic of How to Determine Delay

The EOI can be regarded as one form of an extended development theory (EDT) of SLI. The EDT can be compared to the notion of delayed language that has been in the literature for some time. "Delay" carries with it several notions. One is the idea that, typically, language emerges at a later age for children with SLI than for nonimpaired children. This observation raises the possibility of a methodological refinement, in which there are two control groups, one matched for chronological age and the other for general language development (for preschool children, this is usually the mean length of utterance). This design, as employed in the study previously described, allows for determination of whether or not certain aspects of the grammatical systems of children with SLI lag behind their general language development, that is, if there is a difference beyond the delayed emergence of grammar relative to age peers (recall that these children are clinically identified because they do not show the grammatical competencies of their age peers). In the logic of the language-control-group design, no differences between the SLI group and the language control group are taken as evidence that the grammar of SLI children does not differ from normally developing children in any way other than delayed emergence. More specifically, when the two groups do not differ on targeted grammatical morphemes, it is said that SLI children do not have "extraordinary difficulty" learning grammatical morphemes (Lahey, Liebergott, Chesnick, Menyuk, & Adams, 1992). On the other hand, if differences are found between the SLI group and the language control group, they are taken as evidence for ways in which the grammar of children with SLI is different from younger nonimpaired children. More specifically, it is evidence that certain grammatical morphemes are extraordinarily difficult for children with SLI (e.g., Leonard, 1989; Rice & Oetting, 1993; Watkins & Rice, 1991).

Interpretation of Grammatical Differences
Not Accounted for by a Delay

A finding of a difference reveals a more interesting problem than a simple delayed onset of language acquisition. The problem is how to account for the fact that some aspects of the grammar of children with SLI pose selective and extraordinary difficulty, and how to characterize these difficulties in terms of the grammar and language acquisition mechanisms of nonaffected children. Leonard (1989) invoked surface properties of the grammar as an explanatory construct, proposing that the grammatical morphemes that differentiate children with SLI from language-control children are ones that are unstressed and phonetically unsalient. This interpretation has been challenged by evidence that forms similar in surface structure, but different in underlying grammatical function, do not pattern together in differentiating children with SLI from their controls (Rice & Oetting, 1993; Watkins & Rice, 1991).

Delay as Something to Be Outgrown

Another aspect of the notion of delay is the assumption that children will outgrow their initial symptomology. Presumably, whatever factors that account for the initial extended period of time before the children's language emerges can be overcome or are no longer operative and then language acquisition can proceed unhampered. Recent studies of children with delays in expressive language only, sometimes referred to as late talkers or specific expressive language delay (SELD), suggested that about half of the children with this diagnosis as preschoolers may catch up with their nonimpaired peers by age 4 years (cf. Paul & Alforde, 1993). Notice that the children in the study reported here were children whose receptive language milestones, as well as expressive language, were significantly below age expectations. Children with such receptive limitations are less likely to outgrow their initial language impairment (cf. Thal, Tobias, & Morrison, 1991), perhaps because their underlying linguistic representations are less fully developed than those of children who perform within normative range on receptive tests.

Recent evidence suggests that affected individuals may not, in fact, outgrow the kinds of grammatical differences evident during the preschool years. Two recent studies—Marchman and Weismer (1994) and Oetting, Horohov, and Costanza (1995)—reported that children with SLI (defined according to criteria similar to the ones used in Rice, Wexler, & Cleave, 1994), who were 7 to 9 years old, performed below their comparison groups on an elicitation task of tense marking. The children with SLI were distinctive for their overuse of zero-marked stems. This kind of error was observed for regular and irregular past tense forms. Tomblin (1994) reported that in a sample of young adults with a positive history for childhood language disorders, their spontaneous samples yielded more errors with tense marking than did samples from control subjects. In a detailed case study report of affected adults in a family with a high incidence of language disorders, Ullman and Gopnik (1994) documented that these individuals had not mastered tense marking on either regular or irregular verbs. The conclusion is that tense-marking, in particular, may remain as a way in which the grammar of affected persons does not fully approximate the expected adult levels of performance.

Accounts of Deviant Language

Some investigators following models of normative language acquisition have concluded that the grammars of individuals with SLI are different from those of nonimpaired speakers. Clahsen (1989, 1991) argued that German-speaking children with SLI are missing agreement relationships whereas Grimm and Weinert (1990) identified the problem as one of word order rules. Gopnik and Crago (1991) proposed a missing features model of SLI, in which affected individuals are thought to be missing features of the systems that guide the morphological acquisition of nonimpaired children, including features of number, tense, agreement, gender, aspect, animacy, person, and mass/count (cf. Oetting & Rice, 1993, and Rice & Oetting, 1993 for evidence that the number feature, counter to the predictions of the missing features model, is available to children with SLI).

EOI AS EXTENDED DEVELOPMENT OF LANGUAGE

What is needed is a way to capture how the grammar of children affected with SLI can be delayed in emergence, show selective ways in which certain grammatical morphemes differentiate the performance of children with SLI from language-level-control children, and, at the same time, maintain strong parallels with the grammars of nonaffected children. This account should also accurately predict long-term outcomes.

Under an EOI account, impairment can be thought of as having three manifestations. One is a later-than-expected emergence of the targeted grammatical forms. In this sense, language is delayed. The second manifestation is evident in the period after emergence, when there is a lower-than-expected optional use of finite forms in contexts where the adult grammar requires finiteness. In this sense, certain grammatical forms are extraordinarily difficult. At the same time, there are deep underlying similarities between children with SLI and their nonimpaired peers in that both groups adhere to the linguistic principles that guide normative acquisition of morphosyntax. The third manifestation is an extended period of time in which an optional use of finite forms is evident. In the EOI account, the variation evident in SLI is attributable to a long, drawn-out period of development for certain very specific linguistic principles, principles that appear late in the morphosyntax of nonimpaired children, and that apply to a specific set of surface morphemes. In this way, normative and non-normative variations are linked.

Within this view, language acquisition could be likened to a train. At the outset, the train is at the station with a certain configuration of engines, cars, potential for acceleration, and tracks for guidance. For most children, the train leaves the station at a particular time, with the cars in alignment and the coupling between them carefully synchronized. In the case of children with SLI, the train seems to depart later. Yet the train follows the same physical laws and constraints that govern the functioning of the train of non-SLI children. On closer inspection, it is apparent that the coupling of the cars for the children with SLI is not the same as for the other children. Instead of tight coupling, some cars are attached in a more elastic way, as if made of material like a bungee cord. This allows for the train to maintain its configuration, but some cars fall behind the momentum of others. Thus, the SLI children's morphosyntax can be characterized as having a delayed emergence, an extended period of development, adherence to the same principles as the morphosyntax of non-SLI children, and at the same time a general configuration not quite the same as other children by virtue of the fact that not all aspects of the linguistic system are fully synchronized. What is not known is whether or not the language train of the individuals with SLI ever comes to fully align in the same way as the train of nonaffected individuals.

To summarize, 5-year-old children with SLI, who optionally mark tense in contexts where the adult grammar requires tense, share linguistic characteristics that also apply to much younger children. An EOI is not deviant in the sense of "abnormal" or "missing" principles or structures that are attested to in younger children. At the same time, an EOI is unlike normative acquisition in a lower-than-expected use of finite forms over an extended period of time. There is an extended development of the acquisition of tense. The full duration of this period is not known. The recent reports of adults with SLI who demonstrate less-than-expected use of tense markers suggest that the period may persist into adulthood.

POSSIBLE REASONS FOR THE EOI

Explanations have been put forward for grammatical limitations of SLI of the kind described here. Possible explanations fall into two broad categories. One category carries the theme of missing components of the grammar. In a sense, this category of explanations regards the grammar of children with SLI, the underlying linguistic representations, as deviant in some way. This category focuses on linguistic structures or processes. Examples of this category are the missing features account of Gopnik and Crago (1991) and the missing agreement account of Clahsen (1989). Another category of explanations focuses on psychological processing mechanisms thought to be at least somewhat independent of linguistic representation. On models of this sort, observed differences in grammatical forms are attributable to problems in processing the input language in a way that restricts a child's ability to fill out linguistic representations or to problems with capacity constraints. This line of explanation emphasizes the surface properties of grammatical forms, such as salience, patterns of stress, and frequency of input. Examples of this category are Bishop (1992), Leonard (1989), Johnston (1994), and Tallal, Stark, Kallman, and Mellits (1980).

Explanations from either of these two camps, at least as currently formulated, encounter significant obstacles when applied to the phenomena associated with EOI. With regard to missing components of morphosyntax, it is obvious from available evidence that any missing elements must be quite constrained and are constrained in the same way that younger, normally developing children's language is constrained. It is imperative that this issue be examined from a highly specified model of the adult grammar. What can seem to be missing, such as agreement, may instead be confounded with other dimensions. An example is the observation that English-speaking SLI children tend to use a bare stem form of the verb in contexts that require -s (in utterances such as (1), previously mentioned). One possible interpretation is that this phenomenon is attributable to missing agreement (an interpretation examined in Rice, 1994). What now seems to be operative is a limitation in tense-marking that is also marked by the third person singular present tense -s morpheme. Another contribution of a highly specified model of the adult grammar is that it can identify contexts where agreement is unambiguously marked. In English, *be* offers such a context.

When examined for evidence of agreement (Hadley & Rice, in preparation; Rice, Wexler, & Cleave, 1995), the incorrect form choice predicted by a missing agreement account is not evident.

If there is a candidate for a missing element of the morphosyntax at some early and perhaps later stages, it might be knowledge of the need to mark tense in a main clause. But much more work needs to be done to flesh out what this may mean. It may be attributable to underspecification of tense features, or limited mechanisms for movement of the verb to the functional category of INFL, or other, as yet unspecified, factors (cf. Wexler, 1994 for the original discussion of possible analyses of the OI stage in normally developing children. Also see Hyams, 1994; Rizzie, 1994). A more complete understanding will require evidence from cross-linguistic studies that allow for observation of different parts of the finiteness marking systems of the grammar. The EOI stage manifestations depend upon the particular linguistic properties of a given language (cf. Crago & Allen, this volume; Leonard, this volume; Wexler, 1994, this volume). Further evidence from particular contexts of English will also continue to play an important role in the search for illumination.

Processing or limited capacity accounts also encounter challenges when applied to the phenomena of the EOI. At the outset, it is evident that children with SLI do not make haphazard mistakes. They do not misapply morphemes across lexical or grammatical categories, nor do they formulate odd word orders. Any processing or limited capacity account must be able to capture the fact that performance is constrained by underlying linguistic structures and processes (cf. Leonard, this volume, who stipulates that the underlying linguistic representations for children with SLI do not differ from those of nonimpaired children, a position similar in that respect to that of the EDT previously sketched out).

Linguistically constrained processing accounts seem to be best suited to accounting for omissions of expected surface forms. Intuitively, there is something quite appealing about the notion that the missing forms characteristic of English-speaking children with SLI correspond to the small, unstressed parts of the surface grammar and are, therefore, missing because they are small, unstressed forms. This characterization has been in the literature from the beginning. There are some important caveats that apply, however. One is that the salience account is empty unless there is an independent definition of salience. If there is not, then there is no way to disprove a salience theory; one simply defines elements that are omitted to have low salience. Second, a low salience account makes overly broad predictions. For example, surface properties alone would predict that different morphemes that share the same surface form should be equally vulnerable. One such example is Leonard's (this volume) prediction that regular plural -s should be vulnerable for English-speaking children with SLI, as well as -s to mark third person singular present tense on verbs. This prediction is not upheld, however, in studies of SLI children's use of plurals (cf. Oetting & Rice, 1993; Rice, 1994; Rice & Oetting, 1993). In these studies, plural usage is surprisingly robust in samples of SLI children, with levels of accuracy at or near mastery in spontaneous speech, whereas the -s verbal inflection is at low levels of accuracy. The point to be emphasized here is that a salience or processing account of the obtained evidence for -s, -ed, be, and do would also predict bare forms of the noun where regular plural -s is required, a prediction that does not hold for the children in the study reported here. In a parallel vein, Watkins and Rice (1991) showed that children with SLI were more likely to omit forms such as in and on when they were used as particles than when they were used as prepositions, a difference not predicted by a surface processing account.

Further counterevidence is found in recent reports that irregular past tense forms are also vulnerable for affected individuals (cf. Marchman & Weismer, 1994; Oetting et al., 1995; Ullman & Gopnik, 1994).[4] On low-salience accounts, the regular past tense suffix -ed is usually regarded as low salient whereas the stem-internal changes of irregular verbs, such as ride/rode, are not. If both categories of verbs are likely to be missing tense-marking, then the explanation must shift from that of the surface phonetics to the underlying grammatical function.

Equally problematic for a processing account, however, is the new finding that agreement is evident for be. The available models of processing limitations (e.g., Bishop, 1994; Gathercole & Baddeley, 1990; Leonard, 1989; Tallal & Stark, 1981) focus on the input available to a child: Perhaps the original phonetic form is not detected in the input or, if detected, is sufficiently demanding to create a processing load that exceeds a youngster's capacity, or jeopardizes a child's ability to recognize and place a grammatical form in the appropriate paradigm, such that sufficient instances of verb +ed, for example, are not entered into a past tense paradigm to establish a general rule. Under these scenarios, memory of stored forms is vulnerable, as is the formulation of full paradigms (i.e., knowledge of the morphological forms that mark the grammatical distinction).

A processing account would readily predict the fact that the SLI children in the study previously described, as a group, omitted be forms in 55% of the contexts where they were required. On this kind of explanation, these small, unstressed forms would not be detected in the input or, if detected, would require additional instances or redundancy of use before they would finally be entered into a grammatical paradigm. One would predict, then, during the period of high probability of omission, confusion about which surface forms appear with which grammatical subjects, a confusion reflective of an incomplete or inaccurate paradigm, and/or limited memory of the surface forms of the morphemes. What would not be predicted is the highly consistent choice of correct forms of be in the 45% of the contexts in which be could appear, as observed in the study reported here. This high level of accuracy suggests that the forms of

---

[4]Current analyses underway in our lab also show that irregular past tense forms are affected. In our data, children with SLI are much more likely to use a bare stem form than are the nonaffected children, a tendency that persists from the preschool years into elementary school.

be are available in the linguistic representations, hence, they have passed through the input mechanisms and entered the grammar. There does not seem to be any obvious way to predict that input processing limitations will lead to a high rate of omissions of a set of different surface forms that are also used with a high degree of accuracy when they do appear. Accurate use implies underlying representations and sufficient memory for recall.

Exactly the same point can be made with regard to regular third person singular *-s*. If *-s* is nonsalient and, thus, not entered correctly into a paradigm, if it appears at all, it should appear randomly with regard to subjects. However, we have already seen that *-s* is almost always used correctly, that is, with third person singular subjects, and almost never with other subjects. This consistent pattern of use of *-s* indicates that its grammatical properties are known, counter to the salience hypothesis.

We conclude that neither of the two kinds of available explanations of the limited morphosyntax of children with SLI offers an immediate and compelling explanation of the observed phenomena of the EOI stage. Wexler (1994) commented on the possibility of a maturational mechanism at work in the OI stage of young nonimpaired children, such that a part of the grammar that necessitates tense in main clauses would not be available early on, but would "kick in" at some later time, probably around 3 years of age. Clearly, the tense-marking system is fully operative by 5 years of age. This implies that, for children with SLI, there may be a problem with the activation of a time-referenced onset switch of some sort in the grammar. This onset switch may be set late, or may be only partially operational once activated, or may never be switched on. Answers to these possibilities will depend upon the collection of evidence that documents the longitudinal trajectory of the EOI stage of SLI, and its ultimate resolution or outcome. That is the direction of our current investigations of the EOI stage.

IMPLICATIONS FOR A PHENOTYPE OF SLI

In the previous sections of this chapter, we described an EOI stage that is likely to be characteristic of English-speaking children with SLI. Now we consider the implications of this finding for the specification of a phenotype for genetic contributions to language acquisition. We examine several properties that enhance the promise of candidate phenotypes of language acquisition and use the EOI as an illustration of issues that arise.

Variation Appears Where Variation Is Not Expected

As noted at the outset, a good phenotype is one that demonstrates variation across individuals, and allows for grouping of individuals on the basis of the behaviors associated with the phenotype. In the case of the EOI, variation appears where, for theoretical and empirical reasons, variation is not expected. Tense-marking should not be optional in individuals over 5 years of age. It is a fundamental property of the grammar that must be in place for the formulation of sentences. So, speakers should know this property of the grammar. If there is a grammatical trait that is characteristic of the general population that is not evident in some individuals, that variation could plausibly be attributable to inherited variations.

A closely related point is that the obtained variations constitute extreme values. The accuracy of tense-marking of children with SLI is far below the expected levels of the general population, at the age levels observed (available evidence is strongest for the period of 2–7 years). This means that a distributional analysis of children's accuracy of tense-marking at age 5 years would show that most children would fall in the uppermost levels whereas there would be a group of children who would still cluster at the lower end of accuracy. Although such normative information from a randomly drawn sample is not yet available, the evidence from our experimental studies of group differences provides evidence of this distribution (Rice & Wexler, in preparation). If it turns out that an EOI stage is distributionally distinct, that would be relevant to attempts to determine if the condition is also etiologically distinct (cf. attempts to demonstrate that the genes that contribute to reading disabilities are the same as those contributing to variations in reading at higher levels of ability; Gilger, this volume; Gilger, Borecki, Smith, DeFries, & Pennington, 1994; Lefly & Pennington, this volume; Shaywitz, Escobar, Shaywitz, Fletcher, & Makuch, 1992; Smith, this volume). What is at issue is whether or not the individual variation evident in EOI is attributable to the same etiological factors as the variation evident in the OI stage of development. Is there one gene or cluster of genes that influences grammatical ability that causes variation across individuals, and the condition of SLI is but the bottom tail of this variation (cf. Leonard, 1991)? Or is the variation not a matter of one distribution, but two overlapping distributions, such that the clump at the bottom of the distribution of children in an EOI stage is attributable to a different etiology? These are fundamentally important questions that are on the immediate horizon, just within view. If answers are to be found, there must be a way to identify individuals in an EOI stage, and there must be an understanding of the trait involved, and how this trait may or may not be considered to be part of the normative distribution and expected linguistic competencies.

Variations in Observed Behavior Are Interpretable in Terms of Higher Order Cognitive Representations

What differentiates the genetics of higher cognitive abilities in humans from the behaviors of animals is the postulation of underlying representational mechanisms that account for why linguistic behaviors, for example, show the constraints that they do and the observed patterns of contingency relationships in the surface structures. So the challenge is to do more than just link behaviors or surface markers with genes. The full task is to show relationships in a way that corresponds to the underlying grammar. Perhaps it will be possible to demonstrate that tense-marking is associated with familial aggregations of affectedness that correspond to known patterns of genetic transmission and that the affected family members show characteristic allelic variations. That would be a major advancement in what we know, without any doubt. But what would make it really interesting would be the way in which such a demonstrated association could contribute to our understanding of human linguistic capacities. In the case of EOI, tense-marking is not defined in terms of a particular surface instantiation of tense-marking, such as the -s or -ed suffix, although this may turn out to be an efficient way to measure it. Instead, the theoretical framework provides an account of how tense-marking is integrated into the grammatical system. The point is that clinical markers, at best, should have an associated interpretive framework that allows for full explication of the possible genetic contribution.

The Observed Variation May Be Quite Localized

Language can be defined in ways that highlight general characteristics or that delineate carefully differentiated subparts. The best examples of general characterizations are to be found in the normative, omnibus tests of language acquisition. There is usually relatively gross differentiation of what constitutes language. Often the tests are organized according to performance modality (i.e., receptive vs. expressive and test in broadly defined areas of vocabulary and grammar). In contrast, for example, in theoretical terms, morphosyntax can be partitioned from other parts of the linguistic system and further subdivided into phenomena, such as finiteness-marking.

Almost all of what we know about the condition of SLI is at the level of general milestones of language acquisition. Conventional means of clinical identification rely heavily on normative assessment with omnibus tests of language performance. Although the limitations of this procedure are frequently noted, it remains a mainstay of clinical practice. Therefore, any identification of affected individuals is likely to rely on such methods.

What remains obscured by such an approach is the possibility of localized variations in certain areas of the grammar. If these areas are tested only sporadically in the standardized tests, and if other areas of language acquisition are not problematic, individuals may be considered unaffected when in fact there are interesting variations.

There is evidence that an EOI stage may not be detected by a conventional language test. A case study underway (Schuele, 1992, 1995) documented such a possibility with a young girl, AM, age 6½ years at first assessment. This child was interesting because she demonstrated selective impairments of morphosyntax. Her cognitive and reading abilities were at or above age expectations, and she tested within the normal range on general tests of language ability. Her vocabulary levels were within age expectations. Her speech abilities were intact. At the same time, she had not mastered certain grammatical morphemes indicative of an EOI stage: -ed, -s, and be. She made mistakes, such as *he run home* and *yesterday he walk the dog*. She was referred for evaluation because of the persistence of her mother, whose own positive history of speech/language therapy when she was young along with her husband's positive history led her to wonder about her daughter's grammatical abilities. Subsequent evaluation of this child's siblings and cousins reveal that her younger brother is quite delayed in the emergence of language milestones, and two of her cousins demonstrate similar profiles of selective impairment of morphosyntax.

It is not known how many individuals of this kind may be in the general population. They are quite likely to be overlooked by virtue of current practices in speech/language pathology, where SLI is typically defined by performance below age expectations on standardized omnibus tests. A further factor is that this selective impairment may not be recognized if, as in the case of AM, it co-exists with clear speech, insofar as problems with intelligibility are quite likely to be noticed by adults and serve as an impetus for further language evaluation.

The existence of children such as AM has theoretical and empirical implications. Such selective impairment supports the idea of localized parts of the grammar and, more specifically, the psychological validity of the functional grammatical categories that control tense and agreement. Empirically, it will be important to ascertain if such children are not picked up by current identification procedures. If they are not identified, they would contribute to error in the analyses of possible genetic contributions to SLI or EOI.

The Issue of Cognitive Functioning

Traditionally, the condition of SLI has been defined in terms of exclusionary criteria for intellectual impairment. It was necessary to establish an operational definition for research and clinical purposes. That has come to be defined as a standard score of one standard deviation below the mean, or in the neighborhood of 80 to 85 standard score points. Children below this score were excluded from the SLI group. That level has been challenged in recent studies. Fey, Long, & Cleave (1994) found that intervention outcomes are similar for two groups of children: those whose cognitive scores drop below these conventional levels and those whose scores are higher. Bishop (1994) reported that concordance rates of twins are more meaningful if the cognitive score criteria are relaxed to include subjects with lower scores. Miller (this volume) also challenges the conventional criteria, on the basis of a variety of linguistic measures that are not predicted by nonverbal intelligence estimates.

What these arguments suggest is that it is probably overly restrictive to define an SLI phenotype too narrowly. But some caution is necessary. On the one hand, it is quite possible that certain kinds of linguistic knowledge may be relatively independent of general cognitive estimates (cf. the case of AM, observations of children with Williams Syndrome). On the other hand, it is also the case that deficits in general cognitive abilities are known to correlate highly with general measures of language impairment. And the etiology of pervasive developmental delay is likely to be distinct from the etiology of more selective impairment in the speech and language systems.

To return to the example of the EOI, what is not known is how much of the cognitive prerequisites must be in place to support the morphosyntactic system. Presumably, certain fundamentals must be present such as memory, ability to process incoming speech, basic categorization abilities, and fundamental conceptual development. Minimal levels of these fundamentals may suffice. As Wexler (1994) pointed out, very young children (under 2 years of age) have a wide variety of morphosyntactic abilities; children at this age would not usually be characterized as having a particularly large measure of conceptual development. So, the question is how to define the level of cognitive performance that is essential for grammatical acquisition. It seems that the current experimental standards are probably too conservative; evidence suggests that SLI is detectable at even lower levels of conceptual performance, and the minimal conceptual levels necessary for morphosyntactic development may be quite low. Adjustments in the definition of SLI with regard to cognitive levels are quite likely and in need of further evaluation.

Evidence of Affectedness Is Likely to Be Time-Referenced

There are two senses in which time-referencing applies to the possible phenotype of SLI. One is that the surface manifestation of the grammatical difference is likely to change with age. Young children could be identified by an EOI stage. Presumably, this stage will not be as apparent in older individuals, although little is known of how this stage plays out over time. So, there is the empirical issue of how to evaluate grammatical competence in a way that captures age-referenced grammatical expectations and, at the same time, captures interesting individual variations in ability. The second sense of time-referencing bears on the trait itself. The trait may change over time. There are several possibilities. One is that an underlying representation for the obligatoriness of tense is not available early on but comes on-line at a certain time for nonaffected individuals, say around 3 years, but does not ever come on-line for other individuals, or, does so at a much later time, say around 7 years. If obligatory TNS is late, or incompletely specified somehow, it is certainly possible for individuals to develop compensatory strategies, such that their grammatical performance could stay within the range of what they can manage. As in the case of reading disabilities, individuals could be difficult to detect because they compensate. In the case of EOI, for a compensated individual, tense-marking may be nonoptional in highly familiar contexts, or in grammatically simple configurations, but not in more demanding linguistic contexts. What will be needed to sort out these possibilities are careful empirical investigations that provide the necessary descriptive information. In short, much more needs to be known about the clinical manifestations of an EOI stage and how it changes over time.

CONCLUDING REMARKS

In this chapter, we have shown that the condition of SLI can be characterized in young children as an extended optional infinitive (EOI) stage of morphosyntax. The EOI is detectable because of the observed OI stage of normative variation which is, in turn, derived from current theoretical models of adult grammar. The fundamental question is whether or not the EOI stage, and the related OI stage, could be linked to possible variations in genetic mechanisms underlying linguistic representations. Answers to this question are not now available. At the same time, we know that any answers will depend upon the specification of a phenotype. Such specification subdivides into two issues: One is the ability to identify affected individuals and the second is an understanding of the underlying trait.

REFERENCES

Bishop, D. V. M. (1992). The underlying nature of specific language impairment. *Journal of Child Psychology & Psychiatry and Allied Disciplines, 33*, 3–66.

Bishop, D. V. M. (1994). Grammatical errors in specific language impairment: Competence or performance limitations? *Applied Psycholinguistics, 15*, 507–550.

Bishop, D. V. M. (1994). Is specific language impairment a valid diagnostic category? Genetic and psycholinguistic evidence. *Philosophical Transactions of the Royal Society, 346*, 105–111.

Bishop, D. V. M., North, T., & Donlan, C. (1995). Genetic basis of specific language impairment: Evidence from a twin study. *Developmental Medicine and Child Neurology, 37*, 56–71.

Clahsen, H. (1989). The grammatical characterization of developmental dysphasia. *Linguistics, 27*, 897–920.

Clahsen, H. (1991). *Child language and developmental dysphasia: Linguistic studies of the acquisition of German*. Philadelphia, PA: J. Benjamins.

Fey, M. E., Long, S. H., & Cleave, P. L. (1994). Reconsideration of IQ criteria in the definition of specific language impairment. In R. V. Watkins & M. L. Rice (Eds.), *Specific language impairments in children* (pp. 161–178). Baltimore, MD: Brookes.

Gathercole, S. E., & Baddeley, A. D. (1990). Phonological memory deficits in language disordered children: Is there a causal connection? *Journal of Memory and Language, 29*, 336–360.

Gilger, J. W., Borecki, I. B., Smith, S. D., DeFries, J. C., & Pennington, B. F. (1994). The etiology of extreme scores for complex phenotypes: An illustration using reading performance. In C. Chase, G. Rosen, & G. Sherman (Eds.), *Developmental dyslexia: Neural, cognitive and genetic mechanisms*. Timonium, MD: York Press.

Gopnik, M., & Crago, M. B. (1991). Familial aggregation of a developmental language disorder. *Cognition, 39*, 1–50.

Grimm, H., & Winert, S. (1990). Is the syntax development of dysphasic children deviant and why? New findings to an old question. *Journal of Speech and Hearing Research, 33*, 220–228.

Hadley, P. A., & Rice, M. L. (in press). Emergent uses of BE and DO: Evidence from children with specific language impairment. *Language Acquisition*.

Hyams, N. (1994, March). *The underspecification of functional categories in early grammar*. Paper presented at the U of Essex Child Language Seminar, England.

Johnston, J. R. (1994). Cognitive abilities of children with language impairment. In R. V. Watkins & M. L. Rice (Eds.), *Specific language impairments in children* (pp. 107–122). Baltimore, MD: Brookes.

Lahey, M., Liebergott, J., Chesnick, M., Menyuk, P., & Adams, J. (1992). Variability in children's use of grammatical morphemes. *Applied Psycholinguistics, 13*, 373–398.

Lander, E. S., & Schork, N. J. (1994). Genetic dissection of complex traits. *Science, 265*, 2037–2048.

Leonard, L. (1991). Specific language impairment as a clinical category. *Language, Speech, and Hearing Services in the Schools, 22*, 66–68.

Leonard, L. B. (1989). Language learnability and specific language impairment in children. *Applied Psycholinguistics, 10*, 179–202.

Marchman, V. A., & Weismer, S. E. (1994, June). *Patterns of productivity in children with SLI and NL: A study of the English past tense*. Poster presented at Symposium for Research on Child Language Disorders, University of Wisconsin, Madison.

Oetting, J., Horohov, J. E., & Costanza, A. L. (1995). *Influences of stem and root characteristics on past tense marking: Evaluation of children with SLI*. Poster presented at the annual Symposium on Research in Child Language Disorders, Madison, WI.

Oetting, J. B., & Rice, M. L. (1993). Plural acquisition in children with specific language impairment. *Journal of Speech and Hearing Research, 36*, 1236–1248.

Paul, R., & Alforde, S. (1993). Grammatical morpheme acquisition in 4-year-olds with normal, impaired, and late-developing language. *Journal of Speech and Hearing Research, 36*, 1271–1275.

Pierce, A. E. (1992). *Language acquisition and syntactic theory: A comparative analysis of French and English child grammars*. Boston: Kluwer.

Rice, M. L. (1991). Children with specific language impairment: Toward a model of teachability. In N. A. Krasnegor, D. M. Rumbaugh, R. L. Schiefelbusch, & M. Studdert-Kennedy (Eds.), *Biological and behavioral determinants of language development* (pp. 447–480). Hillsdale, NJ: Lawrence Erlbaum Associates.

Rice, M. L. (1994). Grammatical categories of children with specific language impairments. In R. V. Watkins & M. L. Rice (Eds.), *Specific language impairments in children* (pp. 69–88). Baltimore, MA: Brookes Publishing Co.

Rice, M. L., & Oetting, J. B. (1993). Morphological deficits of children with SLI: Evaluation of number marking and agreement. *Journal of Speech and Hearing Research, 36*, 1249–1257.

Rice, M. L., & Wexler, K. (in preparation). *Tense as a clinical marker of specific language impairment in English-speaking children*.

Rice, M. L., Wexler, K., & Cleave, P. L. (1995). Specific language impairment as a period of extended optional infinitive. *Journal of Speech & Hearing Research, 38*, 850–863.

Rizzi, L. (1994, January). *Root infinitives as truncated clausal structures in early grammars*. Paper presented at the Boston University Conference on Language Development, Boston, MA.

Schuele, C. M. (1992, November). *The limitations of standardized testing: A case study*. Poster presented at the Annual Convention of the American Speech-Language-Hearing Association, San Antonio, TX.

Schuele, C. M. (1995). *Specific language impairment: An investigation of morphosyntax across family members*. Unpublished doctoral dissertation, University of Kansas, Lawrence, KS.

Shaywitz, S. E., Escobar, M. D., Shaywitz, B. A., Fletcher, J. M., & Makuch, R. (1992). Evidence that dyslexia may represent the lower tail of a normal distribution of reading ability. *New England Journal of Medicine, 326,* 145–150.

Tallal, P., & Stark, R. (1981). Speech acoustic-cue discrimination abilities of normally developing and language-impaired children. *Journal of the Acoustical Society of America, 69,* 568–574.

Tallal, P., Stark, R., Kallman, C., & Mellits, D. (1980). Developmental dysphasia: The relation between acoustic processing deficits and verbal processing. *Neuropsychologia, 18,* 273–284.

Thal, D., Tobias, S., & Morrison, D. (1991). Language and gesture in late talkers. A one year follow-up. *Journal of Speech and Hearing Research, 34,* 604–612.

Tomblin, B. (1994, February). Family and twin studies of language impairment. In M. Rice (Chair), *Inherited speech and language disorders: In Search of a phenotype.* Session conducted at the American Association for the Advancement of Science annual meeting, San Francisco.

Ullman, M., & Gopnik, M. (1994). *The production of inflectional morphology in hereditary specific language impairment.* McGill Working Papers 10: Special Issue.

Watkins, R. V., & Rice, M. L. (1991). Verb particle and preposition acquisition in language-impaired preschoolers. *Journal of Speech & Hearing Research, 34,* 1130–1141.

Weissenborn, J. (1994). Constraining the child's grammar: Local wellformedness in the development of verb movement in German and French. In B. Lust, J. Whitman, & J. Kornfilt (Eds.), *Syntactic theory and language acquisition: Crosslinguistic perspectives. Vol. 1. Phrase structure* (pp. 215–248). Hillsdale, NJ: Lawrence Erlbaum Associates.

Wexler, K. (1994). Optional infinitives. In D. Lightfoot & N. Hornstein (Eds.), *Verb movement* (pp. 313–350). New York, NY: Cambridge University Press.

COMMENTARY ON
CHAPTER 8

Jeffrey W. Gilger
University of Kansas

Mabel Rice and Ken Wexler present theory and preliminary data on a clearly interesting developmental phenomenon and a potentially useful language impairment phenotype, the extended optional infinitive (EOI). As noted in their chapter, and in other chapters throughout this book, accurate characterization of an SLI phenotype is critical for future genetic work. All too often research has relied on omnibus language tests with little or no theoretical basis. Moreover, it has been common practice to use qualitative diagnoses that are based on relatively arbitrary cutoffs. Both omnibus tests and qualitative diagnoses can result in a "noisy" phenotype and lead to erroneous conclusions regarding the genetics for a disorder, such as SLI.

Although the EOI/SLI phenotype may not be perfect, it does possess several characteristics that make it attractive for future genetic study. Particularly valuable is that the phenotype is well-described and is based in (linguistic) theory. Future research may also show that the phenotype can be used as a semicontinuous scale, thus avoiding some of the statistical and diagnostic problems associated with cut-off scores. If it is not used as a continuous phenotype of language performance or problems, then individual differences in performance may pose a problem when researchers must decide who is and who is not affected. For example, should someone be called SLI (as opposed to language delayed) if he or she reaches 80% correct on the *be* or *do* items, but only 40% correct on -*s*? Although linguistic theory provides hypotheses about what types of problems should and should not cluster together, there will always be some variation and trouble with characterizing a linguistic phenotype quantitatively.

This difficulty aside, the optional infinitive (OI) and EOI stages are more than simple language tests of general ability: They also have specific and delineated developmental qualities (e.g., Wexler, 1994). Under the EOI account of impairment, SLI is manifested in three major ways: First, there is the delayed onset of certain grammatical forms; second, after emergence of finiteness there remains a lower-than-expected use of finite forms where they are required; and third, there is a prolonged period of development of specific linguistic principles that are applied to a specific set of surface morphemes. Rice and Wexler also make the point that the underlying linguistic representations in SLI children do not differ from those of non-SLI children. Thus, there is some continuity between the language structures of normally and nonnormally developing individuals.

In Fig. 1 I have presented hypothetical curves of the OI and EOI stages in normal and SLI children, respectively, based on the research of Rice and Wexler. These curves show the differences in mean performance at any particular age as well as differences in developmental rate and trajectory. The curve of the younger language-matched controls used in the Rice and Wexler studies is not shown here, though it is noteworthy that the data suggest that the SLI children

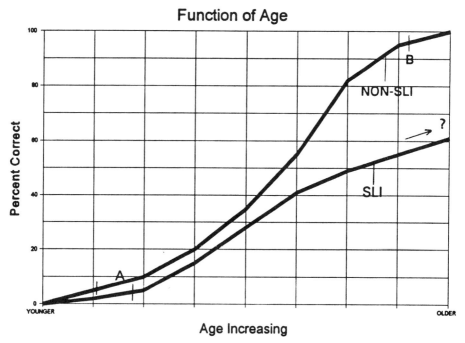

FIG. 1. The question mark for the progression of the SLI children represents our uncertainty as to whether or not they will meet the performance levels of NON-SLI adults. See chapter 8 of this volume.

are not simply delayed developmentally given that they perform significantly less well than the language-matched group.

A developmental phenomenon, such as that pictured in Fig. 1 raises several questions relevant to the genetics of language and SLI. First, there is the question of the role that genes and environment play in the onset of the understanding and use of and the understanding or use of optional infinitives (point A). Second, how do genes and environment interact to modify the rate (i.e., the slope of the lines) of development through the OI stage? Third, to what extent are individual differences in the age of mastery reflective of genetic or environmental differences among children (point B)? Fourth, are the genetic and nongenetic factors operating on language at stages prior to mastery the same as those operating postmastery? And fifth, are the lines drawn for the SLI and normal children in Fig. 1 members of the same distribution of all possible individual developmental trends for the OI stage? That is, is the slower and later developmental trend for the SLI group merely representative of the lower end of the possible range of OI performance? If it is, it is likely that the genetic and nongenetic factors influencing SLI development are the same as those influencing the language of normally developing children. To the extent that it is not, then SLI, and the EOI stage, is an aberration of development, and its etiology may be largely unique from that involved in normal language development (see discussions in Gilger, Rice, & Wexler, this volume).

Although space prohibits a thorough discussion of the behavioral genetic methodologies to address these questions, it is worth noting that the issues they address are relevant to the very essence of the nature of language, and the role of language input in development and remediation (Rice & Schiefelbusch, 1989; Watkins & Rice, 1994). Although I could have plotted the developmental trends of, say, global vocabulary performance, they would not have been so theoretically interesting. A linguistically based phenotype, such as OI and its nonnormal counterpart EOI, provides hypotheses about the underlying structures responsible for the group differences and developmental trends observed. Depending on the degree to which environment or genes affect these structures and various aspects of the child's performance, statements can be made about the nature of the linguistic process and machinery. Moreover, because the EOI phenomenon is theory driven, alternate interpretations of the phenomenon could be tested (e.g., missing components of grammar vs. cognitive processing deficits), by including the appropriate measures in a behavioral genetic design and testing the extent to which, say, variations in short-term memory share the same genes with performance on EOI tasks.

It is likely that our future understanding of normal and nonnormal language development will depend on theory-driven tests of alternative models. Rice and Wexler provide one such model that, by its nature, stimulates many questions about the roles played by genes and environment in language acquisition. As they correctly point out in the last paragraph of their chapter "... answers will depend upon the specification of a phenotype. Such specification subdivides into two issues: One is the ability to identify affected individuals, and the second is an understanding of the underlying trait" (p. 235). Interactions between research and researchers addressing each of these points will eventually lead to success.

REFERENCES

Rice, M. L., & Schiefelbusch, R. L. (1989). *The teachability of language*. Baltimore: Paul Brookes.
Watkins, R. V., & Rice, M. L. (1994). *Specific language impairments in children*. Baltimore: Paul Brookes.
Wexler, K. (1994). Optional infinitives. In D. Lightfoot & N. Hornstein (Eds.), *Verb movement* (pp. 305–350). New York: Cambridge University Press.

9

CHARACTERIZING SPECIFIC LANGUAGE IMPAIRMENT: A CROSSLINGUISTIC PERSPECTIVE

Laurence B. Leonard
Purdue University

Children with specific language impairment (SLI) constitute one of the groups of children for whom the question of a genetics of language is highly relevant. These are children who show significant deficits in language ability but do not meet the criteria for other disability categories. In particular, children with SLI show normal hearing, age-appropriate scores on nonverbal tests of intelligence, no clear signs of neurological impairment, and none of the symptoms (apart from a language problem) associated with autism. Although children with SLI vary in the severity of their language deficits, the problems are often long-standing (e.g., Aram, Ekelman, & Nation, 1984; Weiner, 1985). The familial concentration of this disorder raises the possibility that it has a genetic basis (e.g., Tallal, Ross, & Curtiss, 1989; Tomblin, 1989).

There has been much debate about the proper characterization of the language problems of children with SLI (see reviews in Bishop, 1992; Johnston, 1988). On the one hand, the omission and commission errors reflected in these children's speech are also observed in the speech of younger normally developing children. On the other hand, the language profiles of these children do not closely match those seen at any point in normal development. For example, English-speaking children with SLI are often more limited in their use of grammatical morphology than would be expected given their utterance lengths. An additional factor contributing to the view that these children's language is not merely like that of younger children is the fact that these children's nonverbal cognitive abilities approximate age expectations, thereby raising the possibility that these children make use of compensatory strategies that would not even occur to younger children.

This chapter continues the debate over the proper way to characterize these children's language deficits. Specifically, it is proposed that the underlying grammars of children with SLI are fundamentally normal. Although these children develop language slowly and show a grammatical profile that differs somewhat from the profiles of normally developing children, the claim will be made that these differences have more to do with the interaction between a general processing limitation and the particular language being acquired than with the nature of the children's underlying grammars.

Three types of evidence are used to advance the general argument. First, I show that the grammatical profiles of children with SLI vary according to the particular language being learned, suggesting that there may not be a common component of these children's grammars that is awry. Second, I will offer evidence suggesting that even when some areas of the grammar lag atypically behind other areas, the underlying grammar may show a normal configuration. Finally, I will use evidence from several studies to argue that the weak areas of grammar among children with SLI may simply be those that are most vulnerable to breakdown under a variety of degraded learning or processing conditions. Any genetic factors responsible for specific language impairment, therefore, need not directly affect precise mechanisms.

THE VARYING PROFILES OF SPECIFIC LANGUAGE IMPAIRMENT

I begin with a consideration of how the grammatical profiles of children with SLI can vary as a function of the language being learned. I will focus on evidence from English, Italian, and Hebrew, using data from recent collaborative work of which I have been a part (Dromi, Leonard, & Shteiman, 1993; Leonard, Bortolini, Caselli, McGregor, & Sabbadini, 1992). The studies in each of these languages employed the same procedure. Children with SLI and two groups of normally developing children participated, the first matched with the children with SLI according to mean length of utterance (MLU), the second according to chronological age. The ages of the children with SLI ranged from just under 4 years to 6 years. Their MLUs, as measured in words, ranged from 2.0 to 4.0. In this chapter, I will focus only on the data from the children with SLI and their MLU controls, for this comparison allows us to examine the profiles of the children with SLI relative to that which might be expected from younger normally developing children whose sentences are similarly limited in length. Although not presented here, the data from the age controls showed higher degrees of use than seen for the children with SLI, as expected.

The children were presented with line drawings whose descriptions required the use of particular inflections or freestanding closed-class morphemes. To ensure obligatory contexts for the forms of interest, many of the drawings were accompanied by a question asked by the examiner or a sentence that had to be completed by the child.

A summary of the children's percentages of use of some representative grammatical morphemes in obligatory contexts is provided in Table 9.1. Not all grammatical morphemes examined in these languages are reflected in the table; only those for which comparable data across languages are included. There were at least 10 obligatory contexts for each grammatical morpheme for each child.

To the extent possible, comparable morphemes across languages appear in the same row. Of course, in Italian and Hebrew each morpheme typically expresses a greater number of features than its English counterpart. For example, both Italian and Hebrew noun plurals convey gender as well as number. The morphemes appearing in the same row differ in another important respect. Only in English can an inflection be construed as an addition to the bare stem. In Italian and Hebrew, nouns and verbs (as well as adjectives) are always inflected, even in the singular; it is a matter of selecting the appropriate inflected form, not adding an inflection to a stem that can sometimes stand by itself.

Another comment that must be made about Table 9.1 is that, even with these allowances, morphemes that are highly comparable to those of English cannot always be found. For example, in present tense, Hebrew verbs are inflected for gender and number but not person; therefore, a form that directly corresponds to the English third person singular does not exist. In this case, a separate row was created, and the mean percentages are based on all present verb inflections collapsed across gender and number.

The data from English are consistent with findings from other investigations showing lower percentages of use by children with SLI than by MLU controls.[1] These group differences were seen when percentages were computed across all freestanding closed-class morphemes examined by Leonard et al. (1992) as well as when they were computed across all grammatical inflections. Findings of this type suggest that grammatical morphology is even weaker in these children than their utterance length would lead one to expect. The possible reasons for this are many; however, an inspection of comparable data from Italian and Hebrew allow a few contenders to be eliminated.

As can be seen from Table 9.1, the Italian-speaking children with SLI did not appear to have the same difficulty relative to MLU controls that was seen for the English-speaking children. In particular, grammatical inflections seemed less problematic. The freestanding articles were relatively difficult for these children. When the Italian-speaking SLI children and MLU controls were compared in

---

[1]There is some variation from study to study in the particular grammatical morphemes that reveal significant differences between children with SLI and MLU controls, as well as in the magnitude of the differences reported. For example, Oetting and Rice (1993) and Rice and Oetting (1993) found that their subjects with SLI were quite comparable to MLU controls in their use of noun plurals.

TABLE 9.1
Mean Percentages of Use in Obligatory Contexts by English- and
Italian-Speaking Children[a] and Hebrew-Speaking Children[b]

Morpheme	ESLI	END-MLU	ISLI	IND-MLU	HSLI	HND-MLU
Articles	52	62	41	83		
Noun plurals	69	96	87	89	76	74
Third person present verb inflections	34	59	93	93		
Present verb inflections					76	79
Regular past inflections	32	65				
Past verb inflections					56	29
Adjective inflections			97	99	71	78

[a]Data reported in Leonard et al. (1992).
[b]Data reported in Dromi et al. (1993).

terms of their percentages of use for the same grammatical morphemes examined in English, significant differences favoring the MLU controls were seen only for the freestanding closed-class morphemes, not for the inflections.

The data from Hebrew-speaking children with SLI shown in Table 9.1 reveal a pattern not unlike that seen for Italian. These children used noun, adjective, and verb inflections with percentages similar to those seen for the MLU controls. There were, in fact, no significant differences between the two groups on these inflections. The children's use of one freestanding grammatical morpheme, the definite accusative case marker *et*, was also examined. This form (that has no equivalent in English or Italian) was used with significantly higher percentages by the MLU controls.

Percentages of use in obligatory contexts do not constitute a complete picture of a child's ability. Another indicator is the degree to which these morphemes are applied to inappropriate contexts. As reported in Dromi et al. (1993) and Leonard et al. (1992), the children with SLI only rarely extended these morphemes to inappropriate contexts (e.g., *I draws*), and did not do so any more frequently than the MLU controls. The findings summarized in Table 9.1, then, do not seem to reflect the haphazard use of a surface form that has no function in the child's underlying grammar.

It is possible that the data shown in Table 9.1 are masking other kinds of differences between the children with SLI and their MLU controls. The means for some of the grammatical morpheme categories are the result of collapsing data across several distinct forms of the paradigm. Without further inspection, it cannot be assumed that each form contributed to the overall percentage in the same way in the two groups of children. One way to examine this question is to inspect the children's use of each form within the paradigm. Two examples are provided here. Table 9.2 displays the use of articles by the Italian-speaking

TABLE 9.2
Mean Percentages of Use of Articles in Obligatory
Contexts by Italian-Speaking Children

	Singular			Plural		
	il	la	lo[a]	i	le	gli[b]
ISLI	27	64	21	49	46	9
IND-MLU	83	89	77	78	85	52

Note. Data reported in Leonard et al. (1993).
[a]Masculine singular form used when subsequent word begins with [z], or [s] cluster.
[b]Masculine plural form used when subsequent word begins with vowel, [z], or [s]cluster.

children studied by Leonard et al. (1992) and reported in Leonard, Bortolini, Caselli, and Sabbadini (1993).

The first column in Table 9.2, under both singular and plural, shows the masculine form; the second column shows the feminine form. The third column under both singular and plural shows a masculine form that is used in restricted phonetic contexts. There is also the contracted article form *l'*, which was not examined because young children frequently segment this article as if it were part of the following noun.

As already indicated in Table 9.1, the Italian-speaking children with SLI produced the articles with lower overall percentages than the MLU controls. However, the profiles of use were similar. Statistical analysis revealed that both groups showed greater use of the singular articles than the plural articles, collapsed across gender, and greater use of widely applicable articles than articles with restricted phonetic environments of the same gender, collapsed across number. (The article *il* seemed to represent a special problem for the children with SLI which, based on some additional analyses reported in Leonard et al., 1993, appeared to be attributable to a phonotactic constraint in using a syllable-final consonant followed by another consonant.)

The second example comes from the Hebrew data. Shown in Table 9.3 is the use of each verb inflection type in the Hebrew paradigm by the children studied by Dromi et al. (1993). These data have been reported in Leonard and Dromi (1994). It can be seen from Table 9.1 that the two groups did not differ in their overall percentages for either present or past tense inflections. Table 9.3 reveals that they were also similar in their profiles of use. For both groups, percentages for present feminine singular inflections were significantly higher than those for present feminine plural inflections; present masculine plural inflections showed higher percentages than present feminine plural inflections; present masculine singular inflections were produced with higher percentages than past masculine singular inflections; and present feminine singular inflections were produced with higher percentages than past feminine singular inflections. It can also be

TABLE 9.3
Mean Percentages of Use of Correct Verb
Inflections by Hebrew-Speaking Children

	Present				Past		
	Singular		Plural		Singular		Plural
	Masc	*Fem*	*Masc*	*Fem*	*Masc*	*Fem*	
HSLI	93	88	87	28	64	50	62
HND-MLU	97	94	95	40	41	27	25

Note. Data reported in Leonard and Dromi (1994).

seen that the MLU controls seemed to have somewhat more difficulty with the past inflections than did the children with SLI. I am not certain why this is the case. One possibility is that because verb forms were elicited by means of drawings, the past forms might have placed additional cognitive demands on the children. That is, the actions requiring present forms were actually depicted whereas already-completed actions requiring past forms had to be inferred from the drawings. This might have had the greatest effect on the performance of the MLU controls whose nonlinguistic cognitive abilities were less developed.

Although the paradigm data shown in Tables 9.2 and 9.3 suggest similar acquisition patterns by the children with SLI and their MLU controls, it is possible that these patterns were brought about through different means. Because the children with SLI were older, they might have had the time to acquire these forms through rote learning whereas the control children's use might have reflected the acquisition of a productive system. Fortunately, the materials that we used provided occasions to observe overregularizations. Such productions would seem to reflect the operation of a productive rule.

There were, in fact, a number of instances of apparent overregularization by the children with SLI. For example, the Italian article data contained productions such as **il zaino*, "the backpack," instead of *lo zaino*, reflecting use of the more widely applicable masculine singular article in place of one that has a more restricted phonetic context. It is important to note here that the number and gender of the article in such cases were correct; forms such as **la zaino* were not used.

Overregularization by the children with SLI was also seen in the Hebrew verb inflection data. For example, present feminine singular verb forms can take the inflection *-et* or the inflection *-a*, depending on the verb. Some of the children with SLI used one of these inflections with verbs requiring the other. One such example was **mevashla*, "she cooks," instead of *mevashelet*. There were also instances in which the children used the correct inflected form, but failed to show the phonological alternation in the consonantal root. For example, for the verb *cook*, the pronunciation rules call for the first consonant of the root to be pronounced with [v], as in *mevashelet*. However, the past forms of this verb require that the first consonant of the root be pronounced as [b] not [v]. Several children produced forms, such as *vishel, *vishla, and *vishlu instead of *bishel, bishla,* and *bishlu,* respectively.

Not only were overregularizations seen in the data for the children with SLI in each of these languages, but also such overregularizations were used by these children as frequently as by the MLU controls. Thus, although the possibility that some of the grammatical morpheme use was the product of rote learning cannot be ruled out, the data from the children with SLI provided sufficient evidence of productivity to conclude that the similar profiles in the two groups came about in a similar way.

To summarize thus far, the profiles of Italian- and Hebrew-speaking children with SLI (relative to MLU controls) do not match those of children with SLI acquiring English. English-speaking children with SLI often show lower percentages than MLU controls on both freestanding grammatical morphemes and inflections. For Italian- and Hebrew-speaking children, on the other hand, differences are much more likely to be seen on freestanding morphemes only. Like their normally developing compatriots, the children with SLI in each language only rarely use a grammatical morpheme in an inappropriate context and they do, on occasion, overregularize. For grammatical morpheme types that form a paradigm, these children also resemble normally developing children in terms of which items present the greatest or least difficulty.

There have been several attempts to account for the fact that children with SLI acquiring Italian and Hebrew lag behind MLU controls primarily in the use of freestanding forms whereas children with SLI acquiring English have special problems with inflections as well. These accounts invoke factors such as the acoustic and articulatory differences in the inflections across languages as well as the typology of the language, specifically, whether or not noun, verb, and adjectives must be inflected (see Leonard, 1992). It is assumed that the problems are those of accessing the relevant data in a complete and timely fashion; the underlying grammars of these children are, in other respects, assumed to be unremarkable.[2]

Of course, in principle, crosslinguistic differences, such as those summarized in Table 9.1, might still be due to some fundamental flaw in the underlying grammar. The grammars of different languages are presumably configured differently, and certain configurations might be unavailable to children with SLI. For example, the recent literature on principles-and-parameters theory suggests that, whereas in English there might be separate nodes for T (tense) and Agr (agreement), in Italian and Hebrew tense features might be within Agr (Jaeggli

---

[2]The nature and boundaries of such processing limitations are in need of explication (see Johnston, 1992). Leonard (1994) has recently discussed some of the specific hypotheses that might be tested.

& Safir, 1989). This difference might be crucial in the case of specific language impairment.

However, until there is a consensus on how the grammars of these languages differ, it is difficult to posit a specific locus of difficulty in the grammars of children with SLI that permits the types of crosslinguistic differences that are observed. For now, we must acknowledge the possibility that the underlying grammar is not the source of the problem.

THE COMPREHENSIVE STUDY OF THE GRAMMAR OF ENGLISH-SPEAKING CHILDREN WITH SPECIFIC LANGUAGE IMPAIRMENT

Although crosslinguistic comparisons seem especially useful for testing hypotheses about the nature of the underlying grammars of children with SLI, detailed studies within a single language can also shed light on the matter. Even in English, where atypical profiles are most often noted among children with SLI, careful study might provide evidence that the underlying grammars of these children are basically normal. A recent study by Eyer and Leonard (1995) serves as an illustration. These investigators described the spontaneous productions of one English-speaking child with SLI. When initially seen at age 3;7 (MLU = 1.8 morphemes), the child showed no use of any grammatical morphemes. The principal analyses concerned two subsequent points in time, when the child was 3;9 and 4;5. At age 3;9 (MLU = 2.6 morphemes) his productions were highly reminiscent of those described in the literature on English-speaking children with SLI. For example, his mean percentage of use across all (nonthematic) grammatical morphemes was 21. Based on the data of de Villiers and de Villiers (1973), normally developing children at this MLU level might be expected to use such morphemes with a mean percentage of approximately 70.

Of course, it is one thing to note that the relationship between utterance length and grammatical morpheme use is different in a child with SLI compared to normally developing children and quite another to conclude that the child's underlying grammar is fundamentally abnormal. In fact, a comprehensive look at the child's utterances suggested that there was nothing peculiar about the grammar itself.

The child's utterances were examined according to Radford's (1990a, 1990b) treatment of functional categories, a framework falling within principles-and-parameters theory. In this framework, lexical categories are assumed to be noun (N), verb (V), adjective (A), and preposition (P), with their maximal projections NP, VP, AP, and PP each assumed to be X''. The functional categories are assumed to be determiner (D), complementizer (C), and inflection (I) and their maximal projections DP, CP, and IP.

The rationale for applying this framework to data from a child with SLI is as follows. English-speaking children with SLI appear to be limited in their use of grammatical morphology. Because grammatical morphemes are associated with functional categories, it is highly possible that difficulties with particular grammatical morphemes reflect problems with an entire functional category system. Furthermore, because the elements of one functional category system interact with those of another, it is possible that the underlying grammars of children with SLI are lacking all functional category systems. By examining elements of each of the three functional category systems, one might determine whether the functional categories are indeed absent from the underlying grammars of these children or, alternatively, whether these children's problems are best characterized as a pattern of slow development in which functional categories lag well behind other components of the grammar.

The two columns in Table 9.4 display selected utterances from the two ages: 3;9 and 4;5. In each column, the utterances are grouped according to whether they contain elements from the D-, C-, or I-system.

From the first grouping of the first column, it can be seen that at age 3;9 the child's utterances provided evidence of elements associated with the D-system. Specifically, the child made use of articles, the prenominal determiner *that*, and the case-marked pronominal determiner *my*. The data are not overwhelming, of course. Articles were used in only 30% of their obligatory contexts. Furthermore, the genitive *'s* inflection, which is also viewed as an element of the D-system, was absent from all obligatory contexts.

Evidence for the C-system was even more limited at this age, as can be seen in the second grouping. The first utterance in this grouping seems to contain an overt complementizer that was base-generated in C (glossed as "I know that my guy is right there"). The second utterance, glossed as "I know what are in these," is listed as it has a construction similar to that of indirect question forms, in which there is movement of a *wh*-word to the specifier position in CP, with the head position remaining empty. There were no attempts at *wh*-questions in this sample, and thus no opportunity to observe *wh*-movement.

Finally, the third grouping displays utterances containing morphemes that might be interpreted as elements of the I-system. Except for the modals, and the use of *don't*, evidence was limited. Infinitival *to* was absent from all obligatory contexts, as were regular finite inflections (one example of an irregular past form was observed). Auxiliary *be* was also absent.

From the utterances in the first column of Table 9.4 it can also be seen that this child sometimes failed to produce the proper case in preverbal pronouns. In fact, pronominal forms of nominative case were used in only 47% of their obligatory contexts. Nominative case assignment is an aspect of grammar that is associated with both the I-system and the D-system. That is, although case is assigned by I, a D-system is needed to receive case. Were it not for the presence of several D- and I-system elements in this child's sample, the observed variation between nominative and objective forms would probably be interpreted as evidence that this child's grammar consisted of lexical categories only.

TABLE 9.4
Productions of a Child With SLI Studied by Eyer and Leonard (in press)

Age 3;9	Age 4;5
D-system	*D-system*
Yeah put that on *the* horse	That *a* trashcan
Want *a* long one	Hey, this *a* bump too
Hey, *a* ear!	This go with *the* bus
Me want *that* piece go right there	*This* guy back
Hey, like *that* color there	Me don't have *this* bacon
My new candy	*My* daddy shake him
Mom, me have these at *my* school	*My* hand fit in there
C-system	*C-system*
Me know *that* my guy right there	*What is* that?
Me know *what* in these	*Where's* ketchup?
	Why won't that go?
	How you get up here?
	How you know they napkins?
I-system	*I-system*
Will bite hay in there	That *could* be mom
Yeah, me *can*	Me take a hot dog *to* cook
Can't do it	Need *to* go over
Me *don't* need that	He *didn't* fall off
Don't put mine in there	I *do* too
	I *did*
	I *'m* getting this out
	No, this *is* full
	It *was* up here
	All those *are* done
	Yeah, it *is*
	This park*s* right here
	Hey, it fit*s* on there
	It pop*ped* out
	Me drink*ed* it all

The utterances in the second column of Table 9.4, produced when the child was age 4;5 (MLU = 3.2 morphemes), makes a stronger case that functional categories were represented in his grammar. Although the elements of the D-system resemble those in the corresponding grouping at age 3;9, there was a clear increase in the use of articles in obligatory contexts (from 30% to 67%).

The second grouping contains several instances of *wh*-questions, some with possible copula forms and one with a possible preposed modal auxiliary. A few of these involve *wh*-words that cannot be interpreted to be the subject. Evidence of this type suggests the presence of a C-system.

The final grouping provides examples of elements from the I-system. These include modals in past as well as present tense, the dummy auxiliary *do* as well as *don't* in present and past tense, infinitival *to*, copula and auxiliary *be* forms, and present and past regular verb inflections. The latter seemed to be overregularized. Despite the variety of elements of the I-system that were used, percentages in obligatory contexts were not high. The highest percentage observed was 69%, for infinitival *to*.

As at the earlier age, the child often failed to mark nominative case. However, a greater variety of nominative forms was seen at the later age (*I, he, she, they*). Furthermore, the percentage of nominative forms used in obligatory contexts increased from 47% to 71% at age 4;5.

It should be noted that even at age 4;5 this child's use of nonthematic grammatical morphemes still trailed MLU expectations. His overall mean percentage was only 48 on this measure in contrast to a mean percentage of 82 observed in the de Villiers and de Villiers (1973) data from normally developing children at the same MLU level. It can also be observed that at age 4;5 functional category elements appeared alongside elements typically associated with a grammar consisting solely of lexical categories. The utterance *me drinked it all* is perhaps the most striking example. Here, the child used *me* in preverbal position, suggesting the absence of an I-system that assigns nominative case. Yet, *drinked* appears to contain an overregularization of *-ed*, suggesting the presence of an I-system.

Nevertheless, inspection of the data leads one to conclude that there was nothing aberrant about the grammar itself. The child's grammatical development was slow, with particular grammatical morphemes from functional categories especially slow to develop. However, development proceeded in the right direction. Across time, errors decreased in frequency, and elements of functional categories increased in frequency and variety. The unusual co-occurrences that were observed seemed to be due to the vantage point afforded by the study of development in slow motion, where new forms develop slowly, and remnants from a previous stage are slow to be shaken off.

HOW SPECIFIC MUST THE SOURCE OF THE PROBLEM BE? IMPLICATIONS FOR A GENETICS OF SPECIFIC LANGUAGE IMPAIRMENT

The picture of SLI that seems to emerge is one of a child with slow language development whose grammatical profile represents an exaggerated version of the profile seen in young normally developing children acquiring the same language. For example, it has been noted that young normally developing English-speaking children acquire grammatical inflections more slowly than children acquiring languages such as Italian or Hebrew (e.g., Berman, 1986). English-speaking children with SLI show an even greater disadvantage in the area of grammatical inflections. In the speech of young normally developing Italian-speaking children, freestanding closed-class morphemes are not as well-developed as noun, verb, and adjective inflections (e.g., Caselli, Leonard, Volterra, & Campagnoli, 1993). In the case of children with SLI acquiring Italian, the gap between freestanding forms and inflections is even wider. In short, it appears that grammatical morphemes that are most fragile in a language are especially so for children with SLI. We do not find instances where features that are especially easy for normally developing children are especially difficult for children with SLI acquiring the same language.

It would seem that the findings reviewed here are consistent with genetic factors that influence more general processing mechanisms rather than some highly circumscribed area of the grammar. That is, the factors would seem to be those that limit children's ability to process and acquire details of the language within the typical time span; those details that are most fragile in terms of relative frequency, redundancy, regularity, perceptual salience, and pronounceability (see Peters, 1985; Slobin, 1973), would be the most vulnerable.

Support for the view that limitations in general processing mechanisms are at work can be found in a range of studies involving English. The grammatical profile observed in English-speaking children with SLI can be replicated in normal speakers during tasks in which linguistic material must be produced or comprehended under difficult listening conditions or when cognitive resources must be shared with another task (e.g., Kilborn, 1991). Connectionist studies produce the same profile when either degrading the input (Hoeffner & McClelland, 1993) or severing a proportion of the connections (Marchman, 1993). Locke (1993) has recently proposed a modular theory in which this type of profile is explained by an assumed neuromaturational delay (which can be genetic) that causes a delay in the acquisition of lexical material. According to this theory, when a presumed timelocked encapsulated grammatical analysis mechanism reaches its point of onset, there is insufficient lexical material to activate it. When enough lexical material is accumulated to permit activation, the analytical mechanism is nearing offset, resulting in a limited capacity and a shift of language functions to homologous but less efficient mechanisms in the other hemisphere.

In summary, what the literature seems to reveal is that aspects of a language that are already the most fragile due to any of a number of reasons are the first to show a significant drop under a variety of difficult conditions. Of course, languages differ in the aspects of language that are most fragile. The fact that children with SLI show grammatical profiles consistent with the input language with reductions in these areas of frailty, rather than some universal profile, suggests that it is not the underlying grammar itself that is directly responsible for this disorder.

REFERENCES

Aram, D., Ekelman, B., & Nation, J. (1984). Preschoolers with language disorders: 10 years later. *Journal of Speech and Hearing Research, 27,* 232–244.

Berman, R. (1986). A crosslinguistic perspective: Morphology and syntax. In P. Fletcher & M. Garman (Eds.), *Language acquisition* (2nd ed., pp. 429–447). Cambridge, England: Cambridge University Press.

Bishop, D. (1992). The underlying nature of specific language impairment. *Journal of Child Psychology and Psychiatry, 33*, 3–66.

Caselli, M. C., Leonard, L., Volterra, V., & Campagnoli, M. G. (1993). Toward mastery of Italian morphology: A cross-sectional study. *Journal of Child Language, 20*, 377–393.

de Villiers, J., & de Villiers, P. (1973). A cross-sectional study of the acquisition of grammatical morphemes in child speech. *Journal of Psycholinguistic Research, 2*, 267–278.

Dromi, E., Leonard, L., & Shteiman, M. (1993). The grammatical morphology of Hebrew-speaking children with specific language impairment: Some competing hypotheses. *Journal of Speech and Hearing Research, 36*, 760–771.

Eyer, J., & Leonard, L. (1995). Functional categories and specific language impairment: A case study. *Language Acquisition, 4*, 177–203.

Hoeffner, J., & McClelland, J. (1993, April). *Can a perceptual processing deficit explain the impairment of inflectional morphology in developmental dsyphasia? A computational investigation.* Paper presented at the Stanford Child Language Research Forum, Stanford, CA.

Jaeggli, O., & Safir, K. (1989). The null subject parameter and parametric theory. In O. Jaeggli & K. Safir (Eds.), *The null subject parameter* (pp. 1–44). Dordrecht, The Netherlands: Kluwer.

Johnston, J. (1988). Specific language disorders in the child. In N. Lass, J. Northern, L. McReynolds, & D. Yoder (Eds.), *Handbook of speech-language pathology and audiology* (pp. 685–715). Philadelphia: B. C. Decker.

Johnston, J. (1992). Cognitive abilities of language-impaired children. In P. Fletcher & D. Hall (Eds.), *Specific speech and language disorders in children* (pp. 105–116). London: Whurr.

Kilborn, K. (1991). Selective impairment of grammatical morphology due to induced stress in normal listeners: Implications for aphasia. *Brain and Language, 41*, 275–288.

Leonard, L. (1992). The use of morphology by children with specific language impairment: Evidence from three languages. In R. Chapman (Ed.), *Processes in language acquisition and disorders* (pp. 186–201). St. Louis: Mosby.

Leonard, L. (1994). Some problems facing accounts of morphological deficits in children with specific language impairment. In R. Watkins & M. Rice (Eds.), *Specific language impairments in children: Current directions in research and intervention* (pp. 91–105). Baltimore: Paul H. Brookes.

Leonard, L., Bortolini, U., Caselli, M. C., McGregor, K., & Sabbadini, L. (1992). Morphological deficits in children with specific language impairment: The status of features in the underlying grammar. *Language Acquisition, 2*, 151–179.

Leonard, L., Bortolini, U., Caselli, M. C., & Sabbadini, L. (1993). The use of articles by Italian-speaking children with specific language impairment. *Clinical Linguistics and Phonetics, 7*, 19–27.

Leonard, L., & Dromi, E. (1994). The use of Hebrew verb morphology by children with specific language impairment and children developing language normally. *First Language, 14*, 283–304.

Locke, J. (1993). *The child's path to spoken language.* Cambridge, MA: Harvard University Press.

Marchman, V. (1993). Constraints on plasticity in a connectionist model of the English past tense. *Journal of Cognitive Neuroscience, 5*, 215–234.

Oetting, J., & Rice, M. (1993). Plural acquisition in children with specific language impairment. *Journal of Speech and Hearing Research, 36*, 1236–1248.

Peters, A. (1985). Language segmentation: Operating principles for the perception and analysis of language. In D. Slobin (Ed.), *The crosslinguistic study of language acquisition. Vol. 2: Theoretical issues* (pp. 1029–1067). Hillsdale, NJ: Lawrence Erlbaum Associates.

Radford, A. (1990a). *Syntactic theory and the acquisition of English syntax.* Oxford, England: Basil Blackwell.

Radford, A. (1990b). The syntax of nominal arguments in early child English. *Language Acquisition*, *1*, 195–223.

Rice, M., & Oetting, J. (1993). Morphological deficits of children with SLI: Evaluation of number marking and agreement. *Journal of Speech and Hearing Research, 36*, 1249–1257.

Slobin D. (1973). Cognitive prerequisites for the development of grammar. In C. Ferguson & D. Slobin (Eds.), *Studies of child language development* (pp. 175–208). New York: Holt, Rinehart, & Winston.

Tallal, P., Ross, R., & Curtiss, S. (1989). Familial aggregation in specific language impairment. *Journal of Speech and Hearing Disorders, 54*, 167–173.

Tomblin, J. B. (1989). Familial concentration of developmental language impairment. *Journal of Speech and Hearing Disorders, 54*, 287–295.

Weiner, P. (1985). The value of follow-up studies. *Topics in Language Disorders, 5*, 78–92.

COMMENTARY ON
CHAPTER 9

Elena Plante
The University of Arizona

Leonard's work addresses the fundamental issue of what it means to have a specific language impairment (SLI). Although his work is linguistically driven, it has important implications for those who are interested in the disorder's biologically based brain-language interface. His data begin to sift the effects of a developmentally compromised brain from the independent effects of the child's spoken language. It is only by understanding the language-dependent effects that we will begin to understand what it means to say that the brain's language proficiency is altered in these children.

How strong is the language effect in the brain-language relation? In a cross-linguistic study of aphasia, Bates, Friederici, and Wulfeck (1987) stated, "We can predict more about an individual's performance [on the experimental task] if we know his native language than we can by knowing whether or not the individual has suffered focal brain damage" (p. 44). The native language effect was by far the strongest effect in this particular study, accounting for a remarkable 41% of the total variance. Leonard's data, likewise, support the importance of the role of native language on the presenting signs of developmentally impaired language of SLI. His data illustrate that morphosyntactic dissociations are evident across languages for SLI children, but the specifics of these dissociations can vary across languages. This promotes a reinterpretation of the presenting signs of the language disorder to de-emphasize the role of the impairment on specific linguistic structures to a greater emphasis on general limitations that impair structures that are already "fragile" for other reasons.

This argument is reminiscent of the reasoning behind the competition model of language acquisition and processing developed by Bates and MacWhinney (1987). In fact, there are several parallels between the predictions of the competition model and Leonard's data-based conclusions. The competition model is a general theory of language processing that has been used to explain language acquisition and language breakdown from a cross-linguistic perspective. The model predicts that the relative information load carried by the morphosyntactic, phonological, and lexical elements of a given language influence their acquisition during childhood and normal performance during adulthood. The more information conveyed by the linguistic elements, the earlier they are acquired and the more resistant they are to breakdown under stressed conditions. Conversely, elements that are redundant or weak in terms of their contribution to overall meaning tend to be more fragile. These are the elements predicted to be most affected under adverse processing conditions (e.g., degraded listening, general illness, aging, and neurological damage).

It is interesting, in light of the competition model, that the relative impact of SLI on morphology varies with the "morphological richness" of the language. In English, for which word order is a much stronger cue to sentence meaning than inflectional morphology, bound morphemes are hit relatively hard. In contrast, in Italian, for which bound morphemes carry substantially more of the information load, it is the lower information, freestanding articles that are harder hit.

How does one interpret the nature of an impairment that appears to affect the components of a given language that are already weakest? The idea of normal-but-delayed does not seem appropriate because these relative impairments of morphology are already couched within the context of a non-normal MLU-morphology dissociation; the children with SLI in Leonard's work are compared with younger MLU matches. However, it might be argued that the MLU-morphology associations themselves should not be assumed to be standard across languages. Therefore, the MLU-morphology dissociations that characterize SLI in English may not be indicative of non-normal processes in all languages.

The nature of an MLU match may also obscure smaller, but nonetheless illuminating, effects across languages. Specifically, the MLU matches promote only those effects associated with impaired language that exceed those associated with the age difference inherent in the MLU match (Plante, Swisher, Kiernan, & Restrepo, 1993). This is a problem for just the reason Leonard raises to explain conditions where SLI children outperform their younger language matches: Older children have cognitive resources that are unavailable to, or are underused by, younger children. This same situation can mitigate the degree of apparent impairment in SLI as well, so that no difference is found between children with SLI and MLU matches, but both perform below the normal age-matched subjects. I raise this issue because, if general processing difficulty is central to the expression of SLI, then the degree of impact on various linguistic elements ought to be graded with their relative difficulty, as predicted by a strong form of the competition model. The optimal comparison to examine graded effects is with age-matched normal language controls.

If we assume, however, that further study continues to bear out Leonard's insight that linguistic deficits may be a reflection of broader processing limitations, what then are the implications for the nature of the underlying brain-language relation? Must it be the case, as Leonard has suggested elsewhere (Leonard, 1991, this volume) that SLI is the result of normal variation that makes some children linguistically untalented? Leonard's data suggest a pattern of deficits that has more to do with linguistic vulnerability than grammatical character. Although this pattern parallels normal variation (e.g., later acquisition), it does not guarantee that the behavioral pattern reflects normal variation in the underlying brain-based mechanism.

To return to our example from adult aphasia, Bates et al. (1987) provide a striking illustration of how patient groups that are neurologically distinct can produce similar patterns of performance on linguistic tasks. In this example, several groups of Italian-speaking subjects were asked to act out sentences that contained a verb and two nouns. The response patterns seen in normal young adults were maintained in subject groups that included three types of aphasic patients and two hospitalized control groups. The sole difference between these various groups was in the degree of attenuation of the response pattern, not a qualitative change in the pattern. If we did not know that some of these groups of patients had suffered brain damage, we might conclude that their attenuated, but normal, response pattern reflected brain functioning that was poor, but generally normal. However, these data illustrate that poor, but generally normal, response patterns can occur in a pathologically damaged brain.

This example promotes an alternate interpretation of poor language skills that may apply both to acquired aphasia and SLI. Low, but essentially normal, patterns of performance do not guarantee that the underlying neurological system is normal. It might instead demonstrate the degree to which the brain tends to conserve those processing biases that are most useful even when its neurological condition has been compromised. If so, then SLI has the potential of serving a contrastive population to acquired aphasia to test the degree to which these brain-language principles are conserved within the constraints of a very different, but still suboptimal, neurological system. This is an intriguing possibility.

REFERENCES

Bates, E., Friederici, A., & Wulfeck, B. (1987). Comprehension in aphasia: A cross-linguistic study. *Brain and Language, 32,* 19–67.

Bates, E., & MacWhinney, B. (1987). Competition, variation, and language learning. In B. MacWhinney (Ed.), *Mechanisms of language acquisition* (pp. 157–193). Hillsdale, NJ: Lawrence Erlbaum Associates.

Leonard, L. (1991). Specific language impairment as a clinical category. *Language, Speech, and Hearing Services in Schools, 22*, 66–68.

Plante, E., Swisher, L., Kiernan, B., & Restrepo, M. A. (1993). Language matches: Illuminating or confounding? *Journal of Speech and Hearing Research, 36*, 772–776.

10

BUILDING THE CASE FOR IMPAIRMENT IN LINGUISTIC REPRESENTATION

Martha B. Crago
McGill University

Shanley E. M. Allen
Max-Planck-Institut für Psycholinguistik

Causal explanations for the morphological vulnerability observed in specific language impairment (SLI) have attributed the problem to (a) the lack of phonological salience of morphological inflections (Leonard, 1988, 1989), to (b) differential typologies of languages with sparse and complex morphology (Dromi, Leonard, & Shteiman, 1993; Leonard, Bortolini, Caselli, McGregor, & Sabbadini, 1992; Lindner & Johnston, 1992), as well as to (c) various deficiencies in the underlying grammar, including a featureless grammar (Gopnik 1990a, 1990b) and an impairment in rule construction (Gopnik, in press-a, in press-b), (d) the extended use of infinitives for inflected forms (Rice, 1993, 1994; Rice & Wexler, 1993), and to (e) problems in agreement checking relationships that differentially affect agreement across the grammar (Rice, 1993, 1994).

In this chapter, these various explanatory hypotheses for SLI are described and related to data on SLI in Inuktitut. None of the five hypotheses can fully account for the Inuktitut findings. In light of this, the overall enterprise of building the case for impairment in linguistic representation will be discussed with linguistic theory as well as with information from the literature on agrammatism in aphasia.

CHARACTERISTICS OF INUKTITUT

Inuktitut is the language of some 25,000 Inuit in northern Canada and is part of the Eskimo-Aleut family extending across the circumpolar regions from Siberia to Greenland. It exhibits a high degree of polysynthesis with prolific verbal and nominal inflections. This means that a nominal or verbal root is followed by from zero to eight morphemes corresponding to the Indo-European independent verbs, auxiliaries, deverbals, denominals, adverbials, adjectivals, and so on; then, an obligatory inflectional affix; and finally optional enclitics. In addition, there are over 1,000 verb- and noun-internal productive morphemes that serve as nominalizers, verbalizers, valency-changers, and modifiers.

Nominal inflection represents eight cases and three numbers, and the possessive paradigm encompasses four persons and three numbers. The following example shows how nominal elements may include a variety of modifiers suffixed to the root:

(1) *Quttukallakutaatsiaraapimmut.*
    qut-juq-kallaq-kutaaq-tsiaq-apik-mut
    be.funny-NOM-DIM-tall-nice-handsome-ALL.SG
    'To a nice tall handsome cute funny person.'

Verbal inflection agrees with both subject and object for four persons, three numbers, and 10 verbal modalities. Verbal elements typically show a greater degree of polysynthesis, as in this example:

(2) *Annuraarsimalukatsitipaujaaluumijuq.*
    annuraaq-sima-lukat-siti-paujaaluk-u-mi-juq
    clothe-PERF-unusually-well-very-be-also-PAR.3sS
    'She also often dresses up very unusually.'

Inuktitut also has an ergative case marking system, SOXV word order, and ellipsis of both subject and object. Furthermore, Inuktitut has no uninflected infinitival form.

In northern Quebec, Inuktitut is spoken by 95% of the population (Dorais, 1986). It is used on a routine, almost exclusive, basis by people of all ages in a number of the communities in this region.

DESCRIPTION OF THE STUDY

This study focused on one monolingual Inuk girl aged 5;4. This child, LE, lived in a settlement of approximately 350 people 1,000 miles north of Montreal. Inuktitut was the exclusive language of her everyday life. LE was selected for this case study from among several subjects on the basis of her Inuit parents', her Inuit teachers', the special education consultants', and healthcare workers' reports, one of the author's clinical impressions, and the clarity of her disorder. LE's clinical profile was congruent with the usual criteria for SLI (see Table 10.1). Despite a difficult start in her first adoptive family where she did not remain, this child was judged to be normal socially and emotionally with no frank signs

TABLE 10.1
Subject Characteristics

	Subject		
Characteristic	LE (SLI)	Language (MLU)	Age (CA)
Age	5;4	2;1	5;4
Intelligence	normal	normal	normal
Hearing	normal	normal	normal
Neurological indicators	none	none	none
Social/emotional status	normal	normal	normal
MLU	2.48	2.27	4.28
Utterances*	261	200	386

*Total utterances in 34 minutes (for SLI and CA only).

of neurological impairment, and was considered by virtue of her play and school achievement (she was in kindergarten at the time of the study) to be intellectually at the level of her age-matched peers. LE had hearing within the normal limits despite the fact that she had recurrent otitis media in the first year of her life. However, 25% of all children in northern Quebec have also had chronic otitis media (Julien, Baxter, Crago, Ilecki, & Therien, 1987). The number and severity of LE's episodes of otitis media were equivalent to those experienced by these peers whose language is not considered to be impaired. Of interest in the study is the fact that this child was a member of a large extended family in which there are more than three first-degree relatives with SLI spread over three generations, including a maternal uncle and three cousins.

No language tests are presently available for providing normative data in Inuktitut. Instead, a nonstandardized procedure was used to determine the nature and extent of LE's language abilities. LE was taped in her home where she was engaged in a free play situation with a chronologically age-matched normally developing friend who was four days older than LE. The session lasted approximately 1½ hours and was taped in its entirety by one of the authors of this paper. Of this hour and a half, two sections of approximately 20 minutes in length were selected for transcription based on the audibility of the utterances and representativeness of the overall sample. A sample matched for mean length of utterance (MLU) of similar length and taped in similiar circumstances was taken from data videotaped for a separate study on the acquisition of Inuktitut in normally developing Inuit children (Allen, 1994). MLUs for data presented in this chapter were based on counts of productively used morphemes. Frozen forms (e.g., *Merry Christmas*) and portemanteau morphemes (i.e., morphemes that encode more than one thing) were counted as only one morpheme. Additionally, nominals, some of which were composed of more than one morpheme were, nevertheless, counted as single morphemes. This was done because the children did not show full productivity of the various components that made up such synthesized lexical items.

Data from LE, from an MLU-matched normal, and from her age-matched normal friend were then transcribed by native speakers. Reliability was established by consensus verification between both the transcriber and a native Inuktitut-speaking colleague who was an expert in Inuktitut grammar and Inuktitut child language, and the two authors of this paper. All data were then entered into a computerized database following the CHAT conventions of the CHILDES project (MacWhinney & Snow, 1990) and were coded morphologically. The findings reported in this chapter are based on the first 200 utterances of each of the three children. Self-repetitions, exact imitations of others, exclamations, and routines were not included in the 200 utterances (see Appendix).

LEXICON

LE's lexicon was considerably less developed than her age-matched friend (see Table 10.2). It also differed somewhat from her MLU match. Although LE's type-token ratio for noun and verb roots was not very different than her MLU match, her age-matched playmate, on the other hand, had almost twice as many actual types and tokens of verb roots. The type-token ratio for all morphological units, including both inflections and roots, was quite similar for all three of the children. However, again, the age-matched friend had approximately twice as many actual types and tokens of morphemes as LE did. LE also showed two behaviors that neither her age match nor her MLU match showed. She experienced a number of word-finding difficulties evidenced by frequent use of *um* throughout sentences, and by making multiple attempts at different lexical items in her search for the right word or by attempting to imitate a word just spoken by her friend. Although these types of difficulties occur in normal speakers for a variety of extralinguistic reasons, lexical problems have been described as characteristic of children with developmental language impairment (Rescorla, 1989). LE also had an abnormally high frequency of the all purpose word,

TABLE 10.2
Type-token Ratios for Vocabulary Classes

	Subject		
Vocabulary class	SLI	MLU	CA
General vocabulary*	151/496	143/454	283/856
	(.304)	(.315)	(.331)
Verb roots	22/45	17/29	50/87
	(.489)	(.586)	(.575)
Noun roots	27/56	34/73	35/103
	(.487)	(.466)	(.301)
imaittumik ('thing like this')	12	0	0

*General vocabulary includes roots and inflections in all utterances.

imaittumik, meaning 'thing' or 'one like this' when she was unable to find the correct lexical item as in the following examples:

(3) *una au au uumunga au imaittumi*
u-na au au u-munga au imaittumik
this.one-ABS.SG um um this.one-ALL.SG um one.like.this
'this one um um with this one um one like this'

(4) *aulla aullangami imittumi*
aullaq aullaq-nnguaq-MI imaittumik
leave leave-pretend-MI one.like.this
'leave pretend to leave one like this'

(5) *qimmiralu qimmira vulli imittumi*
qimmiq-ga-lu qimmiq-ga vut-li imaittumik
dog-my-and dog-my our-and one.like.this
'and my dog, my dog, and our one like this'

Furthermore, in Inuktitut, limitations in the lexicon can represent either, or both, a lack of lexical items and a lack of mastery of synthesizing devices used to handle the morphological complexity necessary to construct certain lexical items. (See examples (1) and (2) in the previous section of this chapter for illustrations of such devices.)

GRAMMATICAL MORPHOLOGY

General Characteristics of LE's Grammatical Morphology

In several categories of grammatical morphology for which frequency counts were made, LE performed quite differently than her age match (see Table 10.3 and the Appendix). She had strikingly fewer tokens of verbal inflections that

TABLE 10.3
Number of Tokens of Certain Grammatical Morphology in 200 Utterances*

	Subject		
Grammatical morphology	*SLI*	*MLU*	*CA*
Verbal inflections	32	26	136
Verbal modifiers	30	42	144
Verbalizers	5	6	44
Nominal inflections	46	51	61
Nominalizers	4	11	11
Passives	0	2	9

*See Appendix for list of all morphemes used by the three children.

mark person, number, and modality. LE also had fewer tokens of verb-verb affixes, those morphemes that mark causative, passive and adverbials, and she used fewer verbalizers with which to change nouns into verbs. These kinds of morphemes are verb and noun internal. Despite such differences with her age-matched friend, LE's performance in these same grammatical categories was not much different than her MLU match if we consider only the absolute number of tokens. Nominal inflections and nominalizers showed a different pattern. LE had fewer instances of them than either her age or MLU match. It is interesting to note that LE and her MLU match had more nominal inflections than verbal inflections. The age-matched child had the reverse, that is, more verbal than nominal inflections.

Allen and Crago (in press) have previously documented the early and frequent use of passives in very young Inuit children. Normally developing 2- and 3-year-old Inuit children productively used 2.8 passives per hour in comparison with 0.4 passives per hour used by American English-speaking children of the same age (Pinker, Lebeaux, & Frost, 1987). The passive in Inuktitut is signaled by the use of the single syllable -jau-, which never occurs in a final position. Comparisons across the three children in this study showed that the 2-year-old MLU match used two passive constructions in his 200 utterances, and the 5-year-old age match used nine in her 200 utterances. LE, on the other hand, used no passives in her 200 utterances or, for that matter, in her entire 1½ hours of taped spontaneous speech. The lack of passives in LE's language was clearly unusual for Inuit children of either her age or MLU (Allen & Crago, in press).

Omitted Inflections

To explore LE's use of verbal inflections more fully, potential grammatical contexts for the use of inflections on verbs and locatives were identified. In Inuktitut, there are various optional ways that a verbal root can be followed. For instance, a word beginning with a verbal root can either take a verbal inflection immediately following the root or it can be passivized or nominalized or have a variety of other affixes added before the final inflection. Without knowing the speaker's intent, it is difficult to determine what obligatory form an utterance should take. For this reason, the term *potential* rather than *obligatory* has been used to refer to contexts for the use of particular verbal inflections. However, this does not mean that verbal inflections are optional. The optional element is what is added to the root in addition to the verbal inflection. LE used verbal inflections correctly in 46% of the potential contexts that were identified (see Table 10.4). This was less than her MLU match and greatly less than her age match whose inflections were correct nearly 100% of the time. LE's correct inflections included correct use of person and number. It is important to note, however, that spontaneous naturalistic elicitations, as opposed to experimentally designed elicitations, allow the speaker not to use certain grammatical forms that may be difficult.

TABLE 10.4
Number of Tokens of Verbal Inflection Used Correctly
or Incorrectly in 200 Utterances

	Subject		
Verbal/locative inflection	SLI	MLU	CA
Percentage use in potential situations	46%	59%	95%
	(32/69)	(26/44)	(136/142)
Correct uses	32	26	136
Incorrect omissions on verbal stems	10	5	3
Incorrect omissions on locative stems	8	2	0
Potential omissions on locative stems	5	11	3
Insertion of filler inflection -mi	10	0	0
Overt pronoun in place of inflection	4	0	0

Verbal Omissions. Ten of LE's incorrect representations of inflection consisted of complete omission of the inflection. These omissions included person and number inflections that were sometimes multisyllabic, such as -nama (CSV.1sS) or -junga (PAR.1sS), and sometimes stressed, as in the interrogative form. LE's MLU and age match omitted many fewer inflectional endings than she did.

Locative Omissions. The Inuktitut locative system is quite complex, having a variety of somewhat subtle distinctions of location in space relative to the speaker and different forms of reference for dynamic and static elements. Locatives require a verbalizer and then a verbal inflection in certain obligatory pragmatic and semantic contexts, but not in all contexts. For this reason, two separate counts of locative contexts were made, those in which inflections were clearly obligatory and those in which they were optional. LE had more omissions of obligatory inflectional endings on locatives than either her age or MLU match. The following is an example of this sort of omission:

(6) LE: *maani*
    ma-ani
    here-LOC
    'here'

which contrasts with correct constructions used by her age match,

(7) CA: *maaniittuq*
    ma-ani-it-juq
    here-LOC-be-PAR.3sS
    'it is here'

and

(8) CA: *maunngaruk*
    ma-unga-aq-gu
    here-ALL-go-IMP.2sS.3sO
    'put it here'

LE's use of the bare stem locative in obligatory contexts is highly irregular in Inuktitut.

Additional Irregularities in Grammatical Morphology

Two additional irregularities in LE's use of inflections were most surprising and noticeable to our Inuit colleagues. They were LE's use of the inflection *-mi* and her use of overt pronouns.

-Mi Insertions. On 10 different occasions, LE used the inflection *-mi* in a highly irregular way. She inserted *-mi* as a filler inflection on verbs and locatives where it does not normally occur. The inflection *-mi* is most frequently used on nouns as the singular locative ending ('in'), the singular modalis case ending, and occasionally on verbs as an internal morpheme meaning ('also'), but it can never appear on the ends of verbs or locatives as it does in LE's language. Examples of her irregular use of *-mi* and comparisons of correct use by her CA match are as follows:

(9) LE: *ummali pi- sininnguami*
    u-mma-li pi- sinik-nnguaq-MI
    this.one-ERG.SG-and pi sleep-pretend-MI
    'how about this pi- pretend to sleep -MI'
   CA: *imaa sinisijunga*
    imaak sinik-si-junga
    like.this sleep-PRES-PAR.1sS
    'I'm sleeping like this'
(10) LE: *unali maanimi*
    u-na-li ma-ani-MI
    this.one-ABS.SG-and here-LOC-MI
    'and this one right here -MI'
   CA: *unali maanituq*
    u-na-li ma-ani-it-juq
    this.one-ABS.SG-and here-LOC-be-PAR.3sS
    'and this one is here'

Neither LE's age- nor MLU-match ever made this type of error, and it is highly irregular for normally developing Inuktitut speakers of any age.

Overt Pronouns. Due to the prolific verbal agreement in Inuktitut, individual pronouns that stand alone, such as *ivvit* ('you') or *uvanga* ('I'), are typically reserved for extreme emphasis or responses to questions that require only the pronoun as an answer. However, LE produced four instances in her 200 utterances of individual pronouns in place of the appropriate verbal inflection. These are illustrated as follows:

(11) LE: *maani ivvit*
       ma-ani ivvit
       here-LOC you
       'here you'
   Correct form:
       maaniigit
       ma-ani-it-git
       here-LOC-be-IMP.2sS
       'be here'
(12) LE: *taku ivvit uumaa*
       taku ivvit u-mma
       see you this.one-ERG.SG
       'see you this one'
   Correct form:
       uumunga takujauvutit
       u-munga taku-jau-vutit
       this.one-ALL.SG see-PASS-IND.2sS
       'you were seen by this one'

The use of overt pronouns is so unusual at any age in normally developing Inuit children's language acquisition that it flabbergasted our Inuit colleagues and made a big enough impression on LE's family members that they affectionately teased her by calling her the nickname *ivvit–uvangaraapik* meaning 'adorable little you–me' or by mimicking the sentences in which she used overt pronouns. Her use of overt pronouns left bare verb stems. LE's use of the bare verb stem in this situation, as well as her use of the bare locative stem, essentially contradicted the basically polysynthetic nature of her language and was highly irregular in a pro-drop language, such as Inuktitut.

Outstanding Properties of LE's Language

In summary, then, this particular child's language problems were manifested in the following ways:

1. She had a restricted lexicon with evidence of word-finding difficulties.

2. Her grammatical morphology included (a) a paucity of verbal inflections on both verbs and locatives, (b) the irregular insertion of the largely nominal inflection -mi in the final position on verbs and locatives, (c) the irregular use of overt pronouns, (d) the existence of bare verbal and locative stems, and (e) a lack of passive constructions.

3. On the other hand, LE showed excellent interactional skills at play with her friend.

In conclusion, this description of a single child has been made with full awareness that specific language impairment has considerable heterogeneity in its expression (Korkman & Haakkenen-Rihu, 1994; Lahey, 1988).

BUILDING THE CASE WITH LINGUISTIC THEORY

In this section of the chapter, the Inuktitut data will be related to various explanatory hypotheses and linguistic theories pertinent to specific language impairment.

Surface Hypothesis

Leonard in his surface hypothesis (Leonard, 1988, 1989) claimed that individuals with SLI have difficulties with grammatical morphology that are directly related to the phonological salience of the particular morpheme or morphemes in question. Leonard defined phonological salience in terms of relative stress, relative position within the word or utterance, relative vowel duration, and relative vulnerability to syllable or consonant deletion in production. He claimed that morphemes that are low in phonological salience are more vulnerable to difficulty for SLI subjects, but morphemes that are high in phonological salience are less vulnerable.

The language learning of normally developing English-speaking children reflected difficulty with these unsalient morphemes in terms of the children's relative tardiness in the acquisition sequence (deVilliers & deVilliers, 1985). English-speaking SLI children seemed to have even more difficulty with these morphemes as compared to unimpaired children. However, Leonard and his colleagues claimed in their crosslinguistic research that individuals with SLI experience significantly less difficulty with morphological inflections in languages in which the morphological elements are more phonologically salient.

In Inuktitut, verbal and nominal agreement morphemes are all word final (with the exception of those followed by occasional enclitics) and can be monosyllabic (CV or CVC), bisyllabic (CVCV, CVCVC, CVCCV, CVCCVC), or trisyllabic (CVCVCV), but are never purely consonantal. Utterance-final syllables are often lengthened, particularly in questions. Word-final consonant deletion is quite common in Inuktitut, but deletion of word-final inflectional units is limited to a constrained set of situations in colloquial speech.

Results from the production data in this case study are not consistent with the surface hypothesis. LE omitted more of what might be called *salient* morphological inflections, ones that are multisyllabic and stressed, than her MLU match did. She also had no instances of consonant-only deletions, with all omissions being complete inflectional units. Furthermore, her use of overt pronouns and the insertion of an irregular inflection like *-mi* are not congruent with this explanation because LE added the morpheme *-mi* rather than dropping it as the surface hypothesis would predict.

Sparse Morphology Hypothesis

The sparse morphology hypothesis (Dromi, Leonard, & Shteiman, 1993; Leonard et al., 1992; Lindner & Johnston, 1992) was based on the idea that children pay most attention in their language-learning process to the structural mechanisms that convey the most useful information. These mechanisms include grammatical morphology, word order, and animacy among others. Forms that have the most communicative relevance will be acquired first, with others following in more or less descending order (Lindner & Johnston, 1992; MacWhinney, Bates, & Kliegl, 1984). Thus, both unimpaired and impaired children learning languages in which grammatical morphology carries relatively little information are later in learning or have more difficulty with this morphology (Brown, 1973; Leonard et al., 1992) whereas children learning languages, such as Italian in which grammatical morphology plays an important role, are relatively earlier and more proficient in their learning of the morphology (Hyams, 1986; Leonard et al., 1992).

Inuktitut is a language in which word order, being variable, conveys relatively little information. Animacy also plays little role. Grammatical morphology, however, is extremely prolific and plays a large role in determining structural relationships in Inuktitut.

In LE's production data, her utterances differed from those of both her MLU match and her CA match in her omissions of inflections of various kinds, a number of which result in bare stem forms that are ungrammatical in Inuktitut. These results cannot be accounted for by the sparse morphology hypothesis.

Missing Features and Impaired Morphological Rule Construction Hypotheses

Missing Features. In the missing features hypothesis, Gopnik (1990a, 1990b) claimed that SLI grammar is unusual in that it is missing the notion of obligatory marking of grammatical features. This theory also claims that such grammatical features are not represented underlyingly in Universal Grammar in conjunction with the rule-governed behavior of unimpaired speakers. These features include number, gender, animacy, mass/count, tense, and aspect. This claim does not entail that grammatical features will never be marked but, rather, that their marking will be perceived as totally optional by SLI individuals. Thus, marked forms will be interspersed with unmarked forms in their speech, and they will not be reliable in judging the absence of these morphemes in grammaticality judgment tasks. This theory was perhaps wrongly named and might have been more accurately represented as a theory of missing markings.

Inuktitut has a large number of grammatical features that must obligatorily be marked morphologically. Thus, it is a candidate language in which such a hypothesis may be observed.

Production data from LE showed a pattern of optional use of grammatical features in that the same verbal inflection appears in some obligatory contexts and not in others. Thus, although this hypothesis seems to be descriptively adequate, it is not sufficient to predict or explain deficits in LE's grammar, such as -*mi* insertion, use of overt pronouns, and lexical searching.

Impaired Morphological Rules. Gopnik's more recent hypothesis (Gopnik, 1994), the missing rule hypothesis, stated that the ability to construct implicit rules is impaired in individuals with SLI. This hypothesis postulated that such individuals are able to compensate for this kind of deficit by learning the forms in question as unanalyzed lexical items and by using explicitly learned rules. This prediction meant that individuals with SLI will not recognize that inflectional markings are obligatory and that they will have problems with producing the correctly inflected forms of nonsense words. These inabilities should be reflected in their on-line processing. Furthermore, Gopnik's hypothesis predicted that characteristic errors in the misapplication of explicit rules will occur. For instance, forms that encode conceptually tangible information, such as plural, will be easy to learn lexically, but forms that mark less conceptually tangible information, like agreement, will be much more difficult to learn. It is, therefore, hypothesized that when the semantic information carried by these morphological markers is not obligatorily represented, individuals with SLI are likely to use explicit words to carry the important semantic meanings. In this way, the missing rule hypothesis would account for LE's use of overt pronouns to mark person, her pattern of lexical searching, and her omission of obligatory inflections on both nouns and verbs. It does not, however, account for her -*mi* insertion.

Optional Infinitive Hypothesis

The optional infinitive hypothesis (Rice, 1993, 1994; Rice & Wexler, 1993) was developed by Wexler (1992; this volume) to explain normal first language acquisition. It arises out of the observation that very young children typically pass through a stage in which verbal inflection is not marked consistently (Pierce, 1992; Poeppel & Wexler, 1993; Verrips & Weissenborn, 1992). This is linked to a related set of phenomena, including placement of negation and presence of overt subjects in the utterance. Wexler claimed that these children pass through a period in which they do not recognize the obligatoriness of verbal inflection and optionally permit the presence of infinitive forms in place of finite inflected forms. Wexler links this behaviour with these children's lack of awareness that formal features are required for the expression of tense. Because they do not recognize the difference between finite and nonfinite tense, children consider both these verbal forms interchangeable. The crucial point here is that children are not neglecting to mark verbal inflection and, thus, leaving a bare root but, rather, they are substituting an infinitive form in place of a finite form. Because in English the bare root and the infinitive are homophonous, Wexler provided data from French, German, Dutch, Swedish, Danish, and Norwegian that show the infinitive and inflected forms being interchanged by children. He then showed how a similar analysis, using the same tests, easily extends to English. Depending on the language involved, children tend to pass out of this stage by about 2;6, once they realize the features of tense and, thus, realize the obligatoriness of reflecting tense in verbal inflection.

Rice (this volume) and Rice & Wexler (1993) have adapted this analysis as an explanation for difficulty with tense marking in SLI individuals. Their idea was that the optional infinitives stage in these individuals extends significantly beyond the normal range for acquisition. Individuals with SLI, then, continue to use infinitival and inflected forms optionally well into childhood.

Two problems seem evident for the optional infinitive hypothesis in relationship to Inuktitut. First, tense is not implicated in the Inuktitut verbal inflection system, as argued in some detail in Shaer (1990). This is supported by at least two pieces of evidence. First, Fortescue (1984), among others, has pointed out that "[u]nmarked [for tense] indicative verb forms may be interpreted as either past or present ... depending on the stem and the context" (p. 272). Thus, it does not seem that the feature [+/- tense] is obligatorily represented within the inflection. Second, the representation of tense or, rather, time on the verbal stem is completely independent of the presence of person and number inflection on the verb. Time is represented by verbal adverbs, some even interpreted as verbs, which are affixed to the verb stem. Because Inuktitut verbal inflection is not tied to tense, any hypothesis of unimpaired or impaired acquisition based on the role of the feature [+/- tense] within the grammar cannot apply to the acquisition of Inuktitut verbal inflection.

Second, and probably closely tied to the first reason, Inuktitut has no infinitive form. Inuktitut does have a gerundive form that appears in both intransitive and transitive conjugations with inflection only for object, but it does not have a clear infinitival form. Thus, there is no possibility for Inuit children to go through a stage of optional infinitives, because there is no infinitive. This means that there are two possibilities. Either children would optionally produce a bare root, or they would produce a citation form, such as third person singular, in place of the correctly inflected form. Bare roots are ungrammatical in adult Inuktitut.

In the spontaneous speech of both LE and normally developing Inuit children (Crago, Allen, & Hough-Eyamie, in press), the bare root was sometimes produced in place of a correctly inflected form. The normally developing Inuit children did this primarily at the one-word stage. At the two-word stage, their utterances were already considerably more inflected than those of English-speaking children at a similar stage of language development. Preliminary analyses of normally developing Inuit children's data do show some substitutions of incorrect inflections for correct inflections. These are likely to represent performance errors because the substitution pattern seems to be random, rather than showing consistent replacement of a variety of inflections with one citation form. In the impaired data, however, no cases of substitution were observed apart from the incorrect use of the -mi inflection on verbal stems.

The resulting SLI pattern of omission of inflection (see Table 10.4), resulting in ungrammatical bare roots, also seems somewhat surprising in light of recent crosslinguistic research from agrammatism that claims that inflectional substitution is always observed in cases in which omission would produce an ungrammatical form (Menn & Obler, 1990). Obviously, the present Inuktitut data sample is small and restricted to only one subject. However, preliminary findings point to the need for further research into this hypothesis in languages, such as Inuktitut, that handle tense and inflection differently from the Romance and Germanic languages addressed by Wexler. It is important to note that the Inuktitut findings do not directly contradict the optional infinitives hypothesis. Simply put, this hypothesis cannot be the full explanation across all languages because certain languages do not have the infinitive form.

Differential Agreement Checking Relationships Hypothesis

The differential agreement checking hypothesis (Rice, 1993, 1994; Rice & Wexler, 1993) is based on the differing types of agreement relationships and methods of representing and checking them in current versions of the Principles and Parameters approach (Chomsky, 1992). Basically, two types of agreement-checking relationships are currently expounded within Principles and Parameters literature. The first is a SPEC-HEAD relationship. A typical verbal clause is generated as in the tree in Fig. 10.1. The subject is base-generated in SPEC,VP and subsequently moved to SPEC,AGR-S. The object is base-generated in NP,VP and subsequently moved to SPEC,AGR-O. The verb, fully inflected, is base-generated in V,VP and moved sequentially through all the heads up the tree. In each head position, it checks the features in its inflection with the relevant features of the element in SPEC related to that head. If the checking reveals a nonmatch, the derivation crashes, and the sentence is not uttered. If the checking reveals a

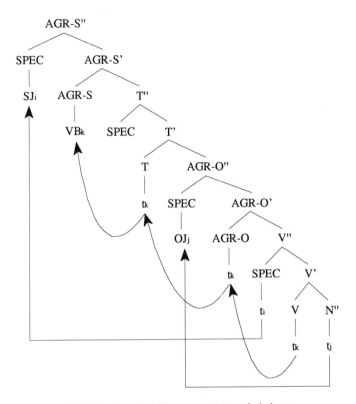

FIG. 10.1. SPEC–HEAD agreement in verbal clause.

match, the verb moves up to the next head. Because the checking of features occurs between items in a SPEC and in a head, this form of checking is known as SPEC-HEAD agreement. It is generally assumed that SPEC-HEAD agreement is invoked for verbal agreement and for agreement between quantifiers and nouns (Bittner, 1994; Branigan, 1992; Johns, 1992; Murasugi, 1992).

The second form of checking is known as HEAD-HEAD agreement. Consider the tree in Fig. 10.2. In this situation, it is assumed that the primary head, here the noun, is base-generated fully inflected. Additionally, the relevant features are marked in the head positions of the respective phrases up the tree. In this case, the entire NP moves up through the SPECs of the respective phrases. In each position, the head N checks its features against those of the respective head of the phrase. If the check reveals a nonmatch, the construction crashes and is not uttered. If the check reveals a match, the NP moves to the next SPEC. Because checking in this system is between two heads, it is referred to as HEAD-HEAD agreement. It is generally assumed that HEAD-HEAD agreement is invoked for agreement in the nominal system (e.g., Ritter, 1992; Travis, 1992; Valois, 1991).

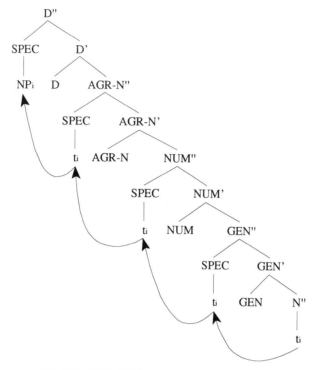

FIG. 10.2. HEAD–HEAD agreement in nominal phrase.

Rice (1993, 1994) had applied this notion to explain the parts of the grammar at risk for SLI individuals. Rice derived her evidence from the observation that English-speaking individuals with SLI perform much better on determiner-noun agreement than on either quantifier-noun agreement or verb-noun agreement (Oetting & Rice, 1993; Rice, 1993, 1994; Rice & Oetting, 1993). This is quite consistent with the different patterns of agreement checking, because the former is checked by HEAD-HEAD agreement and the latter two by SPEC-HEAD agreement. Thus, the claim is that SLI subjects have particular difficulty with SPEC-HEAD agreement relationships.

Rice's differential agreement checking hypothesis is most intriguing in that it represents a creative approach to the explanation of SLI within Chomsky's Minimalist Program (Chomsky, 1992). However, one must beware that these theories of agreement are relatively new, still hotly debated within the theory, and not even wholly accepted by their proponents. In addition, it is not yet clear how these configurations work out across languages, and it is not yet clear which phrasal categories appear in which languages. For instance, in some languages, negation is treated as a phrasal category and, in others, solely as a head without a phrasal projection (de Freitas, 1993). There are currently at least four ideas as to the construction of Inuktitut within the Minimalist Program (Bittner, 1994;

Bobaljik, 1992; Johns, 1992; Murasugi, 1992), only one of which uses the notion of agreement checking in the way that Rice does. So, although an agreement relationship analysis seems an interesting explanation for English data, it is not clear that it can be directly applied crosslinguistically at this early stage. A further caution is posed by new analyses of Rice and Wexler (in progress) reported elsewhere in this volume. In these analyses, there is evidence that SLI children do show agreement checking in the VP as well as in the *be* and *do* systems of English. This evidence suggests that the apparent difficulty with the English third person singular present tense marker, -*s*, may be attributable to the tense feature it carries instead of to agreement.

Putting these theoretical and empirical issues aside, there are a variety of issues in Inuktitut that would be interesting to address in association with Rice's hypothesis. First, Inuktitut does not have overt determiners, such as articles or demonstratives.[1] Although it would not be surprising for there to be agreement features in the head of DP position, the agreement between these and the NP cannot be checked in transcripts because they are not overtly represented. Second, nominal ellipsis in Inuktitut for both subjects and objects is extremely common. This makes it difficult to check for noun-verb agreement in the transcripts in other than a discourse sense. Third, case is a very important feature of Inuktitut although it is not in English. Thus, a theory of agreement marking that would serve crosslinguistically must include reference to case agreement. Because case is presumed to be checked within the verbal paradigm in a SPEC-HEAD configuration, we would predict substantial difficulty with case marking on behalf of SLI children. Fourth, Inuktitut is an ergative language, different than the accusative languages that have been addressed so far in SLI literature. It is quite likely that the agreement marking in ergative languages is mediated somewhat differently than in accusative languages, and it would be important to address this. Fifth, Inuktitut inflects on the verb for both subject and object, unlike the subject-only agreement present in the languages in which SLI has been studied so far. Therefore, it would be interesting to see if these two agreements were somehow dissociated in SLI production in Inuktitut, something that has not been attested to by the present data. Despite the fact that the differential agreement hypothesis does not yet account for these varying aspects of grammar related to Inuktitut, it appears to have promise for capturing certain essential features of SLI.

Summary

In this section of the chapter, we focus on various theoretical analyses of SLI data in light of linguistic theory and with reference to Inuktitut data. The primary implication, resulting from this discussion, concerns the importance of continu-

---

[1]Demonstratives occur in Inuktitut, but only as full NP pronominals, never as modifiers in determiner position.

ing crosslinguistic research in SLI. Crosslinguistic research has been of substantial importance in research on language acquisition of normally developing children over the last two decades (e.g., Slobin, 1985, 1992, and the vast body of literature cited therein). Theories derived from one language or typological group have been substantially broadened and enhanced by undergoing application to other languages or typological groups. Wexler's (1991) optional infinitive hypothesis is but one example.

In the field of SLI research, crosslinguistic work has been undertaken primarily in the area of phonological and stress saliency by Leonard and his colleagues (1992) in their work on Italian and Hebrew, with most interesting results. Discussion in this section of our chapter has alluded to some of the interesting issues in crosslinguistic SLI research that can be addressed in terms of morphological and syntactic concerns across languages, and to the potential significance of such research, as has been seen in the work of Clahsen (1989) in German, Dalalakis (1994) in Greek, Fukuda and Fukuda (1994) in Japanese, and the forthcoming work of Rice and LeNormand in French. Work in languages of increasingly diverse structures, such as Inuktitut, will continue to aid in fitting together pieces of the puzzle of specific language impairment and provide evidence for theories of normal language as well.

BUILDING THE CASE WITH INFORMATION FROM RESEARCH ON AGRAMMATISM

In the previous section of this chapter, the various hypotheses and linguistic theories used to explain deficits in the grammatical morphology of individuals with SLI were reviewed. There is an uncanny resemblance between them and certain hypotheses and theories that have been used to explore the nature of agrammatism in aphasia. There are a number of similarities as well as certain differences between the way that the case for problems in linguistic representation has been built for agrammatism and the present efforts to build such a case for specific language impairment.

In the 1970s, Goodglass conducted research into his version of the surface hypothesis by investigating the relationship of phonological, stress, and other prosodic patterns with difficulties in morphology. He described what he referred to as a new concept, that of saliency. According to his research (Goodglass, 1973), patterns of stress predicted, among other things, deletion of unstressed initial function words. However, he also noted as early as 1958 (Goodglass & Berko, 1960; Goodglass & Hunt, 1958) that performance with the same phonological form, namely /s/, varied according to its grammatical role. In specific, he found that individuals with agrammatism had more difficulty with the possessive -s and the third person singular -s than they did with the plural -s. A parallel finding for children with SLI has been recently reported by Rice and Oetting (1993).

Researchers into agrammatism have also pursued crosslinguistic investigations into languages with more complex morphology than English (Bates & Wulfeck, 1989). Such studies in Hebrew (Druks & Marshall, 1991; Grodzinsky, 1984) and Italian (Miceli & Mazzuchi, 1990; Miceli, Silveri, Romani, & Caramazza, 1989) were less motivated by questions of how different linguistic typologies related to a surface theory of morphological impairment than they have been in SLI research. Instead these studies were motivated by questions of how omissions of verbal inflections related to the optional infinitive. In fact, agrammatic findings for Hebrew and Italian are quite interesting in view of the work of Leonard and his colleagues (Dromi et al., 1993; Leonard et al., 1992) on those same languages and this present work on Inuktitut. The findings for agrammatism show patterns of substitutions rather than omissions. In other words, fully inflected forms were used, but used incorrectly for the syntactic context, in somewhat the same way that LE used the -*mi* form. Such findings underline the importance of reporting the nature of errors in crosslinguistic studies. The Italian agrammatic studies also show a dissociation between the use of bound and free morphemes.

Parallel to efforts in SLI, agrammatism research has attempted to explain the nature of the morphological deficit from within a grammatical framework, including a government-binding (GB) framework. This work by Grodzinsky (1984) centered around issues of antecedent-trace relationships.

Given the parallelism of these explanations for agrammatism and SLI, it is informative to track other issues and debates that have emerged in research on agrammatism that have relevance to building the case for impairment in linguistic representation. For instance, the heterogeneity of performance by individuals with agrammatism has led to claims that no single unitary account can predict the nature of this disorder. This, in turn, has led Caramazza and his colleagues (Badecker & Caramazza, 1985; Caramazza, 1986; McCloskey & Caramazza, 1988) to espouse the importance of the case study method. Use of this method was motivated, in part, by the early elaboration of criteria for the classical syndromes of Broca's and Werneicke's aphasia. Such criteria have since become targets for findings from individuals that do not fit into them. In SLI, where criteria have been largely exclusionary and less polarized into specific syndromes, there has been less motivation to find cases that are misfits to the overall category. However, an important query for research into the genetic bases of language impairment is whether the nature of the neurological deficits underlying the language impairment will have a more uniform effect than has been shown for acquired lesions.

Other research into agrammatism has explored issues associated with heterogeneity of performance, issues that we have just begun to touch on in SLI research. One of these is the notion of adaptation or compensating strategies. Research and theory in this area (Byng, 1988; Gandour, Marshall, & Windsor, 1989; Kolk & Van Grunsven, 1985) have been inconclusive, but emphasize the importance of analyzing the nature of errors and applying learning paradigms to the study of compensating strategies (Seidenberg, 1988; Shallice, 1988). The definition of phenotypes will need to account for adaptive strategies and their consequences in the language performance of SLI. Considering these sources of heterogeneity in a clinical population, certain researchers in the area of aphasia have argued for the concept of symptom complexes to replace more rigid clinical syndromes (Caplan, 1991) and for statistical procedures that rely on regression analysis and models of best fit like Bates' mean likelihood estimate (Bates, McDonald, MacWhinney, & Appelbaum, 1991). Furthermore, reports of group studies of individuals with agrammatism now feature individual subject scores as well as theoretical accounts for what may cause such variability. Such accounts and accountability for SLI will be meaningful to the construction of phenotypes.

Finally, research into agrammatism has featured both the assessment of comprehension as well as grammaticality judgement tasks (Bates, Friederici, & Wulfeck, 1987; Caplan & Hildebrandt, 1988; Linebarger, Schwartz, & Saffran, 1983; Saffran, Schwartz, & Marin, 1980; Shankweiler, Crain, Gorrel, & Tuller, 1989). In aphasia, comprehension studies have been motivated by the classical syndrome approach in which individuals with Broca's aphasia were not expected to have comprehension difficulties. In SLI, because hypotheses about the relationship of comprehension and production underlie a number of clinical assessment tools, research criteria, and clinical categories, investigation into the comprehension-production link have ongoing importance. Furthermore, grammaticality judgements and real-time processing approaches to the study of agrammatism have yielded very crucial information in the understanding of that disorder. In brief, despite the difficulty with inflectional morphology revealed in the production data on agrammatism, individuals in this population have been shown to have a preserved ability to make grammaticality judgements and to show differential real-time processing effects for grammatical and ungrammatical sentences. These findings have provided information that has opened up the possibility for changing the characterization of agrammatism from one of impairment in linguistic representation to one of impairment in accessing the linguistic representation. Moreover, such findings highlight the importance of building the case for impairment in linguistic representation by using this particular experimental paradigm.

Summary

As investigations of SLI continue, it is important that research includes converging information from comprehension, grammaticality judgements, and real-time tasks. Both for the purposes of crosslinguistic investigation and for understanding adaptation strategies, studies also need to include error analysis. Furthermore, the heterogeneity of the disorder can be represented in both case studies and by reporting individual data within group studies. Individual variation in well-selected subjects can also be accounted for in a theoretically principled way (e.g., Waters, Rochon, & Caplan, 1992) and group tendencies explored through more varied statistical approaches to the data analysis. Case studies, for their part, need to include converging and ample evidence. Finally, both individual variation and symptom complexes should be reflected in phenotypic profiles. In short, there is a case for including not only other languages and normal acquisitional theory into studies of SLI, but also for including issues from adult impairment literature into our collective thought processes and research on the genetic basis for language.

LIST OF ABBREVIATIONS

Nominal Case

ABL	ablative
ABS	absolutive
ALL	allative
EQU	equalis
ERG	ergative
LOC	locative
MOD	modalis
VIA	vialis

Nominal Inflection (e.g. ERG.SG)

SG	singular
DU	dual
PL	plural

Possessed Nominal Inflection (e.g. ERG.1Ssg)

1	first person possessor
2	second person possessor
3	third person possessor (disjoint with referent in main clause)
4	fourth person possessor (coreferent with referent in main clause)
S	singular possessor
P	plural possessor
sg	singular possessum
pl	plural possessum

Verbal Modality

CND	conditional
CSV	causative
CTM	contemporative
ICM	incontemporative
IMP	imperative (including optative)
IND	indicative
INT	interrogative

PAR   participial (functionally equivalent to indicative in northern Quebec)

Verbal Inflection (e.g. CSV.3sS.2pO)

1	first person
2	second person
3	third person (disjoint with referent in main clause)
4	fourth person (coreferent with referent in main clause
X	person variable
s	singular
d	dual
p	plural
x	number variable
S	subject
O	object

Word-Internal Morphology

ANTP	antipassive
CAUS	causative affix
DIFF.SUBJ	different subject
DIM	diminutive
EXCL	exclamation
FUT	future
HAB	habitually
INTR	intransitivizer
NEG	negative
NOM	nominalizer
PASS	passive
PAST	past
PERF	perfective
POL	politeness affix
PRES	present
TR	transitivizer

APPENDIX

GRAMMATICAL MORPHEMES USED

LE	MLU	CA
Demonstrative Inflections		
-*a* 'ABS.DUPL'	-*a* 'ABS.DUPL'	-*a* 'ABS.DUPL'
-*ma* 'ERG.SG'	-*ma* 'ERG.SG'	
		-*minga* 'MOD.SG'
-*munga* 'ALL.SG'	-*munga* 'ALL.SG'	
-*na* 'ABS.SG'	-*na* 'ABS.SG'	-*na* 'ABS.SG'

(Continued)

LE	MLU	CA
Locative Inflections		
-ani 'LOC'	*-ani* 'LOC'	*-ani* 'LOC'
-unnga 'ALL'	*-unnga* 'ALL'	*-unnga* 'ALL'
-uuna 'VIA'		*-uuna* 'VIA'
Nominal Inflections		
-ga 'ABS.1Ssg'	*-ga* 'ABS.1Ssg'	*-ga* 'ABS.1Ssg'
		-ganik 'MOD.1Ssg'
-it 'ABS.2Ssg'	*-it* 'ABS.2Ssg'	*-it* 'ABS.2Ssg'
-it 'ABS.PL'	*-it* 'ABS.PL'	*-it* 'ABS.PL'
-kka 'ABS.1Spl'		
-kkut 'VIA.SG'	*-kkut* 'VIA.SG'	*-kkut* 'VIA.SG'
-mi 'LOC.SG'	*-mi* 'LOC.SG'	*-mi* 'LOC.SG'
-mik 'MOD.SG'	*-mik* 'MOD.SG'	*-mik* 'MOD.SG'
	-minut 'ALL.4Ssg'	*-minut* 'ALL.4Ssg'
	-mit 'ABL.SG'	*-mit* 'ABL.SG'
		-mma 'ERG.1Ssg'
-mut 'ALL.SG'	*-mut* 'ALL.SG'	*-mut* 'ALL.SG'
-nga 'ABS.3Ssg'	*-nga* 'ABS.3Ssg'	*-nga* 'ABS.3Ssg'
		-ngagut 'VIA.3Ssg'
-ngani 'LOC.3Ssg'	*-ngani* 'LOC.3Ssg'	*-ngani* 'LOC.3Ssg'
-nganut 'ALL.3Ssg'	*-nganut* 'ALL.3Ssg'	*-nganut* 'ALL.3Ssg'
	-ni 'LOC.PL'	
		-nik 'MOD.PL'
		-nnut 'ALL.2Ssg'
-nut 'ALL.PL'	*-nut* 'ALL.PL'	*-nut* 'ALL.PL'
		-si 'ABS.2Psg'
		-tit 'ABS.2Spl'
		-titut 'EQU.SG'
	-up 'ERG.SG'	
-vut 'ABS.1Psg'	*-vut* 'ABS.1Psg'	*-vut* 'ABS.1Psg'
Verbal Inflections		
-gakku 'CSV.1sS.3sO'		
-gama 'CSV.1sS'		*-gama* 'CSV.1sS'
		-gani 'CTM&NEG.4sS'
		-gannuk 'CSV.1dS'
		-gatta 'CSV.1pS'
		-gattik 'CSV.2dS'
-git 'IMP.2sS'	*-git* 'IMP.2sS'	*-git* 'IMP.2sS'
	-guk 'IMP.2sS.3sO'	*-guk* 'IMP.2sS.3sO'
		-jagit 'PAR1sS.2sO'
	-jait 'PAR.2sS.3sO'	*-jait* 'PAR.2sS.3sO'
		-jara 'PAR.1sS.3sO'
		-jasi 'PAR.2pS.3sO'
		-javuk 'PAR.1dS.3sO'
		-javut 'PAR.1pS.3sO'
		-jugut 'PAR.1pS'

(Continued)

LE	MLU	CA
-junga 'PAR.1sS'	*-junga* 'PAR.1sS'	*-junga* 'PAR.1sS'
-juq 'PAR.3sS'	*-juq* 'PAR.3sS'	*-juq* 'PAR.3sS'
		-jusi 'PAR.2pS'
		-kkit 'IMP.2sS.3sO'
		-lagit 'IMP.1sS.2sO'
-lagu 'IMP.1sS.3sO'		*-lagu* 'IMP.1sS.3sO'
		-lavut 'IMP.1pS.3sO'
-li 'IMP.3sS'		*-li* 'IMP.3sS'
		-lit 'IMP.3pS'
		-lugu 'ICM.XsS.3sO'
	-luk 'IMP.1dS'	*-luk* 'IMP.1dS'
	-lunga 'ICM.1sS'	*-lunga* 'ICM.1sS'
		-luni 'ICM.4sS'
		-lunuk 'ICM.1dS'
		-lusi 'ICM.3pS'
		-luta 'ICM.1pS'
		-lutit 'ICM.2sS'
	-mmat 'CSV.3sS'	*-mmat* 'CSV.3sS'
		-mmata 'CSV.3pS'
	-nnga 'IMP.2sS.1sO'	
	-paaq 'EXCL.1sS'	*-paaq* 'EXCL.1sS'
		-ppat 'CND.3sS'
-ta 'IMP.1pS'		*-ta* 'IMP.1pS'
		-tsunga 'CTM.1sS'
		-tsuni 'CTM.4sS'
		-tsuta 'CTM.1pS'
		-tsutit 'CTM.2sS'
	-va 'INT.3sS'	
-visi 'INT.2pS'		
-vit- 'INT.2sS'		*-vit* 'INT.2sS'
	-vuguk 'IND.1dS'	
-vunga 'IND.1sS'		*-vunga* 'IND.1sS'
-vuq 'IND.3sS'		

Noun–Noun Morphemes

LE	MLU	CA
		-kutaaq- 'long'
-lik- 'item having'		*-lik-* 'item having'
		-limaaq- 'all of'
		-ngaaq- 'other'
-nnguaq- 'imitation'	*-nnguaq-* 'imitation'	*-nnguaq-* 'imitation'
	-qauti- 'item containing'	
		-siuti- 'item for'
		-suuq- 'HAB'
-tsiaq- 'nice'		
	-viniq- 'former'	*-viniq-* 'former'

(Continued)

LE	MLU	CA
Verb–Verb Morphemes		
		-gi- 'TR'
		-gialik- 'must'
		-giallak- 'a bit'
	-giaqtuq 'in order to'	*-giaqtuq* 'in order to'
-guma- 'want'	*-guma-* 'want'	*-guma-* 'want'
		-gunnaq- 'can'
		-it- 'NEG'
		-jaaq- 'resemble'
		-jaq- 'often'
	-jau- 'PASS'	*-jau-* 'PASS'
		-ji- 'ANTP'
		-juri- 'think'
		-kainnaq- 'PAST'
	-kainnaq- 'for a while'	*-kainnaq-* 'for a while'
-laaq- 'FUT'		
		-langa- 'FUT'
		-lauju- 'PAST'
-laukat- 'for a while'		*-laukat-* 'for a while'
-lauq- 'POL'	*-lauq-* 'POL'	*-lauq-* 'POL'
-lauq- 'PAST'		
-linga- 'PERF'		*-linga-* 'PERF'
-liq- 'POL'		*-liq-* 'POL'
		-liq- 'PRES'
		-luuq- 'do'
		-ma- 'PERF'
-mi- 'also'	*-mi-* 'also'	*-mi-* 'also'
		-naq- 'CAUS'
-niaq- 'FUT'		*-niaq-* 'FUT'
	-niq- 'PAST'	
-nngit- 'NEG'	*-nngit-* 'NEG'	*-nngit-* 'NEG'
-nnguaq- 'pretend'	*-nnguaq-* 'pretend'	*-nnguaq-* 'pretend'
		-qajaq- 'can'
		-qatauti- 'mutual'
		-qattaq- 'HAB'
	-qqau- 'PAST'	
		-qu- 'want'
		-rataaq- 'PAST'
	-ruaq- 'might'	*-ruaq-* 'might'
-si- 'PRES'		*-si-* 'PRES'
	-sima- 'PERF'	*-sima-* 'PERF'
	-taq- 'repeatedly'	*-taq-* 'repeatedly'
	-tit- 'CAUS'	*-tit-* 'CAUS'
		-tit- 'DIFF.SUBJ'
	-tsaq- 'really'	*-tsaq-* 'really'
		-tsaq- 'INTR'
	-tsi- 'ANTP'	*-tsi-* 'ANTP'

(Continued)

APPENDIX
(Continued)

LE	MLU	CA
-tsiaq- 'well'		
		-tuinnaq- 'merely'
		-vatsuaq- 'very'
Denominal Morphemes		
		-gi- 'have as'
		-ijaq- 'remove'
-it- 'be'	*-it-* 'be'	*-it-* 'be'
		-liuq- 'make'
		-ngaaq- 'rather'
		-qaq- 'have'
		-taaq- 'acquire'
	-tuq- 'consume'	
		-tuu- 'be the only'
-u- 'be'	*-u-* 'be'	*-u-* 'be'
-uq- 'arrive at'	*-uq-* 'arrive at'	*-uq-* 'arrive at'
Deverbal Morphemes		
-juq- 'NOM'	*-juq-* 'NOM'	*-juq-* 'NOM'

ACKNOWLEDGMENTS

In closing, we would particularly like to thank Annie Tukkiapik for her transcriptions and Lizzie Ningiuruvik for the major role she played in data collection and advice. We also thank Betsy Annahatak for her help and guidance. Three of our McGill colleagues, Myrna Gopnik, Shari Baum, and José Bonneau, have been particularly helpful in discussions of issues related to this chapter. Our greatest debt, of course, is to LE and her family; we are most grateful to them. Funds for the research were provided by grant #410-90-1744 to Drs. Gopnik, Crago, and Paradis, and doctoral fellowships #452-89-1120, 453-90-0005, 752-91-0543 to Shanley Allen from the Social Sciences and Humanities Research Council of Canada as well as by generous funding for the normal acquisition study from the Kativik Research Council to Martha Crago and Shanley Allen.

REFERENCES

Allen, S. E. M. (1994). *Acquisition of some mechanisms of transitivity alternation in arctic Quebec Inuktitut.* Unpublished doctoral dissertation, McGill University, Montreal.

Allen, S. E. M., & Crago, M. B. (in press). Early passive acquisition in Inuktitut. *Journal of Child Language.*

Badecker, W., & Caramazza, A. (1985). On considerations of method and theory governing the use of clinical categories in neurolinguistics and cognitive neuropsychology: The case against agrammatism. *Cognition, 20*, 97–126.

Bates, E., Friederici, A., & Wulfeck, B. (1987). Sentence comprehension in aphasia: A cross-linguistic study. *Brain and Language, 32*, 19–67.

Bates, E., McDonald, J., MacWhinney, B., & Appelbaum, M. (1991). A maximum likelihood procedure for the analysis of group and individual data in aphasia research. *Brain and Language, 40*, 231–265.

Bates, E., & Wulfeck, B. (1989). Cross-linguistic studies of aphasia. In B. MacWhinney & E. Bates (Eds.), *The cross-linguistic study of sentence processing* (pp. 328–371). Cambridge: Cambridge University Press.

Bittner, M. (1994). *Case, scope and binding*. Dordrecht, The Netherlands: Kluwer.

Bobaljik, J. D. (1992). Nominally absolutive is not absolutely nominative. In J. Mead (Ed.), *Proceedings of the Eleventh West Coast Conference on Formal Linguistics* (pp. 44–60). Stanford, CA: CSLI.

Branigan, P. (1992). *Subjects and complementisers*. Unpublished doctoral dissertation, Massachusetts Institute of Technology, Cambridge.

Brown, R. (1973). *A first language: The early stages*. Cambridge, MA: Harvard University Press.

Byng, S. (1988). Sentence comprehension deficit: Theoretical analysis and remediation. *Cognitive Neuropsychology, 5*, 629–676.

Caplan, D. (1991). Notes and discussion: Agrammatism is a theoretically coherent aphasic category. *Brain and Language, 40*, 274–281.

Caplan, D., & Hildebrandt, N. (1988). Conclusions and discussion. In D. Caplan & N. Hildebrandt (Eds.), *Disorders of syntactic comprehension*, (pp. 269–304). Cambridge, MA: MIT Press.

Caramazza, A. (1986). On drawing inferences about the structure of normal cognitive systems from the analysis of patterns of impaired performance: The case for single-patient studies. *Brain and Cognition, 5*, 41–66.

Chomsky, N. (1992). A minimalist program for linguistic theory. *MIT Occasional Papers in Linguistic Theory*. Cambridge, MA: Massachusetts Institute of Technology.

Clahsen, H. (1989). The grammatical characterization of developmental dysphasia. *Linguistics, 27*, 897–920.

Crago, M. B., Allen, S., & Hough-Eyamie, W. (in press). Exploring innateness through cultural and linguistic variation. In M. Gopnik (Ed.), *The biological foundations of language*. Oxford: Oxford University Press.

Dalalakis, J. (1994). Familial language impairment in Greek. *McGill Working Papers in Linguistics, 10*, 216–228.

de Freitas, L. (1993). *The syntax of sentential negation: Interactions with case, agreement and (in)definiteness*. Unpublished doctoral dissertation, McGill University, Montreal, Canada.

deVilliers, J. G., & deVilliers, P. (1985). *Language acquisition*. Cambridge, MA: Harvard University Press.

Dorais, L. J. (1986). La survie et le developpement de la langue des inuit [The survival and development of the Inuit language]. *Revue de l'Universite Laurentienne, 18*(1), 89–103.

Dromi, E., Leonard, L. B., & Shteiman, M. (1993). The grammatical morphology of Hebrew-speaking children with specific language impairment: Some competing hypotheses. *Journal of Speech and Hearing Research, 36*, 760–771.

Druks, J., & Marshall, J. C. (1991). Agrammatism: An analysis and critique, with new evidence from four Hebrew-speaking aphasic patients. *Cognitive Neuropsychology, 8*(6), 415–433.

Fortescue, M. D. (1984). *West Greenlandic*. London: Croom Helm.

Fukuda, S., & Fukuda, S. (1994). Familial language impairment in Japanese: A linguistic investigation. *McGill Working Papers in Linguistics, 10*, 150–177.

Gandour, J., Marshall, R., & Windsor, J. (1989). Idiosyncratic strategies in sentence production: A case report. *Brain and Language, 36*, 614–624.

Goodglass, H. (1973). Studies on the grammar of aphasics. In H. Goodglass & S. Blumstein (Eds.), *Psycholinguistics and aphasia*, (pp. 183–215). New York: Academic Press.

Goodglass, H., & Berko, J. (1960). Agrammatism and inflectional morphology in English. *Journal of Speech and Hearing Research, 3,* 257–267.

Goodglass, H., & Hunt, J. (1958). Grammatical complexity and aphasic speech. *Word, 14,* 197–207.

Gopnik, M. (1994). Impairments of tense in a familial language disorder. *Journal of Neurolinguistics.*

Gopnik, M. (1990a). Feature blindness: A case study. *Language Acquisition, 1*(2), 139–164.

Gopnik, M. (1990b). Feature-blind grammar and dysphasia. *Nature, 344,* 715.

Grodzinsky, Y. (1984). The syntactic characterization of agrammatism. *Cognition, 16,* 99–120.

Hyams, N. (1986). *Language acquisition and the theory of parameters.* Dordrecht, The Netherlands: Reidel.

Johns, A. (1992). Deriving ergativity. *Linguistic Inquiry, 23,* 57–87.

Julien, G., Baxter, J. D., Crago, M., Ilecki, H. I., & Therien, F. (1987). Chronic otitis media and hearing deficit among Native children of Kuujjuaraapik (Northern Quebec). *Canadian Journal of Health, 78,* 57–61.

Kolk, H., & Van Grunsven, M. (1985). Agrammatism as a variable phenomenon. *Cognitive Neuropsychology, 2,* 347–384.

Korkman, M., & Haakkenen-Rihu, P. (1994). A new class of developmental language disorders. *Brain and Language, 87,* 96–116.

Lahey, M. (1988). *Language disorders and language development.* New York: Macmillan.

Leonard, L. B. (1988). Lexical development and processing in specific language impairment. In R. L. Schiefelbusch & L. Lloyd (Eds.), *Language perspectives: Acquisition, retardation and intervention* (Vol. 2), (pp. 69–87). Austin, TX: Pro-ed.

Leonard, L. B. (1989). Language learnability and specific language impairment in children. *Applied Psycholinguistics, 10,* 179–202.

Leonard, L. B., Bortolini, U., Caselli, M. C., McGregor, K. K., & Sabbadini, L. (1992). Morphological deficits in children with specific language impairment: The status of features in the underlying grammar. *Language Acquisition, 2*(2), 151–179.

Lindner, K., & Johnston, J. R. (1992). Grammatical morphology in language-impaired children acquiring English or German as their first langauge: A functional perspective. *Applied Psycholinguistics, 13,* 115–129.

Linebarger, M. C., Schwartz, M., & Saffran, E. (1983). Sensitivity to grammatical structure in so-called agrammatic aphasics. *Cognition, 13,* 361–393.

MacWhinney, B., Bates, E., & Kliegl, R. (1984). Cue validity and sentence interpretation in English, German and Italian. *Journal of Verbal Learning and Verbal Behavior, 23,* 127–150.

MacWhinney, B., & Snow, C. (1990). The Child Language Data Exchange System: An update. *Journal of Child Language, 17,* 457–472.

McCloskey, M., & Caramazza, A. (1988). Theory and methodology in cognitive neuropsychology. *Cognitive Neuropsychology, 5,* 583–623.

Menn, L., & Obler, L. K., (1990). Cross-language data and theories of agrammatism. In L. Menn, L. K. Obler, & H. Goodglass (Eds.), *Agrammatic aphasia: A cross-language narrative sourcebook* (pp. 1368–1389). Philadelphia, PA: John Benjamins.

Miceli, G., & Mazzucchi, A. (1990). The speech production deficit of so-called agrammatic aphasics: Evidence from two Italian patients. In L. Menn, L. K. Obler, & H. Goodglass (Eds.), *Agrammatic aphasia: A cross-language narrative sourcebook* (pp. 717–816). Philadelphia, PA: John Benjamins.

Miceli, G., Silveri, M. C., Romani, C., & Caramazza, A. (1989). Variation in the pattern of omissions and substitutions of grammatical morphemes in the spontaneous speech of so-called agrammatic patients. *Brain and Language, 36,* 447–492.

Murasugi, K. (1992). *Crossing and nested paths: NP movement in accusative and ergative languages.* Unpublished doctoral dissertation, Massachusetts Institute of Technology, Cambridge.

Oetting, J., & Rice, M. (1993). Plural acquisition in children with specific language impairment. *Journal of Speech and Hearing Research, 36*(6), 1236–1248.

Pierce, A. (1992). *Language acquisition and syntactic theory: A comparative analysis of French and English children's grammars.* Dordrecht, The Netherlands: Kluwer.

Pinker, S., Lebeaux, D., & Frost, L. A. (1987). Productivity and constraints in the acquisition of the passive. *Cognition, 26,* 195–267.

Poeppel, D., & Wexler, K. (1993). The full competence hypothesis of clause structure in early German. *Language, 69*(1), 1–33.

Rescorla, L. (1989). The language development survey: A screening tool for delayed language in toddlers. *Journal of Speech-Language and Hearing Disorders, 54,* 587–599.

Rice, M. (1993, October). *Extended optionality or missing constraints: Two models of the morphosyntax of children with SLI.* Paper presented at McGill University, School of Communication Sciences and Disorders Seminar Series, Montreal.

Rice, M. (1994). Grammatical categories of specifically language-impaired children. In R. Watkins & M. Rice (Eds.), *New directions in specific language impairment* (pp. 69–89). Baltimore, MD: Brookes.

Rice, M., & Oetting, J. (1993). Morphological deficits of children with SLI: Evaluation of number marking and agreement. *Journal of Speech and Hearing Research, 36,* 1249–257.

Rice, M., & Wexler, K. (1993, July). *Types of agreement in SLI children.* Paper presented at the Sixth International Conference for the Study of Child Language, Trieste, Italy.

Ritter, E. (1992). Cross-linguistic evidence for number phrase. *Canadian Journal of Linguistics, 37*(2), 197–218.

Saffran, E., Schwartz, M., & Marin, O. (1980). The word order problem in agrammatism: Production. *Brain and Language, 10,* 263–280.

Seidenberg, M. S. (1988). Cognitive neuropsychology and language: The state of the art. *Cognitive Neuropsychology, 5*(4), 403–426.

Shaer, B. M. (1990). *Functional categories, reference, and the representation of tense.* Paper presented at the 13th GLOW Colloquium, Cambridge, England.

Shallice, T. (1988). *From neuropsychology to mental structure.* Cambridge, England: Cambridge University Press.

Shankweiler, D., Crain, S., Gorrell, P., & Tuller, B. (1989). Reception of language in Broca's aphasia. *Language and Cognitive Processes, 4*(1), 1–33.

Slobin, D. (1985). *The cross-linguistic study of language acquisition* (Vol. 1). Hillsdale, NJ: Lawrence Erlbaum Associates.

Slobin, D. (1992). *The cross-linguistic study of language acquisition* (Vol. 3). Hillsdale, NJ: Lawrence Erlbaum Associates.

Travis, L. (1992). Inner tense with NPs: The position of number. *Cahiers de Linguistique de l'UQAM, 1*(1), 327–345.

Valois, D. (1991). *The internal syntax of DP.* Unpublished doctoral dissertation, UCLA.

Verrips, M., & Weissenborn, J. (1992). Routes to verb placement in early German and French: The independence of finiteness and agreement. In J. Meisel (Ed.), *The acquisition of verb placement* (pp. 283–331). Dordrecht, The Netherlands: Kluwer.

Waters, G., Rochon, E., & Caplan, D. (1992). The role of high level speech planning in rehearsal: Evidence from patients with apraxias of speech. *Journal of Memory and Language, 31,* 54–73.

Wexler, K. (1992). *Optional infinitives, head movement, and the economy of derivations in child grammar* (Cognitive Science, Occasional Paper No. 45). Cambridge, MA: MIT Center for Cognitive Science.

COMMENTARY ON
CHAPTER 10

David Poeppel
University of California, San Francisco

CRAGO AND ALLEN'S CONTRIBUTION

Crago and Allen's (CA) chapter is an important contribution to the study of language acquisition in several respects. First, their work extends the empirical basis of research on normal and pathological language acquisition by considering Inuktitut, a polysynthetic language with ergative case marking. The data that have shaped our present understanding of language acquisition—at least from the perspective of research basically sympathetic to linguistic theory—have come primarily from English, a handful of Romance languages, and several Germanic languages. Acquisition studies of languages with the morphological richness of Inuktitut (see CA's appendix) are rare. Importantly, less intensively studied languages like Inuktitut or Quiche Mayan (cf. Pye, 1983) can provide subtle evidence that speaks to current debates.

Second, the data that CA contribute add to a current acquisition discussion focused on very young children's use of default, nonadult forms. In my next section, I take up one of the analyses reported in CA's chapter and show how the results can extend a recent acquisition account in useful ways.

Finally, CA illustrate how the same theoretical considerations that shape an interpretation of acquisition data can also shape our understanding of certain aphasias. Their argument underscores how important explicit psycholinguistic and linguistic models are for clear accounts of language-related phenomena.

291

ELABORATING ON CURRENT ACQUISITION THEORY: AN INTERPRETATION OF SOME INUKTITUT ACQUISITION DATA

A number of studies of early child language demonstrate that children up to approximately 2;6 often produce infinitival verbal forms in contexts in which adults supply finite, inflected forms. Extensive crosslinguistic data speak to this phenomenon and have led to the development of an account called the optional infinitive hypothesis (also often called the root infinitive hypothesis). The three leading ideas that underlie the hypothesis are as follows: First, children at this stage have available a fully articulated phrase structure representation; consequently, their phrasal productions can conform to the adult grammar. Second, although many of the utterances produced by children are correct by criteria of the adult language, many utterances are nonadult: Specifically, the children produce infinitival verbal forms in root (matrix) contexts that require finite forms. Third, the language feature that putatively leads to the frequent use of a "default" infinitival form is the representation (or computation) of tense. (See Wexler, 1994, this volume, for a summary of evidence from a number of different languages and an extensive discussion of these issues.)

Young children's full representational competence has been demonstrated for many languages. In contrast, children's use of overt infinitivals has been shown for some languages and not for others. Although in French, Dutch, German, English, and Swedish, for example, infinitives are used systematically by learners in matrix clauses, there is no clear evidence of root infinitives in languages such as Italian and Spanish. Similarly, there is no evidence for the use of infinitives in Inuktitut, as shown by CA.

Because the use of root infinitives by children is such a pervasive phenomenon in many languages, one might expect to find analogous phenomena in those languages that do not show the overt use of infinitives in finite contexts. The absence of root infinitives in Inuktitut and other languages notwithstanding, one can imagine that, given the fundamental similarities of languages and language learners, there are systematic production phenomena across learners that call for a unified account. In what follows, I use CA's data to suggest a way of thinking about these phenomena that might capture some underlying similarities.

In response to the optional infinitive hypothesis proposed by Wexler (1994) and argued for by Poeppel and Wexler (1993) for normal acquisition and by Rice and Wexler (in press) for the case of specific language impairment, CA emphasize that Inuktitut has no uninflected infinitival form. Consequently, a theory that relies explicitly on the existence of an infinitival form cannot be adequate, at least with respect to Inuktitut. Suppose one wants to extend the reasoning underlying the root infinitive account to include the phenomena of Inuktitut. One will want to abstract from the specific role that infinitives play in a root infinitive framework to include other forms (i.e., bare stems or any other forms that may function as default forms). I propose here an extension to the root infinitive notion that captures crucial aspects of the Inuktitut data and that possibly extends to other languages as well.

Consider CA's Table 10.4 in which they quantify the morphological properties of 200 of their three subjects' utterances.

For the subject LE, CA identify 69 contexts in which they expected some inflection (the 131 other utterances presumably did not qualify for analysis with respect to verbal or locative inflection). Note the pattern of inflections supplied by the subject:

- 32 of 69 forms were correctly inflected.
- 10 instances showed omission of obligatory inflection, yielding either a stem or a form that was insufficiently inflected.
- 8 times LE omitted the necessary inflection on a locative stem, presumably yielding incorrect locative stems or otherwise underspecified forms.
- 5 times the full inflection was not supplied on locative stem; however, in those cases the omission did not generate an incorrect form.
- In 4 cases, LE produced a pronoun instead of the appropriate inflection; again, a stem is left over.
- 10 of LE's 69 relevant utterances had the idiosyncratic filler inflection -*mi*.

For the MLU-matched subject, CA identify 44 relevant contexts. In those contexts:

- 26 of 44 forms were correctly inflected.
- 5 instances showed omission of obligatory inflection.
- 2 times the necessary inflection on a locative stem was omitted.
- 11 tokens were potential omissions on locative stems.

Note that, for both subjects, the omissions yield either stems or forms that bear some affix but are, nevertheless, morphophonologically impoverished from the adult grammar's point of view. The data from the child that was chronologically age-matched to the subject LE are consistent with the view that this child was past the stage at which default forms are optionally used.

Although CA do not emphasize this point, one aspect of the grammar of both the SLI and the MLU-matched subjects that these data strongly show is that they know deep properties of the agreement system. The data that demonstrate this are implicit in the table: Although obligatory affixes are not supplied almost in half of the relevant contexts, inflectional affixes are evidently never misattached. Either an obligatory morpheme is correctly supplied or it is dropped, yielding a stem or an otherwise insufficiently affixed (inflected) form. However, in no (reported) case does the child (or the controls) attach a morpheme in a place wrongly (for example, by supplying first person inflection in a third person context). This implies that the child knows the requirements of the agreement system to the extent that a wrong affix is not produced. Rather, the child will, in many cases, produce an (inflectionally) underspecified form. Note that the child does know the specific morphemes required. As CA point out, ". . . the same verbal inflection appears in some obligatory contexts and not in others" (CA, p. 272). This is particularly interesting in that it shows that even though the child may know a particular verbal form, she does not obligatorily supply it (cf. Poeppel & Wexler, 1993, for an analogous finding in German).

How to Think About Infinitives and Stems

The use of a stem form is not correct in the adult grammar of Inuktitut. But this is true for the infinitives produced in root contexts in other languages as well. However, the characterization as "incorrect" is perhaps not as constructive as possible. Suppose we think of the stems and other nonadult forms shown by CA not as incorrect but as underspecified. From that point of view, children are not choosing an incorrect form. Rather, the form that is expressed is the one that is less specified with respect to features that are morphologically and phonologically obligatory for the adult. If one thinks of these underspecified forms as defaults (though the specific default form chosen depends on various morphophonological properties of a given language), one can consistently characterize the form that the child optionally chooses as featurally underspecified. Although the production of a stem does not satisfy the requirements of the adult grammar, stems (and other underspecified forms) have a different status in child language; depending on the language, specific underspecified forms may be acting as defaults.

If one adopts this view, the systematic underlying similarity between acquisition data from German, French, or Inuktitut can be captured: In all cases, the children produce correctly inflected forms in alternation with forms that can be characterized as underspecified, and that act as defaults. The fact that, in some languages, an infinitival form functions as default but in others a stem form is the default is a consequence of independent properties of these languages.

We, thus, can generalize to the following pattern of results for very young children: In some languages, there is an alternation between infinitives and finite, correctly inflected forms (e.g., French, German, English). In Inuktitut (and, by prediction, in other languages like Inuktitut), there is an alternation between stems and finite, correctly inflected forms. The predicted pattern for still other languages is that there will be an alternation between other forms that are reasonably described as featurally underspecified and finite, correctly inflected forms. The central idea is that what children express are either correctly inflected lexical items or underspecified forms. What children almost never produce are forms in which there is (overt) misattachment of an inflectional affix to a stem.

The logical possibilities of what a learner produces are summarized in (2). If the target is a stem that bears three inflectional affixes, as in (1), the learner might produce any of the forms in (2). Which form is chosen is conditioned by language-specific properties.

(1) Adult target:
      stem - affix 1 - a2 - a3
(2) What the child produces:
      a. form 1: stem - a1 - a2 - a3
      b. form 2: stem - a1 - a2 -
      c. form 3: stem - a1 -
      d. form 3: stem - 0

Importantly, there are theories of morphology that have motivated accounts for phenomena of the type discussed. Distributed Morphology (Halle & Marantz, 1993), for example, provides a theoretical framework in which the ideas previously sketched can be formulated.

Insofar as the concept of underspecified default forms is plausible, one will want to investigate why particular forms are chosen as defaults. Here we must turn to language-specific considerations. It is presumably no accident what a chosen default form is. For German, the fact the *-en*, the adult infinitival form, is chosen as the default morpheme is natural insofar as *-en* is nonspecific with respect to tense. Similarly, in French *-er/-ir* are natural default choices insofar as those affixes correspond to phrases nonspecific for tense. Perhaps the bare stem seen in Inuktitut (and, interestingly, English, where one can think of the uninflected form in *he eat* as a stem or as an infinitive) is also a form that is expressed when tense features remain underspecified. In any case, a generalization that has consistently emerged (cf. Wexler, 1994) is that what is distinct in early child language is the representation/computation of tense.

SUMMARY

The discussion of abstract issues in syntax acquisition issues would not even be possible without the detailed information that CA provide in this case study. It is this type of rich quantitative characterization that allows for the discovery of underlying systematic phenomena. First, insofar as acquisition studies only provide example utterances, possible underlying systematicities that require some abstraction could easily be overlooked. Second, the research underscores the value of crosslinguistic work. A discussion of how stems and infinitives may act in similar ways is not possible unless one takes into account languages that actually have these properties in child or adult language. Third, from the perspective developed here, the data from subject LE provide evidence consistent with the extended optionality account argued for by Rice and Wexler. However, it is not the use of optional infinitives that is extended in SLI ontogenesis but the use of whatever default form is the natural form for a specific language. Finally, the work suggests that accounts that make no reference to linguistic representations cannot capture or describe the phenomena with the same degree of accuracy and richness. Without concepts like *agreement* or *inflection* that have carefully worked out properties and are embedded in larger theories of language, the systematicities in the data cannot be stated.

ACKNOWLEDGMENTS

I thank Alec Marantz, Colin Phillips, and Ken Wexler for their helpful comments. I was supported by the McDonnell-Pew Foundation.

REFERENCES

Halle, M., & Marantz, A. (1993). Distributed morphology. In K. Hale & S. J. Keyser (Eds.), *The view from building 20* (pp. 111–176). Cambridge, MA: MIT Press.

Poeppel, D., & Wexler, K. (1993). The full competence hypothesis of clause structure in early German. *Language, 69*(1), 1–33.

Pye, C. (1983). Mayan telegraphese: intonational determinants of inflectional development in Quiche Mayan. *Language, 59,* 583–604.

Wexler, K. (1994). Optional infinitives, head movement, and the economy of derivations. In D. Lightfoot & N. Hornstein (Eds.), *Verb movement* (pp. 305–382). Cambridge, MA: Cambridge University Press.

11

THE SEARCH FOR THE PHENOTYPE OF DISORDERED LANGUAGE PERFORMANCE

Jon F. Miller
University of Wisconsin – Madison

This chapter focuses on the problem of defining language behavior phenotypes. Language disorders result from a wide variety of etiologies, for example, mental retardation, hearing loss, environmental deprivation, affecting one or more levels of language performance. This broad view of language disorder is contrasted with specific language disorder (SLI), a term commonly understood to refer to deficits in language performance with no known cause, that is, not due to mental retardation, hearing loss, neurological disease, trauma, social-emotional disturbance, or environmental deprivation. The hypothesis asserted here is that SLI should be considered a behavioral classification, describing delays in the onset and rate of language development as well as asynchronies in development across language levels and deviant language performance. The notion of SLI as a unitary construct will be reviewed as well as the evidence supporting the existence of different types of disordered language performance, that is, different phenotypes of SLI. Insight into the genetics of SLI may be gained by reviewing the language outcomes of two syndromes that result from major chromosomal anomalies, Down syndrome and fragile X syndrome. Although both of these syndromes are associated with mental retardation, the language-cognition relationship for each may cast doubt on excluding mental retardation from the definition of SLI. Coming to some agreement on the defining properties of SLI will be central to documenting language phenotypes that may be genetically determined.

Recent research on the language performance of children with specific genetic syndromes documents asynchronies in language development relative to other cognitive skills. The differences in the language phenotype of Down and fragile X syndromes will illustrate the point that mental retardation only describes a general cognitive deficit, but does not predict the course of language development. Mental retardation as an exclusionary category for SLI may be shortsighted, based on antiquated theories of both intelligence and the relationship between language and cognition.

The search for the genetic links to SLI is the search for either a single SLI language phenotype, or several phenotypes that may be associated with biological processes, rather than environmental influences. Although the nature of SLI has remained elusive, I argue that it reflects a variety of language problems, including deficits in morphosyntax, semantic referencing, pragmatics, speaking rate, utterance formulation, and word finding.

DEFINING SLI: IMPLICATIONS FOR ETIOLOGICAL INQUIRY

The Evolving Characteristics of SLI

The concept of a *specific language disorder*, or SLI, was considered by many to be a unitary construct during the late 1970s and early 1980s. In most studies of this period, single aspects of language, such as phonology, vocabulary, syntax, semantics, and pragmatics, were studied. During the second half of the 1980s, several language investigators argued that the majority of the research, focusing on the identification of these individual variables as parsimonious characterizations of disordered performance, was inadequate (Johnston, 1988; Van Kleeck, 1988). Indeed, children with SLI rarely exhibit unique deficits at a single level, but they usually exhibit deficits at several levels of language performance simultaneously.

Research conducted by numerous investigators supports the view that deficits at one level of language performance impact other language levels. For example, Leonard, Camarata, Schwartz, Chapman, and Messick (1985) and Camarata and Schwartz (1985) have found that phonological limitations restrict lexical acquisition. Similarly, syntactic deficits have been found to alter the performance of language-disordered children on a variety of pragmatic tasks (Johnston, 1988; Johnston & Kamhi, 1984; Miller, 1978, 1981). Johnston (1988) argued that children with SLI are communicatively impaired precisely because they lack a command of language form: "Without sufficient grammatical resources, they have difficulty constructing cohesive texts, repairing conversational breakdowns, and varying their speech to fit social situations" (p. 693). The results of these studies clearly indicate a need to understand the interrelationships between linguistic levels, as well as to appreciate the asynchronies within these linguistic levels.

Developmental asynchronies have been observed both within linguistic levels and between processes, that is, between comprehension and production. Examples of within level asynchrony can be found in studies that have documented simple sentence patterns developing in advance of grammatical functions and propositional embedding (Johnston & Kamhi, 1984; Liles & Watt, 1984). Asynchronies between language comprehension and production have been observed for some time in typically developing children (Chapman & Miller, 1975) and in various groups of children, those with language disorders and those with language disorders related to mental retardation (Miller, 1987a, 1988; Miller, Chapman, & MacKenzie, 1981). Asynchronies observed in the language profiles of children with language impairments may continue to change throughout the developmental period. These language profiles may look different over time as a result of development, adaptation, or intervention. This concept of changing profiles of a language disorder questions the view of impaired language as a unitary construct.

Language Disorder Is Multidetermined

Language disorders in children are associated with a wide variety of etiologies including pre- and postnatal trauma, genetic syndromes, inborn errors of metabolism, disease processes, and environmental deprivation. These etiologies result in impairments in sensory, cognitive, and motor performance associated with hearing impairment, mental retardation, emotional disturbance, and brain injury. There remains a significant population of children with language disorders who suffer none of these conditions. These children have been referred to in the literature as specifically language impaired (SLI). Past research characterized children with SLI as a homogenous group, concluding that these children speak less frequently, speak less accurately, process information at a slower rate, and produce more errors than their peers. In general, children with SLI show a late onset of language, their rate of language development is slower, and they may never achieve the language skills of their peers even as adults (Aram, Ekelman, & Nation, 1984; Schery, 1985; Weiner, 1985). Deficits at every level of language, phonological, syntactic, semantic, and pragmatic have been reported in children with SLI; though individual children do not exhibit deficits in all areas at any one point in development (Miller & Klee, 1995). It is possible, however, that SLI is an evolving construct so that an individual child may exhibit deficits in all areas through the developmental period. Children with SLI have also been reported to have word-finding problems (German, 1987), and utterance formulation problems (Dollaghan & Campbell, 1992; MacLachlan & Chapman, 1988). Attempts to identify the specific language features that define SLI among clinical samples of children have been largely unsuccessful, at least where standardized tests have been the primary source of data (Aram, Morris, & Hall, 1993). At the same time, a variety of causal constructs, genetic, cognitive, environmental, and information processing deficits, have been proposed to explain SLI. There appears to be little correspondence between the causes of SLI and the specific language performance to be explained. This may be because research aimed at identifying causal constructs has treated SLI as a unitary construct with a general definition. Clearly, research aimed at documenting causal constructs, genetic or environmental, will benefit from improved precision in describing specific language impairment. Recent research (Rice & Wexler, this volume) shows promise in this regard where a reliable clinical marker of at least one subgroup of children with SLI early in development has been identified.

Types of Specific Language Impairment

The language performance of children with SLI has been categorized in several ways with the primary focus on the differences between the development of language comprehension and production skills. This method of description suggests that children with SLI are not homogenous, but possibly heterogenous with varying impairments related to language processes. The American Psychiatric Association's DSM-IV system identifies two groups of children with language disorders: those children with language comprehension and production deficits and those with only expressive language deficits (American Psychiatric Association, 1994). Tallal (1988) documented three groups of children with SLI: those with comprehension delays relative to production, those with both comprehension and production delays, and those with only production delays. Rapin (1988) also defined three categories for children with SLI: children with expressive language disorders who have normal comprehension, those with combined expressive and receptive language disorders along with impaired articulation, and those with higher order processing disorders.

Recent work suggests that differentiating children with SLI on the basis of only language processes, that is, using language comprehension and language production as the variables to describe these children, in the absence of examining various levels of language, such as phonology, vocabulary, syntax, semantics, and pragmatics, does not adequately describe the heterogeneity found in this group (Bates & Thal, 1991; Bishop, 1992, 1994; Nelson, 1991; Snow, 1991).

In summary, the search for causes of SLI has been limited by the descriptive models, and the descriptive models have been limited by measurement constraints. Variation in describing and defining SLI has led to a number of causal constructs being proposed with the data supporting each less than compelling. Recent reviews of the research literature regarding children with SLI supports the conclusion that there is considerable heterogeneity among children with language impairment (Bishop, 1992, 1994). Bishop suggested that there are several subtypes of language impairment among this population and that different causal constructs will need to be put forth to explain each subtype. To examine this assertion more carefully, I argue that SLI may be analogous to mental retardation in that mental retardation is a behavioral classification that is multidetermined, with each cause resulting in unique language outcome. In support of this argument, I review the language profiles of children with two genetic syndromes, Down syndrome and fragile X syndrome. This contrast will demonstrate that the two syndromes, each with distinct genotypes, result in the same behavioral classification of mental retardation, but very different language phenotypes. Finally, the assertion that SLI is a behavioral classification will be evaluated with new data from a large sample of children (256) with SLI using: (a) a multidimensional model of language performance, (b) language assessment technology that improves the manner in which language performance is quantified, and (c) identify new developmental variables as well as error variables to describe SLI subtypes in children 3 through 13 years of age. These data provide a glimpse of the variation in productive language performance of children with SLI as well as the descriptive detail necessary to identify language phenotypes that may be associated with specific genetic causes.

DISSOCIATIONS OF COGNITION AND LANGUAGE: LANGUAGE PROFILE ANALYSIS OF CHILDREN WITH DIFFERENT GENETIC PHENOTYPES

I now review the language performance profiles of children with chromosomal abnormalities addressing the question: Is there a language phenotype for each of these syndromes? The suggestion that language phenotypes exist for distinct genetic disorders indicates a link between language profile analysis and genetically based communication problems. The language phenotypes of children with Down syndrome and those children with fragile X syndrome will be presented in the context of a larger profile analysis, one that includes basic descriptions of syndrome incidence, neurological characteristics, cognitive function, general behavioral features, and speech-language profiles. This comparison demonstrates that while both syndromes result in the same behavioral classification—mental retardation—their language and communication skills are quite distinct. You will note that these syndromes result in very different central nervous system deficits, and their language abilities are not accounted for by the general label of mental retardation.

CHILDREN WITH DOWN SYNDROME

- Incidence. 1 in 700 on average, with incidence increasing with increasing age.
- Gender issues. Equal numbers of males and females.

- Genetics. Trisomy 21 (96%), Chromosomal translocation (3%), or Mosaicism (1%).

- Phenotype. Persons with Down syndrome are short in stature, with a distinct facial morphology, including a flat facial profile and epicanthal folds. The tongue is relatively larger but only appears so because of a small oral cavity. Motor function is characterized by generalized hypotonia, which may be mild to severe. Joints are hyperextendable. Mental retardation is common, and Down syndrome is the leading genetic cause of mental retardation. For the majority of these children (95%), the causal mechanism is accidental and not inherited. About 50% of these children have hearing defects, and an almost equal number have a variety of gastrointestinal problems or congenital heart disease.

- Neurological differences. A variety of differences have been found in the central nervous system (CNS) of persons with Down syndrome. For example, brain weight that is normal at birth, increases at a very slow rate, resulting in smaller brain size by 2 to 3 years of age. There are also differences found in the size, shape, and function of other CNS structures, such as the cerebellum and brainstem. At a microscopic level, differences have been found in the density of neurons located in cortical layers, in the structure of dendrites, and in the number of synapses. Neurochemical abnormalities have been observed in the neurotransmitter enzyme systems of both the peripheral and central nervous systems. Functionally, neurophysiological investigations have suggested deficits in synaptic transmission. Reports of seizure activity are higher during infancy and adulthood.

- Cognitive skills. Individuals with Down syndrome range in severity of mental retardation from profound to mild, with anecdotal reports of average intelligence. However, historically, it has been reported that the majority of individuals with Down syndrome evidence moderate to severe mental retardation. It has been suggested that the severity of this intellectual deficit is closely related to the institutionalization and education of individuals with Down syndrome. It has been argued by some that children with Down syndrome proceed through the same developmental stages as typically developing children, but this is questionable given the known problems these children exhibit with language comprehension and production. In addition, given the multifaceted nature of cognition, the vast range of talents exhibited by persons with Down syndrome in music, in the visual arts, and in athletics as unexplored areas of cognitive development must be considered.

- Speech skills. Although individual children with Down syndrome may exhibit specific phonological rule deficits or distinct speech sound errors, the most common speech production problem that children with Down syndrome evidence is reduced speech intelligibility.

- Language skills. Children with Down syndrome have particular difficulty acquiring language skills, particularly productive language skills. These children show an asynchrony in the acquisition of language exhibited by their comprehension of language exceeding their production ability. In one sample of 20 children with Down syndrome that I have followed for several years from the onset of first words, only seven children showed appropriate rates of vocabulary acquisition in production but all showed comprehension skills relative to their nonverbal mental age. The onset of multiword utterances was significantly delayed for the group. Mean length of utterance at mental age 30 months was 1.13 on average, with five children not yet combining words. Their rate of progress was very slow for productive syntax, with a predicted MLU of 1.66 for the group by age 8 years, compared to their other cognitive skills averaging 4;2 years of mental age. An MLU of 1.66 is achieved by 24 months in typical children. Given that CA and MA are equivalent in typical children, then the children with Down syndrome are delayed almost two years in acquiring productive syntax. Deficits in acquiring syntax are more pronounced than in vocabulary acquisition, though both are impaired relative to nonverbal mental age. Several investigators have documented the difficulties with syntax (Chapman, in press; Miller, 1987a, 1988), and Fowler (1988, 1990) proposed that syntax is the basis of their language-learning deficit. Deficits in syntactic comprehension appear after cognitive mental age 3 to 4, though vocabulary comprehension is better than nonverbal cognitive skills with increased age.

CHILDREN WITH FRAGILE X SYNDROME

- Incidence. 1 in 1,000 males.
- Gender issues. Males are primarily affected. Females may be affected, but generally to a significantly lesser degree.
- Genetics. The cytogenetic expression of the fragile site is on the long arm of the X chromosome at Xq27.3.
- Phenotype. Males with the fragile X syndrome have a distinctive facial profile characterized by a generally long face, large ears, a prominent jaw, a four-finger palmer crease, and macroorchidism, or large testicles. Motor development is generally delayed and characterized by hypotonia and hyperextendible joints. Mitral valve prolapse is a common feature. Mental retardation is typical, and fragile X syndrome is the leading hereditary cause of mental retardation.
- Neurological differences. There have been very few studies of the neuroanatomical or neurophysiological differences between persons with fragile X syndrome and typically developing individuals, though in all of these studies both young males and females have been studied. Findings indicate cerebral ventricular enlargement in approximately 30% of individuals, decreased size of the posterior cerebellar vermis and enlargement of the fourth ventricle, and abnormal brain stem auditory-evoked potentials in approximately 40% of individuals.

- Cognition. Children with fragile X syndrome have cognitive deficits that range from normal intelligence to severe mental retardation, and males are more severely affected than females. Males with fragile X syndrome generally evidence cognitive impairments that are consistent with mild to moderate mental retardation. Females with fragile X syndrome generally have normal to mildly retarded cognitive function with learning disabilities. Males with fragile X syndrome frequently exhibit attention deficits that interfere with learning. In addition, males with fragile X syndrome evidence auditory attention, sequencing and memory problems, math learning difficulties, and deficits in abstract reasoning. These males show relative strengths in visual recognition, memory and learning skills, as well as reading and spelling.

- Behavior. The behavior of male children with fragile X syndrome is generally characterized as hyperactive and impulsive with a restricted attention span. These children often exhibit autistic-like behaviors: hand flapping, hand biting, and other unusual hand mannerisms. Several recent reports have examined the distinction between children with fragile X syndrome and children with autism, suggesting that children with fragile X syndrome are more social and communicative than children with autism; however, about 14% of males with autism also have fragile X syndrome.

- Language. Where children with Down syndrome exhibit particular difficulty with the syntax of the language and strengths in using language for communication, children with fragile X syndrome show particular difficulty with social use of language. There are two primary features of the language performance problems of children with fragile X syndrome: pragmatic communication impairments and utterance formulation deficits. Specific characteristics of the pragmatic language problems include reduced eye contact and poor topic maintenance with inappropriate or tangential language use. Utterance formulation deficits are evidenced by word-finding problems, vocabulary deficits, perseverative word and phrase repetitions, and the frequent use of automatic phrases.

- Speech. The speech production of children with fragile X syndrome is characterized by a fast speaking rate, increased loudness, and verbal apraxia. Speech intelligibility is frequently reduced and cluttering may be observed.

- Genetic language phenotype. Children with each of these chromosomal disorders exhibit distinct language and communication profiles. These language patterns are not explained by the general cognitive deficits associated with the syndromes, supporting a multicomponent model of cognitive-linguistic performance, rather than a cognitive model. Mental retardation, as the common behavioral classification of these two groups, does not describe or explain the language profiles. Instead, these language patterns suggest that different genetic bases influence the differences in behavior and genetic language phenotypes such that, for the children with Down syndrome, unknown genetic factors are targeting syntax whereas for the fragile X syndrome, such factors impact nonsyntactic aspects of language.

A CLINICALLY REFERENCED STRATEGY
FOR IDENTIFICATION OF SLI PHENOTYPES

Clinician-generated responses to questions regarding the categorization of language production disorders suggest six potentially different categories of language problems (Miller, 1987b; Miller, Frieberg, Rolland, & Reeves, 1992). These problems include:

1. Utterance formulation deficits.
2. Word-finding problems.
3. Rate of message transference problems.
4. Pragmatic or discourse problems.
5. Semantic and referencing problems.
6. Delayed language development.

These categories should be thought of as a starting point for the exploration of language disorder phenotypes. These language problem categories represent a bottom up view of disordered language performance for the purposes of differentiating subtypes of SLI. The categories are not to be thought of as mutually exclusive. Children with language disorders may evidence several of these problems and may fit into multiple categories.

To quantify each category it is necessary to use observable dependent measures. Some dependent measures are more accessible than others, for example, mean length of utterance (MLU), which is a measure that can be used for the delayed language development category, can be calculated using computer software when consistent sampling conditions and transcription rules are practiced. Pragmatic performance, on the other hand, requires complex coding techniques because digital algorithms for automated recognition of topic, repairs, and contingency are not available.

Preliminary insight into four of the six language problem categories, delayed language development, word-finding problems, utterance-formulation deficits, and rate of message-transference problems, can be gained by using language sample measures accessible by a computer software program (Miller & Chapman, 1985–1993; SALT). The first group of variables document delayed development; these include MLU in morphemes, number of different words (DW), and number of total words (TW). These measures have been validated as indices of developmental progress for children 3 through 13 years of age by their high, and significant, correlations with chronological age (Miller, 1987b, 1991).

There is a second group of variables that document repetition and revisions of parts of words, words, and phrases. These characteristics have been referred to as *mazes* (Loban, 1976), and are receiving increasing attention as indices of language deficits associated with neuropathology in adults (Garrett, 1992) and in children (Dollaghan & Campbell, 1992; MacLachlan & Chapman, 1988). Mazes as a class of behaviors have been validated as indices of utterance formulation load (Miller, 1987b). Mazes are relevant descriptions for documenting both word-finding problems and utterance-formulation problems. Three maze measures can easily be calculated: the number of utterances containing mazes, the total number of mazes, and the percent of mazed words calculated by dividing the number of words in mazes by the total number of words in the transcript.

A third group of variables document speaking rate, using two different measures, words per minute and pauses. These measures allow linguists to quantify both slow or fast speaking rates. Clinicians have frequently reported that children with language disorders take more time to produce a language sample than typically developing children. Clinicians have also described children who seem to speak very rapidly, but communicate very little. These children are memorable to clinicians because they talk constantly, but provide very little information. Rate of talking, or of message transference, has never been documented for describing language disordered performance in the research literature, but clinicians regard it as having clinical face validity.

Pause rate is a second variable associated with speaking rate, but one that is also associated with maze productions. A *pause* is identified as a period of silence longer than two seconds either within an utterance or between utterances. For this analysis, the total pause time summed over all pauses in the transcript can be used as the dependent measure. Pauses have been studied relative to their discourse function (e.g., as a signal for turn changes), to their relationship to mazes or speech disruptions (Dollaghan & Campbell, 1992), and to their role in word-finding problems (German, 1987).

THE WISCONSIN STUDY OF AN SLI PHENOTYPE

To address the question of whether an SLI phenotype might exist, and to determine if SLI subtypes exist along the lines of the language-production categories described earlier, I undertook an investigation of 256 children with SLI between the ages of 2 and 14. The 256 participants in this study were recruited from the Madison Metropolitan School District (MMSD) (see Table 11.1 for a summary of participants by age and sex). Each of these children produced a conversational sample, approximately 15 minutes in length, resulting in a tran-

TABLE 11.1
Distribution of Language-disordered Participants by Age and Sex

Age in Years	Males	Females	Total
2–5	37	24	61
6–9	81	48	129
10–14	38	28	66

script of at least 100 complete and intelligible utterances. The samples were audio taperecorded by their school speech-language pathologist using the sampling guidelines detailed in Miller and Leadholm (1992). Each sample was transcribed by the MMSD's transcription laboratory using SALT transcription conventions (Miller & Chapman, 1985–1993). The language production of these children was analyzed using SALT to address the following questions:

1. Do all children with SLI exhibit delayed language development as measured by MLU, DW and TW?
2. Do all children with SLI produce significant increases in the number of utterances containing mazes, total number of mazes, and percent mazed words, compared to typically developing children?
3. Do all children with SLI produce significantly less language per unit time than typically developing control participants as measured by the number of words produced per minute?
4. Are significant increases in the number of pauses in a transcript more likely to be associated with delayed development, increased maze production, or fast or slow speaking rate?

To identify performance deficits, it is necessary to understand how typically developing children were performing under the same speaking conditions. To provide control data, I developed a database consisting of language samples recorded and transcribed from 252 typically developing children in two speaking conditions, conversation and narration. These children ranged in age from 3 to 13 years of age and were from Madison and north central Wisconsin. Fairly equal numbers of boys and girls were studied for a wide range of socioeconomic backgrounds (though the range exhibited in Madison is limited). These participants were separated into age groups labelled 3, 4, 5, 6, 7, 9, 11, and 13 years, with 26 to 49 participants in each group (Miller et. al., 1992; Miller & Leadholm, 1992). The control data is referred to as the reference database (RDB).

To answer the research questions, two kinds of analyses were conducted. First, each participant's transcript was analyzed, and scores were calculated for the three developmental variables (MLU, TW, and DW), the three maze variables (number of utterances containing mazes, total number of mazes, and percent mazed words), rate (words per minute), and number of pauses. These measures were compared to the RDB to determine if the value was within one standard deviation, or was plus or minus one standard deviation from the mean for the child's chronological-age peers. One standard deviation was selected as a minimum criterium as evidence for suspected disordered performance. I was also interested in an index where marginal performance on a number of measures could be noted. The second analysis documented the number of children with performance above or below one standard deviation for each combination of variables: developmental, mazes, and rate.

TABLE 11.2
Number and Percent of Participants With Deficits in Each Category

Category	Count	Percent
A: Delay	147	57
B: Mazes	97	38
C: Slow	98	38
D: Fast	19	7
E: Pause	127	50
X: Not A–E	20	8

Because rate could only be rated as fast or slow, five variables were identified: (a) delayed development, (b) increased mazes, (c) slow rate, (d) fast rate, and (e) increased pausing. To be considered deficient in each variable, a subject had to exhibit performance below one standard deviation for any one of the developmental measures, above one standard deviation for any one of the maze measures, below one standard deviation for the words-per-minute to be considered slow, above one standard deviation for the words-per-minute to be considered fast, and above one standard deviation on the number of pauses. The number of participants performing plus or minus one standard deviation on each of the five variables can be found in Table 11.2. These results will be discussed in relation to each of these five questions:

1. Do all children with SLI exhibit delayed language development as measured by MLU, DW, and TW?

As can be seen in Table 11.2, 147 (or 57%) of the 256 participants scored at least one minus standard deviation on one of the three developmental measures: MLU (mean length of utterance), DW (number of different words), or TW (total words). The age range of these participants covered the complete range of the subject sample, and it is clear that not all participants with SLI show developmental delay. Miller (1991) has argued that each of these measures quantifies a different aspect of language performance, MLU (syntax), DW (semantic diversity), and TW (total language proficiency). Each of these measures correlates very highly with age in the RDB sample. As contradictory evidence, each of these measures are highly correlated with each other (r ranges from .85–.92), suggesting that these measures may somehow be measuring the same underlying construct.

2. Do all children with SLI exhibit significant increases in the number of utterances containing mazes, total number of mazes, and percent mazed words, compared to typically developing control participants?

Thirty-eight percent (97 of 256) of the participants were at least plus one standard deviation in one of the three maze categories.

3. Do all children with SLI produce significantly less language per unit time than typically developing control participants as measured by the number of words produced per minute?

A total of 98 children produced fewer words per minute than their typically developing peers. The anecdotal observation that children with language disorders produce less talk per unit time than typically developing children is true for more than one third of the participants. I was somewhat surprised that slow rate was distributed across the entire age range of the sample; it was expected to be most frequent among the younger children. I expect that slow rate will be frequent among children showing delayed language performance when performance across the five variable categories is examined.

4. Do all children with SLI produce significantly more language per unit time than typically developing control participants as measured by the number of words produced per minute?

Only 19 of the 256 participants exhibited rates more than one standard deviation above the mean for their age peers. This is only about 7% of the sample, and it does demonstrate that some children do have rate problems that run contrary to the usual view of language disordered children.

5. Are significant increases in the number of pauses in a transcript more likely to be associated with the delayed development, increased maze production, or fast or slow speaking rate?

Almost 50% of the participants produced more pauses than their age peers. Pauses were noted if they were 2 seconds or longer in duration. They occurred either within utterances or between utterances. I expected that frequent pausing would be associated with slow rates of talking and with mazes where pauses may be a silent version of a maze. The results were surprising in several ways. First of all, pauses seem to be fairly evenly distributed across the categories. Thirty-one percent of the children had increased pause time with delayed development, 10% had increased mazes with pauses, 28% were slow talkers with pauses, 6% had increased pauses alone, and only 1% were fast talkers with pauses.

INTERPRETING THE RESULTS OF THESE ANALYSES

Language samples of 256 children identified by the Madison Metropolitan School District as language impaired were analyzed in a search for common characteristics among these children. The criteria used by MMSD to qualify students for services is performance of minus 1.5 standard deviations on measures of language performance using at least two different instruments. The identification process does not distinguish between deficits in comprehension or production. Qualification may be minimally based on deficits on language production. These data together address the general questions: Is SLI a unitary construct? Is there an SLI phenotype? Are there subcategories of SLI with distinct language phenotypes? Using a diverse set of general analyses of development, rate, and verbal fluency, there is clearly no unitary construct for SLI. The results indicate that as a group, the children with SLI did not all show deficits on any one of the five categories examined. However, these categories do not exhaust the descriptions that are useful for characterizing SLI; for example, there are no measures of pragmatic or discourse performance or detailed measures of any language level. Given the results, however, I would predict that subsequent detailed analyses would result in more differentiation rather than a more homogeneous outcome.

The initial analysis of these data evaluated each variable as if it were independent, much like the past literature in the field, that is, do children with a language disorder do X? The answer provided by these data are consistent with the literature; some children show deficits in one category, some in another area, and others in several categories. The data evaluated here point out that SLI is a diverse construct. Clearly, large numbers of participants with SLI are needed to adequately describe these children. Further, probably only longitudinal data will provide the details necessary to evaluate how this disorder changes over time.

The relationship that exists among variables is of particular interest in describing profiles of language strengths and weaknesses. When evaluating the possible combinations of the five variable sets, how many participants will show a specific deficit on one variable set versus deficits on several variable sets simultaneously? Table 11.3 shows the overlapping deficits for variables A, B, C, and D. For participants showing deficits in one of the four categories, how many show deficits in the other three, individually or in combination? This table reveals participants with specific combinations of deficits, for example, ABC but not D, contains those participants with deficits on any A variable, plus any B variable, plus C, but not D, excluding any participant with deficits in ABC & D.

Several outcomes are worth mentioning. The largest exclusive combination was a deficit on a developmental measure and a slow rate. Twenty-five percent of the sample had this combination only. Surprisingly, this combination occurred throughout the entire age range. The next most frequent outcome listed on Table 11.3 shows deficits only on a developmental variable (15%), and a maze variable (17%). The remaining combinations were very small. It is of interest that children who exhibit developmental delays do not frequently exhibit maze deficits.

CONCLUSIONS

Clinicians and researchers recognize that SLI is not a unitary construct and that multidimensional models of language development and disordered language must be used to document different disorder types (Fletcher, 1991; Miller, 1987b,

TABLE 11.3
Groups A, B, C, and D in Combination Where A = Delayed Development,
B = Mazes, C = Slow Talker, and D = Fast Talker

Category	Count	Percent	Mean age	Age range
ABC (not D)	22	8.59%	7.85	2.58–13.08
ABD (not C)	0	0.00%		
AB (not C or D)	21	8.20%	7.42	4.92–12.17
AC (not B or D)	64	25.00%	8.26	2.67–14.67
AD (not B or C)	1	0.39%	8.00	
BC (not A or D)	1	0.39%	7.67	
BC (not A or C)	10	3.91%	8.09	5.42–11.33
A (not BC or D)	39	15.23%	7.36	3.75–14.58
B (not AC or D)	43	16.80%	8.79	3.25–14.08
C (not AB or D)	11	4.30%	8.50	4.50–13.75
D (not AB or C)	8	3.13%	7.22	4.67–10.83
X (not ABC or D)	36	14.06%	9.44	3.08–14.83

1991; Snow, 1991; Tallal, 1988). Detailed description, that is, defining the language phenotype, is necessary to pursue any genetic account of specific language impairment; this description must be multidimensional, a view supported by the data discussed in this chapter. There is reason to believe that SLI may have genetic links and a specific phenotype (cf. Gilger, this volume). For example:

1. SLI is more common in males than females.
2. Tomblin (1989) reported significant increases in the incidence of SLI in families with males having SLI over families with females having SLI (cf. Tomblin, this volume).
3. Hier and Rosenberger (1980) reported that 63% of their sample of SLI children had a family history of developmental language disorder.
4. Neils and Aram (1986) reported that more than 20% of their sample of SLI children had a positive family history.

The Neils and Aram study is the only study to report different incidence figures for different types of language impairment, suggesting that a rigorous description of the SLI population is a positive step toward the genetic study of SLI.

Progress in documenting genetic links to SLI are limited by description of language phenotypes and by the identification of those language performance variables that are dependent more on biological effects rather than on environmental influences for their expression. This view of the genetics of the language disorder enterprise suggests that linguists should find not one, but several, genetic links to language disorders, each associated with a different phenotype. The work conducted on the language profiles of persons with specific chromosomal anomalies demonstrates that mental retardation as a behavioral classification has little relevance to the genetics of language disorders, except to point out that mental retardation, like SLI, is a behavioral classification and not the result of a specific cause. The data presented here from studies of children with SLI suggest that there may be several different behavioral phenotypes associated with this language disorder.

ACKNOWLEDGMENT

The preparation of this chapter was supported in part by research grant number HD22393, NICHD, The National Institutes of Health.

REFERENCES

American Psychiatric Association (1994). *Diagnostic and Statistical Manual of Mental Disorders: DSM-IV* (4th ed.). Washington, DC: Author.

Aram, D. M., Ekelman, B. L., & Nation, J. E. (1984). Preschoolers with language disorders: Ten years later. *Journal of Speech and Hearing Research, 27*, 232–244.

Aram, D. M., Morris, R., & Hall, N. (1993). Clinical and research congruence in identifying children with specific language impairment, *Journal of Speech and Hearing Research, 36*, 580–591.

Bates, L., & Thal, D. (1991). Associations and dissociations in child language development. In J. Miller (Ed.), *Research on child language disorders: A decade of progress* (pp. 145–168). Boston: College Hill Press.

Bishop, D. (1992). The underlying nature of specific language impairment. *Journal of Child Psychology and Psychiatry, 33*, 3–66.

Bishop, D. (1994, June). *Is specific language impairment a valid diagnostic category? Genetic and psycholinguistic evidence.* Paper presented at the 15th annual symposium on research in child language disorders, University of Wisconsin-Madison, Madison, WI.

Camarata, S. M., & Schwartz, R. G. (1985). Production of object words and action words. *Journal of Speech and Hearing Research, 28*, 323–330.

Chapman, R. (in press). Language development in children and adolescents with Down syndrome. In P. Fletcher & B. MacWhinney (Eds.), *The handbook of child language.* London: Bazel Blackwell.

Chapman, R. S., & Miller J. F. (1975). Word order in early two and three word utterances: Does production precede comprehension? *Journal of Speech and Hearing Research, 18*(2), 355–371.

Dollaghan, C., & Campbell, T. (1992). A procedure for classifying disruptions in spontaneous language samples. *Topics in Language Disorders, 12*, 42–55.

Fletcher, P. (1991). Evidence from syntax for language impairment. In J. Miller (Ed.), *Research on child language disorders: A decade of progress* (pp. 169–187). Boston: College Hill Press.

Fowler, A. (1988). Determinants of rate of language growth in children with Down syndrome. In L. Nadel (Ed.), *The psychobiology of Down syndrome* (pp. 217–246). Cambridge: MIT Press.

Fowler, A. (1990). Language abilities of children with Down syndrome: Evidence for a specific syntactic delay. In D. Cicchetti & M. Beeghly (Eds.), *Children with Down syndrome: A developmental perspective* (pp. 302–328). Cambridge: Cambridge University Press.

Garrett, M. (1992). Disorders of lexical selection. *Cognition, 11*, 29–46.

German, D. J. (1987). Spontaneous language profiles of children with word-finding problems. *Language, Speech and Hearing Services in the Schools, 18*, 217–230.

Hier, D., & Rosenberger, P. (1980). Focal left temporal lobe lesions and delayed speech acquisition. *Developmental and Behavioral Pediatrics, 1,* 54–57.

Johnston, J. (1988). Specific language disorders in the child. In N. Lass, L. McReynolds, J. Northern, & D. Yoder (Eds.), *The handbook of speech pathology* (pp. 685–715). Philadelphia: Saunders.

Johnston, J., & Kamhi, A. (1984). The same can be less: Syntactic and semantic aspects of the utterances of language impaired children. *Merrill-Palmer Quarterly, 30,* 65–86.

Leonard, L. B., Camarata, S., Schwartz, R. G., Chapman, K., & Messick, C. (1985). Homonymy and the voiced-voiceless distinction in the speech of children with specific language impairment. *Journal of Speech and Hearing Research, 28,* 215–224.

Liles, B., & Watt, J. (1984). On the meaning of "language delay." *Folia Phoniatr, 36,* 40–48.

Loban, W. (1976). *Language development: Kindergarten through grade twelve* (Research Report No. 18), Urbana, IL: National Council of Teachers of English.

MacLachlan, B., & Chapman, R. (1988). Communication breakdowns in normal and language learning-disabled children's conversation and narration. *Journal of Speech and Hearing Research, 53,* 2–7.

Miller, J. F. (1978). Assessing children's language behavior: A developmental process approach. In R. L. Schiefelbusch (Ed.), *The basics of language intervention* (pp. 269–318). Baltimore: University Park Press.

Miller, J. F. (1981). *Assessing language production in children: Experimental procedures.* Baltimore: University Park Press.

Miller, J. F. (1987a). Language and communication characteristics of children with Down syndrome. In S. M. Pueschel, C. Tingey, J. E. Rynders, A. C. Crocker, & D. M. Crutcher (Eds.), *New perspectives on Down syndrome* (pp. 233–262). Baltimore: Brookes.

Miller, J. F. (1987b). A grammatical characterization of language disorder. In A. Martin, P. Fletcher, P. Grunewell, & D. Hall (Eds.), *First international symposium: Specific speech and language disorders in children* (pp. 100–114). London: AFASIC Press.

Miller, J. F. (1988). The developmental asynchrony of language development in children with Down syndrome. In L. Nadel (Ed.), *The psychobiology of Down syndrome* (pp. 167–198). Cambridge: MIT Press.

Miller, J. F. (1991). Quantifying productive language disorders. In J. Miller (Ed.), *Research on child language disorders: A decade of progress* (pp. 211–220). Austin, TX: Pro-Ed.

Miller, J. F., & Chapman, R. S. (1985–1993). *Systematic Analysis of Language Transcripts (SALT): A computer program designed to analyze free speech samples*—DOS Versions 1.0 through 3.0. [computer program]. Language Analysis Laboratory, Waisman Center, University of Wisconsin-Madison.

Miller, J. F., Chapman, R. S., & MacKenzie, H. (1981). *Individual differences in the language acquisition of mentally retarded children.* Paper presented at the second international congress for the study of child language, University of British Columbia, Vancouver, B.C., Canada.

Miller, J. F., Frieberg, C., Rolland, M.-B., Reeves, M. (1992). Implementing computerized language sample analysis in the public school. In C. Dollaghan (Ed.), *Topics in Language Disorders, 12*(2), 69–82.

Miller, J. F., & Klee, T. (1995). *Quantifying language disorders in children.* New York: Cambridge University Press.

Miller, J., & Leadholm, B. (1992). *Language Sample Analysis: The Wisconsin Guide.* Madison, Wisconsin: The Wisconsin Department of Public Instruction.

Neills, J., & Aram, D. (1986). Family history of children with developmental language disorders. *Perceptual and Motor Skills, 63,* 655–658.

Nelson, K. (1991). Event knowledge and the development of language functions. In J. Miller (Ed.), *Research on child language disorders: A decade of progress* (pp. 125–142). Austin, Texas: Pro-Ed.

Rapin, I. (1988). Discussion. In J. Kavanagh & T. Truss (Eds.), *Learning disabilities: Proceedings of the national conference* (pp. 181–272). Parkton, MD: York Press.

Schery, T. K. (1985). Correlates of language development in language-disordered children. *Journal of Speech and Hearing Disorders, 50,* 73–83.

Snow, C. (1991). Diverse conversational contexts for the acquisition of various language skills. In J. Miller (Ed.), *Research on child language disorders: A decade of progress* (pp. 105–124). Austin, TX: Pro-Ed.

Tallal, P. (1988). Developmental language disorders. In J. Kavanagh & T. Truss (Eds.), *Learning disabilities: Proceedings of the national conference* (pp. 181–272). Parkton, MD: York Press.

Tomblin, B. (1989). Familial concentration of developmental language impairment. *Journal of Speech and Hearing Disorders, 54,* 287–295.

Van Kleeck, A. (1988). Language delay in the child. In N. Lass, L. McReynolds, J. Northern, & D. Yoder (Eds.), *The handbook of speech pathology* (pp. 655–684). Philadelphia: Saunders.

Weiner, P. (1985). The value of follow-up studies. *Topics in Language Disorders, 5*(3), 78–92.

BRAIN SCIENCE

12

PHENOTYPIC VARIABILITY IN BRAIN-BEHAVIOR STUDIES OF SPECIFIC LANGUAGE IMPAIRMENT

Elena Plante
The University of Arizona

The underlying assumption of a brain-behavior study of developmental language disorders is that the combination of neurobiological and behavioral information may provide insights into the nature of the disorder beyond that provided by behavioral data alone. This approach serves to expand the phenotype of the disorder to include neuroanatomical as well as behavioral signs. To examine brain-language relations, my colleagues and I concentrate on children identified as specifically language impaired. These children have impaired language skills in the presence of documented normal hearing, normal overall nonverbal skills, and parent report of no other handicapping conditions. At the school ages, affected children also meet standard definitions for a learning disability or dyslexia, as poor language skills impact reading and academic skills. Regardless of their current label, the individuals we study all share a common core of impaired language skills.

Because our subjects provide us with both behavioral and neuroanatomical information, we have an opportunity to assess behavior within a neurobiological context, and conversely, to assess neuroanatomical variation within a behavioral context. This can be important for how we eventually understand both aspects of the disorder. For example, many investigators, when faced with different presenting behaviors across subjects, suggest that their subjects may represent subgroups within the language disorders spectrum. When behavior is paired with knowledge of neuroanatomy, we have an opportunity to assess the degree of variability that might reasonably arise from a shared neuroanatomical profile. This gives us an anchor against which to evaluate whether different behavioral signs may plausibly arise from one, or more than one, biological source.

Although there is great potential for the neuroanatomical and behavioral components to inform each other, there is also the opportunity for each, independently, to reveal variability. Variability occurs both across and within brain-behavior studies. Although some variability is biologically driven, other sources of variability, or pseudo-variability, are attributable to the methodological choices of the investigator. In this chapter, I present examples of both types of variability as they affect neuroanatomical studies of developmental language and learning disorders.

BACKGROUND

The pursuit of neuroanatomical correlates of language and learning disorders has been heavily influenced by Galaburda's autopsy results for developmental dyslexia. In 1985, he and his colleagues described four brains of male dyslexics (Galaburda, Sherman, Rosen, Aboitiz, & Geschwind, 1985). In all cases, cortical ectopias, which are clusters of misplaced neurons, were found in the right and left hemispheres, predominantly in the areas surrounding the sylvian fissure. One case had polymicrogyri, a cortical malformation, in the posterior region of the sylvian fissure. These structural anomalies were accompanied by a symmetry of the plana temporale. This symmetry was described as a left planum temporal of typical size paired with a right planum that was atypically large.

These findings were important for several reasons. First, these neuroanatomical findings clustered within the regions of the brain classically associated with language functioning. These language-related areas appear in the left hemisphere of the brain and surround the sylvian fissure. They include the inferior frontal gyrus and frontal operculum anteriorly, the posterior aspect of the superior temporal gyrus, including the planum temporale, and the supermarginal gyrus and operculum of the parietal lobe (see Fig. 12.1).

Second, these brain anomalies point to a disturbance of brain development. Cortical ectopias result when clusters of newly generated neurons become misplaced during prenatal brain development. Polymicrogyri occur when the development of the cell layers in the cortex (surface layer) is severely disturbed. In this case, the resulting malformation takes on the appearance of many small gyri where fewer, larger gyri should appear. Even planum temporale symmetry can be a reflection of prenatal events. Brain asymmetries emerge with brain development during the third prenatal trimester (Chi, Dooling, & Gilles, 1977a). The typical pattern of asymmetry may be altered by prenatal effects during this time.

Finally, the bilateral nature of these brain findings in dyslexia are consistent with a global effect on brain development. This is distinctly different from the findings typically associated with acquired language disorders. Acquired language disorders, or aphasia, in adults or children typically result from localized damage to perisylvian structures within the left hemisphere. Thus, the profile of developmental alterations in dyslexia shares some neuroanatomical regions with acquired language disorders, but is unique in its bilateral nature.

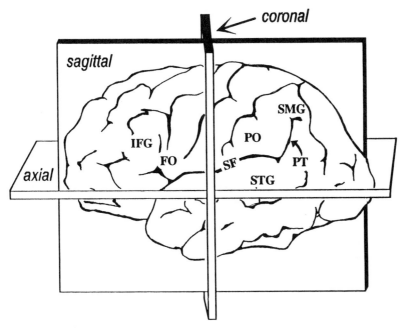

FIG. 12.1. The left hemisphere of the brain, bisected in each of the major planes of view (axial, sagittal, coronal). Language-related structures are labeled as follows: FO = Frontal Operculum, IFG = Inferior Frontal Gyrus, PO = Parietal Operculum, PT = Planum Temporal (internal, lower bank of the sylvian fissure), SF = Sylvian Fissure, SMG = Supramarginal Gyrus, STG = Superior Temporal Gyrus.

These perisylvian findings, along with the importance of these regions for language functioning, convinced us that this was the place to look for structural differences in children with documented language impairments. We decided to look for structural differences in this area by measuring the regional anatomy from the magnetic resonance imaging (MRI) scans of normal and language-impaired individuals. This approach, called *morphometric analysis*, provides the obvious advantage of being able to study the brains of living individuals whose behavioral profile can be well documented.

Morphometric analysis requires accurate identification of target brain structures and a systematic method of quantifying the characteristics (e.g., size, asymmetry) of those structures. However, the shape and location of the planum temporale make it a difficult structure to measure directly from MRI scans. For example, the difference in the course of the sylvian fissure in the left and right hemispheres can result in it being optimally viewed in one hemisphere and less than optimally viewed in the other within any given slice. This poses special challenges in obtaining accurate measures of this particular structure. The difficulties inherent in measuring this area were illustrated by Larsen and colleagues (Larsen, Odegaard, Grude, & Hoien, 1989). They scanned autopsied brains and measured the planum from coronal MRI slices and on the actual brains and found that one third of the planum asymmetries had been misclassified based on the MRI measurement protocol.

I adopted an alternate strategy to counter such problems by adopting an axial-based, volumetric measure of structures surrounding the sylvian fissure. This approach minimized several problematic sources of error variance that affect morphometric analysis of MRI scans (cf. Plante & Turkstra, 1991). The resulting neuroanatomic region of interest included portions of the frontal operculum anteriorly, and the planum temporale posteriorly (see Fig. 12.2 for an

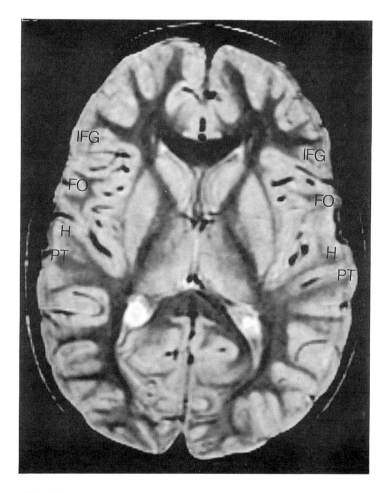

FIG. 12.2. An axial slice of the brain of a language- and reading-disabled subject. Note the relative symmetry of language-related structures of the left hemisphere (on right side of image) compared with the same structures in the right hemisphere.

axial view of these structures). Both of these areas have a typical left-greater-than-right asymmetry in the general population (e.g., Albanese, Merlo, Albanese, & Gomez, 1989; Chi, Dooling, & Gilles, 1977a; Geschwind & Levitsky, 1968; Wada, Clarke, & Hamm, 1975; Witelson & Pallie, 1973). Consequently, this measure produced similar rates of asymmetry in normal individuals as originally reported for the planum temporale by Geschwind and Levitsky (1968).

My colleagues and I (Plante, Swisher, Vance, & Rapcsak, 1991) applied this protocol in a study of eight boys, ranging in ages from 4 to 9 years, with developmental language impairment. Six out of eight of the language-impaired boys had an atypical pattern of asymmetry of the perisylvian structures. As in the autopsy cases, this consisted of an average-sized left perisylvian area paired with an atypically large right perisylvian area.

I extended this inquiry in a study of the families of four of these language-impaired children (Plante, 1991). Consistent with other familial studies (e.g., DeFries, Singer, Foch, & Lewitter, 1978; Neils & Aram, 1986; Tallal, Ross, & Curtiss, 1989; Tomblin, 1989), I also found that when there was one language-impaired child in a family, there tended to be other affected family members. In addition, the atypical asymmetry that was characteristic of specifically language-impaired boys was also found among their siblings and among their parents. However, there was an imperfect correspondence between the behavioral status and the brain findings as they were measured in this study. For most of the subjects, atypical asymmetries co-occurred with evidence of a developmental language impairment. Occasionally, no atypical asymmetry was detected in behaviorally impaired individuals. Atypical asymmetries were also found for individuals in whom we failed to find evidence of the behavioral disorder.

NEUROANATOMICAL VARIABILITY

Measurement Characteristics

We conclude that morphometric analysis of MRI scans is sensitive enough to detect brain differences in a majority of cases. However, for a minority of cases, no effect is found. What might account for this intersubject variability? The first issue is the relative sensitivity of the various methods of assessing neuro-anatomy. Neuroanatomical effects occur on a continuum, ranging from the sub-cellular to the level of gross anatomy. Effects on autopsy can be documented at any of these levels. Although autopsy work offers the finest detail, findings in these studies are strongly influenced by a priori hypotheses. For example, the four brains of male dyslexics (Galaburda et al., 1985) have twice been reanalyzed with additional findings noted each time (Humphreys, Kaufmann, and Galaburda, 1990; Livingstone, Rosen, Drislane, and Galaburda, 1991). In contrast to autopsy analysis, morphometric MRI analysis is only sensitive to the most extreme end of the neuroanatomical continuum. A cellular anomaly must be so widespread that it changes gross anatomy in order to be detected by measurement of regional anatomy from MRI scans. Therefore, morphometric analysis is not sensitive when cellular level disturbances are relatively mild.

On the other hand, clinical readings of MRI scans, as typically performed by a neuroradiologist, are much less sensitive than either autopsy or morphometric analysis. During clinical readings, the neuroradiologist typically visually examines the MRI films, looking for signs of either developmental alterations or acquired lesions on the films. In the eight cases I reported, not one was identified as atypical on a standard clinical reading. This is because deviations must be sufficiently large and visually distinct to be seen on an image that is only inches large. Size or volume differences in regional anatomy, as measured with morphometric analysis, will not necessarily be visually distinct on any given slice of the brain image. Therefore, the relative sensitivity of clinical review, MRI measurement, and autopsy analysis is a strong example of why negative findings in neuroanatomy are not evidence of normalcy.

For imaging studies, differences in measurement protocols across studies are also a potential source of variability. So far, however, differences have reflected more surface, rather than substantive, differences. To date, various measures of perisylvian structures have been obtained from axial, sagittal, and coronal views (see Fig. 12.1) using three-dimensional or volumetric measures (e.g., Jernigan, Hesselink, Sowell, & Tallal, 1991; Plante, 1991; Plante et al., 1991) and two-dimensional or area measures (e.g., Larsen, Hoien, Lundberg, & Odegaard, 1990; Leonard et al., 1993), and one-dimensional or linear measures (Hynd, Semrud-Clikeman, Lorys, Novey, & Eliopulos, 1990). To the extent that these measures include similar anatomical regions, the general outcomes have been remarkably similar. For example, those studies that have examined patterns of asymmetry found deviations from the typical patterns in language-impaired, learning-disabled, and dyslexic subjects. Rather than viewing the protocol differences as a source of variance, I have seen them as providing converging evidence for the nature of the neuroanatomical effect. Indeed, what variability occurs often sheds light on the relative contribution of spatially close regions to the observed neuroanatomical effect (cf. Leonard et al., 1993).

Subject Characteristics

Various subject characteristics are a potential source of biologically-based phenotypic variability. However, the role of subject variables in neuroanatomical studies can be very different from their role in behavioral studies. To illustrate this, I examine three subject traits that are frequently addressed in behavioral investigations: age, ethnicity, and gender.

Age. Age does not seem to affect the perisylvian findings, although it may well affect other types of neurobiological signs (e.g., myelinization). Both autopsy and imaging studies to date have included individuals from ages 4 to 88, with a

TABLE 12.1
Asymmetry Findings Across Ages in Individuals With Developmental
Language Impairments or Dyslexia

Autopsy cases	
Atypical symmetry of the plana temporale	
Cohen, Campbell, & Yaghmai, 1989	Age 7
Galaburda, Sherman, Rosen, Aboitiz, & Geschwind, 1985	Ages 14–32
Humphreys, Kaufman, & Galaburda, 1990	Ages 20–88
Imaging studies	
Atypical asymmetries of perisylvian structures	
Leonard, Voeller, Lombardino, et al., 1993	Ages 15–65
Plante, 1991	Ages 5–57
Plante, Swisher, Vance, & Rapcsak, 1991	Ages 4–9

remarkable similarity in neuroanatomical effect across this age range. Table 12.1 summarizes the findings of several studies that examined asymmetries of perisylvian structures in impaired subjects over a range of ages. This stability across ages is not surprising, given that the neuroanatomical measures are anchored to the brain's gyral configuration and asymmetry, which are products of prenatal development. Gyri appear during the third trimester and the brain's gyral pattern and asymmetries are set during this time (Chi, Dooling, & Gilles, 1977a, 1977b). Autopsy studies indicate the rate of asymmetry does not shift remarkably from birth across the life span (Chi, Dooling, & Gilles, 1977a; Wada, Clarke, & Hamm, 1975; Witelson & Pallie, 1973). Likewise, signs of atypical development of gyral size and configuration in language-impaired and dyslexic individuals remain constant from the preschool to geriatric ages.

This stability across the life span allows for the possibility of tracking atypical asymmetries across generations within families (Plante, 1991). To the extent that the measurement protocols depend on these early appearing and stable neuroanatomical landmarks, the neuroanatomical correlates of the disorder will also be demonstrable across generations. Thus, the stability of this feature allows it to serve as a marker of neurobiologically "affected" individuals within a family, who may or may not also show a behavioral impairment.

Ethnicity. There is a potential effect of ethnicity on the rate of atypical asymmetries in the population. It is reasonable to suppose that there might be genetic variation among ethnic populations that would affect brain morphology. In 1984, McShane, Risse, and Rubens found that the rates of asymmetry of the occipital petalia measured on computerized tomography (CT) scans differed for White, Black, and Native American subjects. Although these results are certainly suggestive, there are several limitations inherent to the method. The resolution of CT at the time these measures originated was such that the asymmetry actually measured was that from the longitudinal fissure to the edge of the skull, not necessarily the edge of the brain itself. These linear measures are remarkably sensitive to head tilt, which will be affected by the shape of the skull as it lays in the scanner. So the difference measured on CT may reflect racial difference in skull shape more so than brain asymmetry.

This technique for measuring brain asymmetries did not transfer well to MRI, as single linear measures are too easily influenced by slight gyral variations and minor deviations in slice angle. Chu, Damasio, and Tranel (1991) reported poor correspondence between CT- and MRI-measured asymmetries within the same subjects. Given the poor correspondence between CT- and MRI-measured asymmetries, it is an open question as to whether an ethnic effect on asymmetries would be replicated with MRI analysis.

Gender. There is emerging evidence that suggests a possible gender effect in the neuroanatomical correlate of language- and learning-disabilities. This possibility is highlighted by a series of autopsies of females identified as language-impaired or dyslexic (Cohen, Campbell, & Yaghmai, 1989; Humphreys et al., 1990). An atypical symmetry of the plana was described in these cases, but neural ectopias were much less frequent than in the earlier-described male cases. In addition, the myelinated and unmyelinated glial scars have appeared more frequently in female cases to date.

Although the available cases are still few, the neuroanatomical findings suggest the brains of males and females were affected most strongly during different periods of development. Early brain development is characterized by cell division, through which new neurons are generated. Newly generated cells then migrate towards the brain's surface where they form the cortical layers. The period of cell migration extends into the third trimester and corresponds to the appearance of the brain's gyri and sulci. Brain asymmetries emerge with gyral development during the third trimester. In contrast, myelinization of cell fibers is largely a postnatal process. When this developmental progression is altered, the resulting signs vary according to the events that were occurring when the effect of development occurred.

The appearance of relatively higher numbers of cortical ectopias in the male cases suggests a heavier effect occurred during the period of cell migration in males than in females. This may explain the higher numbers of neural ectopias reported in the males cases to date. Cells arrested early in migration appear as subcortical ectopic neurons; cells arrested later appear as cortical ectopias and brain warts. In contrast, myelinated glial scarring, more prominent in the female cases, is associated with the late prenatal or early postnatal period, prior to the regional completion of myelinization. This neuroanatomical pattern suggests a somewhat later effect in females. The variation in the neuroanatomical findings in the cases to date may well be a relative timing difference or a difference in severity for the two genders.

This potential effect appears, so far, to be gender-related. This could mean that the effect may be directly mediated by sex-linked variables, or it may be a side effect of maturational differences indirectly associated with gender. For example, in a study of a set of fraternal twins (Plante, Swisher, & Vance, 1989), we suggested that anatomical and subsequent functional differences in a brother and sister discordant for language impairment may reflect an interaction between the time of a neuroanatomical effect and the relative maturation of the two fetuses in utero. In this case, the male twin had a symmetry of the perisylvian areas and impaired language skills. His normally developing twin sister had a reversed (right > left) asymmetry without evidence of impaired language. The female was reported to have reached developmental milestones before her brother, suggesting that she was maturationally somewhat ahead of her brother. If she was also developing faster than her brother in utero, this would result in a timing difference for any prenatal effect that may have altered fetal brain development. Under this scenario, the effect would be indirectly linked to gender by virtue of the relative difference in maturation rates.

VARIABILITY IN CAUSE

At this early stage of investigation, there is the distinct possibility of variability in the factors or events that cause these disorders. We know that, at some level, children with known biological differences can present similar behaviors. This is true, to some extent, with acquired and developmental language impairments. Both types of children will test in the low range on norm-referenced tests of language skills, and in the low-average range on tests of nonverbal abilities (see Aram & Whitaker, 1988, for a review of behavioral profiles associated with unilateral and bilateral lesions in children). A similar phenomenon has been described for reading disabilities. Decker and Bender (1988) reported comparable reading test scores could be obtained from individuals with Klinefelter's syndrome (a genetic disorder) and from individuals with idiopathic reading disabilities.

The similarities in the surface presentation of children with known biological differences are likely a function of the restricted range of behaviors assessed with a given standardized test or research protocol. Children given a particular test or set of tests can only show deficits in the subset of skills sampled. Therefore, biologically-based differences that might emerge with task demands of a different nature can be masked by the tests administered. It may be that different tasks could detect fine distinctions within and between members of biologically-distinct groups.

Neuroanatomical information can, in some cases, be helpful in determining when biologically-based differences in cause exist. Certainly, the distinction between acquired and developmental language impairments is an easy one for MRI to determine; the characteristic signs of brain damage are quite different from the characteristics of altered prenatal development. Conversely, MRI can also be helpful in assessing the extent to which variation in behavior is associated with unique neuroanatomical profiles. One example of such an approach was taken by Hynd and colleagues (1990). They examined brain morphology in subjects with dyslexia and in subjects with attention deficit disorder. Because these two conditions frequently co-occur, they may be biologically related. Although the two conditions did share some atypical neuroanatomical features, other features differed for the two behavioral conditions. Specifically, atypical measures of perisylvian structures were common only for the dyslexic subjects.

My colleagues and I have tested a series of hypotheses concerning the nature of the biological effect(s) that give rise to language impairment. In order to be considered a correlate of the disorder, this biological sign must be consistent with what is known about the behavioral components of the disorder. The first studies (Plante et al., 1989, 1991) linked impaired language skills with atypical neuroanatomy in regions long associated with language functioning. A study of families affected by language impairment established that the neurobiological sign appeared to be familial (Plante, 1991). This led to hypotheses concerning causal agents/events that were both potentially heritable and were capable of producing the types of anatomical signs seen in the language-impaired subjects and their family members.

One such potential agent is the action of testosterone on the developing brain (cf. Geschwind & Behan, 1982). To test the viability of increased testosterone levels as a causal agent, we tested a group of children with a rare genetic disorder, congenital adrenal hyperplasia (CAH), that leads to increased testosterone production. This disorder involves a genetic abnormality located on Chromosome 6 that is recessively transmitted. The atypical perisylvian asymmetries documented first in language-impaired children were also found in children with CAH. Some of the CAH subjects also had poor language skills. In addition, a majority of their families contained at least one individual who was identified as speech/language impaired or learning disabled. Therefore, it is possible that both direct and indirect factors may be linking the presence of altered brain morphology and impaired language skills to the genetic anomaly (Plante, Boliek, Binkiewicz, & Erly, in press).

VARIABILITY IN THE BEHAVIORAL CORRELATES

Let us turn our attention to the second half of the brain-behavior relation: sources of variability in the behavioral phenotype. The method of identification is perhaps the largest single source of variability across studies. The approach to identification is affected by a number of factors that are not biologically-influenced as well as some that are. These include (a) the investigator's focus, (b) test-generated variability (c) true phenotypic variability, (d) age-related variability, and (e) interactive effects on phenotype.

The Investigator's Focus

The investigator's focus influences every decision that follows. For example, Lewis focused on children with phonological disorders (Lewis, 1990, 1992). Because these children also frequently have language impairments, our actual populations overlap despite differences in the feature that draws the child into our respective studies. The same is true, and perhaps even more so, for studies of learning disabilities and language impairments. I routinely recruit school-age language-impaired subjects from Learning Disabilities (LD) classrooms and programs for dyslexia. In a majority of cases, norm-referenced testing confirms poor language skills in these children. This is not surprising, given that the federal law that originally mandated services for learning-disabled students (PL94-142) defines the disability as involving "a disorder in one or more of the basic psychological processes of language, spoken or written, which may manifest itself in an imperfect ability to listen, think, speak, read, write, spell, or do mathematical calculations." The law provides *dyslexia* and *developmental aphasia* as alternate terms.

In addition to the differences in clinical designations, an investigator's focus has a secondary influence that contributes to phenotypic variability across studies. When an investigator is focused on a subset of the behavioral symptomatology, he or she tends to select the measures that will highlight the skills of interest. Therefore, studies of learning- or reading-disabled children tend to select academic achievement batteries and studies of language-impaired children tend to select tests of language skills. At a surface level, this testing difference will tend to make these children look like independent populations when they may be one and the same.

Test-generated Variability

Given a framework for conceptualizing a disorder, an investigator has many options for identifying children who have or do not have the disorder. My colleague Rebecca Vance and I reviewed 21 tests of language that had norms for 4- and 5-year-old children intended to identify childhood language impairments (Plante & Vance, 1994a), which demonstrates the range of options even within this restricted age group. We went on to examine the diagnostic accuracy of four of these tests. We first administered a hearing screening and the Kaufman Assessment Battery for Children nonverbal scales (Kaufman & Kaufman, 1983), to 20 language impaired and 20 normal language 4- and 5-year-olds. These measures established normal hearing and nonverbal skills for all children who served as subjects. In addition, all the language-impaired children were receiving services for impaired language from speech-language pathologists. All normal language children were attending preschool and neither their parents nor their teachers had any concerns about their speech or language development.

We found that our samples' performance varied widely across the four tests, relative to the tests' own normative sample. For two tests, the normal children approximated or exceeded the average for the test norms; for two tests, they fell well below the average. Thus, the behavior of normal children was not directly comparable across these four tests of language skills. More of a problem was that the tests varied widely in their accuracy in discriminating between normal and impaired children. Discrimination was best, at 90% sensitivity and specificity, for the Structured Photographic Elicited Language Test-II (Werner & Kresheck, 1983). This accuracy level has been replicated in a second sample of children (Plante & Vance, 1994b). The diagnostic accuracy of the remaining three tests was fair to poor. Each test identified a different number of children as normal or impaired. In addition, the cutoff score that best discriminated between the two groups of children was different for every test. Thus, a single cutoff score applied across different tests, as typically used in both clinical practice and research studies, systematically under- or over-identified children across tests.

The same variability occurs with performance on nonverbal intelligence tests.[1] Swisher, Plante, and Lowell (1994) examined the relative performance of SLI and NL children on three nonverbal tests. The estimated intelligence of the children varied with the test given, with the average scores of the SLI group always lower than those of the NL group on all tests. The highest IQ scores were obtained with the Leiter International Performance Scale (Leiter, 1969) and the lowest with the short form of the Matrix Analogies Test (Naglieri, 1985).

In addition to this test effect, large between-group effects occurred when test items reflected skills that the literature indicated were weaknesses for SLI children (e.g., Johnston & Weismer, 1983; Kamhi, Catts, Mauer, Apel, & Gentry, 1988; Savich, 1984). These effects of "pocket" nonverbal deficits on IQ testing are somewhat problematic, given specific language impairment is frequently operationally defined as including nonverbal IQ that is "within normal limits."[2] The variability found across nonverbal tests would place some children in the normal range on one measure, but not on another. Thus, the nonverbal test selected may artificially alter the sample of children selected by virtue of its interaction with the nonverbal components of the disorder. When IQ-language discrepancies are used to select children, the situation is further complicated because the disorder's effect on IQ tests is combined with the sensitivity and specificity problems of the language tests (i.e., Aram, Morris, & Hall, 1992; Cole, Mills, &

---

[1]Nonverbal intelligence testing is typically administered to "rule out" poor cognition as an explanation for poor language skills. This is based on the premise that normal cognition is a prerequisite for normal language development. However, specific language impairment, along with other clinical populations, provides evidence of cognitive-linguistic dissociations. Therefore, normal cognition is not necessarily sufficient for normal language development.

[2]The historical use of nonverbal testing as an exclusionary criterion has had the paradoxical effect of suppressing information on cognitive strengths and weaknesses that could as easily be reinterpreted as part of the behavioral phenotype.

Kelly, 1994). This can result in a spurious restriction of the sample, which can interfere with the generalization of research findings to the broader population of children with specific language impairment.

These examples of test-generated variability emphasize the difficulty inherent in accurate identification and description of the behavioral phenotype associated with the disorder. Frequently, the discriminant accuracy (statistical sensitivity and specificity) of tests used to identify affected individuals is unknown. In these cases, investigators are left to choose tests based on their content, rather than their proven utility, for the purpose of identifying or describing impaired skills. In addition, investigators are left to adopt arbitrarily-derived standards (e.g., the lower first, fifth, or tenth of the normal distribution) to indicate the presence of impaired skills. Although this is a widespread practice, empirical evidence has demonstrated the risk of systematic misidentification that result from such practice (cf. Plante & Vance, 1994a, 1994b). Unfortunately, misidentification of affected individuals is a serious problem faced by genetic studies of language and learning disabilities (see Pennington & Smith, this volume).

True Phenotypic Variability

Despite the current challenges to accurate identification, I do think specifically language-impaired children can be empirically defined, given the right measures. Indeed, current work by others (see Leonard, Crago, Rice, & Tomblin, this volume) will contribute to the refinement of phenotypic description. For example, Rice (1994) has demonstrated that linguistic signs of impaired language in family members previously unidentified as affected can be revealed by examination of their use-specific linguistic features. These types of analyses can reveal deficits that are missed by norm-referenced testing. In addition, they can also indicate the degree to which linguistic signs vary within and between families affected by specific language impairment.

However, even with improved methods of identification, I do not think that specific language impairment necessarily constitutes a unique, biologically-based disorder, independent of other forms of developmental language impairments. Instead, there is probably a fair degree of true phenotypic variability within the population. In fact, we can find language impaired children who do and who do not meet classic definitions of specific language impairment within a single family (Plante, 1991).

Within-family variation. In the first family I studied (Plante, 1991), the three children, all boys, were language impaired. The youngest two children qualified as specifically language impaired, with nonverbal cognitive, socioemotional, motor, and hearing within normal limits and no secondary handicapping conditions. The oldest boy did not meet the definition of SLI. His socioemotional functioning, as measured by the Vineland Adaptive Behavior Scales (Sparrow, Balla, & Cicchetti, 1984), was poor, and he was diagnosed as having attention deficit/hyperactivity disorder for which he was taking Ritalin. Within the pair of specifically language-impaired siblings, there was still further variability. The second son only had some phonological errors in his speech as well as impaired language. This variability in the behavioral signs (e.g., presence or absence of a phonological component, comorbid conditions) occurred despite the fact that these three children were biologically related to each other. This opens the possibility that clinically significant variability can be associated with a single or small set of biological preconditions.

Gender variation. Gender is another potential source of true phenotypic variability. However, its exact role in phenotypic expression is an open question. One potential problem in examining the gender issue is the apparent underidentification of girls in the current clinical samples. This phenomenon was well documented by Shaywitz and colleagues (Shaywitz, Shaywitz, Fletcher, & Escobar, 1990). It has also been my experience that when I am able to document a previously unidentified case of developmental language impairment within a family, it tends to be one of the daughters. When girls are underidentified, it is difficult to sample the full range of phenotypes that represent the female, language-impaired population in order to determine where true gender differences may lie.

Age. A second source of true phenotypic variability within the members of a family is the age of the individual. Scarborough and Dobrich (1990) described the phenomenon of "illusory recovery" in language-delayed children. They suggested that the gains made by these children during a period when language skills reach a plateau in their age-mates gives the illusion that the disordered child has caught up with his normal peers and outgrown the earlier deficits. Unfortunately, according to many practicing clinicians, these children reappear on their caseloads, or on those of the learning disabilities specialist, as they grow into problems with higher level language skills, including reading and writing (Snyder, 1984).

I have been able, on an informal basis, to reassess some of the children who have been through our MRI protocols. I find that these children continue to be identified on norm-referenced testing as impaired. However, their presenting signs in terms of communication, academic performance, and enrollment status in special education services change with time. I should note that test scores do not necessarily predict classroom success. Some of the MRI subjects have been dismissed from therapy and have maintained passing, but usually not exceptional, grades despite the continued presence of impaired language skills documented by norm-referenced testing. Others who continue to test as markedly impaired have been receiving everything from consultative services to continued placement in a private school for dyslexia. This suggests that the norm-referenced testing can, in some cases, serve as an identifier of a language impairment, independent of whether the child is currently handicapped by the disorder.

Interactive Effects on Phenotype

Each of the influences on phenotypic description that I have discussed has the potential to interact under certain conditions. One such example of interaction occurs in the ascertainment of the behavioral status of the parents of language-impaired children. Depending on where and when the parents had been raised, language and learning disabilities may not have been widely recognized. Unlike their children, few parents would have received services for a developmental language impairment during childhood. Identification of affected individuals, perhaps for the first time, as adults will reflect interactions between the variables of the investigators' focus, selection of the identification method, and the parents' age.

A variety of methods have been employed to identify affected parents. All suggest that language-impaired children are much more likely to be born to language-impaired parents than are the rest of the population. However, the rate of impairment varies markedly across studies. The differences in the method of identification may be the major source of variability in the rates reported in these studies. Preliminary data from families I have studied serve to illustrate this point (see Fig. 12.3). I had tested adults who had language-impaired children and adults who lacked a family history for any developmental disorders. The adults completed a case history that allowed us to classify parents by their history of therapy (cf. Tomblin, 1989) and self-report of problems indicative of a developmental language impairment (cf. Tallal, Ross, & Curtiss, 1989). In addition, we administered a test battery documented by Tomblin and colleagues (Tomblin, Freese, & Records, 1992) to discriminate between young adults with and without a history of therapy for a language impairment.

The results of multiple methods of classifying adults are dramatically different (see Fig. 12.2). Under all methods, all of the controls are identified as unaffected. When we classify according to who saw a speech-language pathologist, two of the parents of language-impaired children are identified as affected. When we classify by self-report of difficulty, only one of the two previously identified parents also identified herself as having had problems as a child. In addition, one previously unidentified parent identified himself as affected. When we classified subjects against a data set (see Tomblin et al., 1992), yet another subset of two parents were identified as affected. Finally, when we compared the parents of language-impaired children to our own group of normal adults, a preliminary analysis correctly identified all of the parents as parents of language-impaired children and all of the control subjects as controls on the basis of their test scores. This last analysis suggested that subtle differences in test performance, below the level that would indicate a clinical impairment, differentiated parents from control subjects.

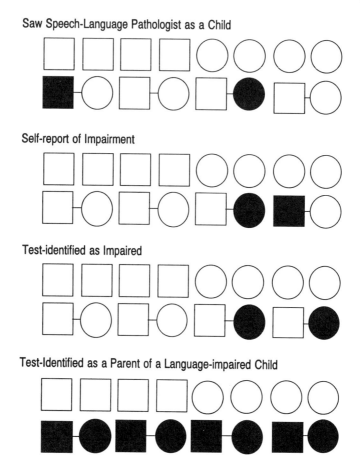

FIG. 12.3. A comparison of four methods of identifying adults affected by developmental language impairments. For each method, the first row depicts control males (squares) and females (circles). The second row depicts male and female parents of specifically language-impaired children. Dark circles indicate individuals identified as "affected."

Despite the preliminary nature of these analyses, there is a clear indication that the method of identification, interacting with age, produces a direct effect on who we identify as an affected adult. These differences across methods do not necessarily mean that one method is superior to another. Rather, it is the case that each method poses a different type of question. History of therapy and self-report of impairment help to triangulate on the issue of functional impairment among adult family members. In contrast, test results can often be sensitive to differences that are below that which might cause clinical concern. Test results may reflect a pattern of behavioral effect that accounts for both strongly (clinically-impaired) and very weakly affected family members. Furthermore, careful linguistic analysis (cf. Rice, 1994) can uncover specific aspects of the behavioral phenotype that appear across generations within particular families that standardized testing would not necessarily reveal.

In conclusion, the issue of subtypes within the language-impaired population is often raised when investigators are faced with unexpected or otherwise unexplained variability. My work with families of language-impaired children suggests that significant variability characterizes this population as a whole. A fair proportion of this variability seems attributable to the questions asked and the method of study. Additional behavioral variability is directly related to the subjects, but not all of it signals biologically distinct subgroups. I suggest that the biological variables may serve as a useful anchor for determining when behavioral variability reflects the population range and when it signals a true subgroup.

REFERENCES

Albanese, E., Merlo, A., Albanese, A., & Gomez, E. (1989). Anterior speech region asymmetry and weight-surface correlation. *Archives of Neurology, 46,* 307–310.

Aram, D. M., Morris, R., & Hall, N. E. (1992). The validity of discrepancy criteria for identifying children with developmental language disorders. *Journal of Learning Disabilities, 25,* 549–554.

Aram, D. M., & Whitaker, H. A. (1988). Cognitive sequelae of unilateral lesions acquired in early childhood. In D. L. Molfese & S. J. Segalowitz (Eds.), *Brain lateralization in children: Developmental implications* (pp. 417–436). New York: Guilford.

Chi, J. G., Dooling, E. C., & Gilles, F. H. (1977a). Left-right asymmetries of the temporal speech areas of the human fetus. *Archives of Neurology, 34,* 346–348.

Chi, J. G., Dooling, E. C., & Gilles, F. H. (1977b). Gyral development of the human brain. *Annals of Neurology, 1,* 86–93.

Chu, C. C., Damasio, H., & Tranel, D. (1991). Intrasubject correlations of occipital asymmetry measurements based on CT and MRI. *Society for Neuroscience Abstracts, 17,* 867.

Cohen, M., Campbell, R., & Yaghmai, F. (1989). Neuropathological abnormalities in developmental dysphasia. *Annals of Neurology, 25,* 567–570.

Cole, K. N., Mills, P. E., & Kelly, D. (1994). Agreement of assessment profiles used in cognitive referencing. *Language, Speech, and Hearing Services in Schools, 25,* 25–31.

Decker, S. N., & Bender, B. G. (1988). Converging evidence for multiple genetic forms of reading disability. *Brain and Language, 33,* 197–215.

DeFries, J. C., Singer, S. M., Foch, T. T., & Lewitter, F. I. (1978). Familial nature of reading disability. *British Journal of Psychiatry, 132,* 361–367.

Galaburda, A. M., Sherman, G. F., Rosen, G. D., Aboitiz, F., & Geschwind, N. (1985). Developmental dyslexia: Four consecutive patients with cortical anomalies. *Annals of Neurology, 18,* 222–233.

Geschwind, N., & Behan, P. (1982). Left-handedness: Association with immune disease, migraine, and developmental learning disorder. *Proceedings of the National Academy of Science, 79,* 5097–5100.

Geschwind, N., & Levitsky, W. (1968). Human brain: Asymmetries in the temporal speech region. *Science, 161,* 186–187.

Humphreys, P., Kaufmann, W. E., & Galaburda, A. M. (1990). Developmental dyslexia in women: Neuropathological findings in three patients. *Annals of Neurology, 28,* 727–738.

Hynd, G. W., Semrud-Clikeman, M., Lorys, A. R., Novey, E. S., & Eliopulos, D. (1990). Brain morphology in developmental dyslexia and attention deficit disorder/hyperactivity. *Archives of Neurology, 47,* 919–926.

Jernigan, T. L., Hesselink, J. R., Sowell, E., & Tallal, P. A. (1991). Cerebral structure on magnetic resonance imaging in language- and learning-impaired children. *Archives of Neurology, 48,* 539–545.

Johnston, J. R., & Weismer, S. E. (1983). Mental rotation abilities in language-disordered children. *Journal of Speech and Hearing Research, 26,* 397–403.

Kamhi, A. G., Catts, H. W., Mauer, D., Apel, K., & Gentry, B. F. (1988). Phonological and spatial processing abilities in language- and reading-impaired children. *Journal of Speech and Hearing Disorders, 53,* 316–327.

Kaufman, A. S., & Kaufman, N. L. (1983). *Kaufman assessment battery for children.* Circle Pines, MN: American Guidance Service.

Larsen, J. P., Hoien, T., Lundberg, I., & Odegaard, H. (1990). MRI evaluation of the size and symmetry of the planum temporale in adolescents with developmental dyslexia. *Brain and Language, 39,* 289–301.

Larsen, J. P., Odegaard, H., Grude, H., & Hoien, T. (1989). Magnetic resonance imaging—a method of studying the size and asymmetry of the planum temporale. *Acta Neurologica Scandinavia, 80,* 438–443.

Leonard, C. M., Voeller, K. K., Lombardino, L. J., Morriss, M. K., Hynd, G. W., Alexander, A. W., Anderson, H. G., Garofalakis, M., Honeyman, J. C., Mao, J., Agge, O. F., & Staab, E. V. (1993). Anomalous cerebral structure in dyslexia revealed with magnetic resonance imaging. *Archives of Neurology, 50,* 461–469.

Lewis, B. A. (1990). Familial phonological disorders: Four pedigrees. *Journal of Speech and Hearing Disorders, 55,* 160–170.

Lewis, B. A. (1992). Pedigree analysis of children with phonology disorder. *Journal of Learning Disabilities, 25,* 586–597.

Leiter, R. G. (1969). *The Leiter international performance scale.* Chicago, IL: Stoelting Co.

Livingstone, M. S., Rosen, G. D., Drislane, F. W., & Galaburda, A. M. (1991). Physiological and anatomical evidence for a magnocellular defect in developmental dyslexia. *Proceedings of the National Academy of Science, USA, 88,* 7943–7947.

McShane, D., Risse, G. L., & Rubens, A. B. (1984). Cerebral asymmetries on CT scan in three ethnic groups. *International Journal of Neurosciences, 23,* 69–74.

Naglieri, J. A. (1985). *Matrix analogies test—Short form.* San Antonio, TX: Psychological Corporation.

Neils, J., & Aram, D. M. (1986). Family history of children with developmental language disorders. *Perceptual and Motor Skills, 63,* 655–658.

Plante, E. (1991). MRI findings in the parents and siblings of specifically language-impaired boys. *Brain and Language, 41,* 67–80.

Plante, E., Boliek, C., Binkiewicz, A., & Erly, W. K. (in press). Elevated androgens, brain development, and language/learning disabilities in children with congenital adrenal hyperplasia. *Developmental Medicine and Child Neurology.*

Plante, E., Swisher, L., & Vance, R. (1989). Anatomical correlates of normal and impaired language in a set of dizygotic twins. *Brain and Language, 37,* 643–655.

Plante, E., Swisher, L., Vance, R., & Rapcsak, S. (1991). MRI findings in boys with specific language impairment. *Brain and Language, 41,* 52–66.

Plante, E., & Turkstra, L. (1991). Sources of error in the quantitative analysis of MRI scans. *Magnetic Resonance Imaging, 9,* 589–595.

Plante, E., & Vance, R. (1994a). Selection of preschool language tests: A data-based approach. *Language, Speech, and Hearing Services in Schools. 25,* 15–24.

Plante E., & Vance, R. (1994b). *Diagnostic accuracy of two tests of preschool language.* Manuscript submitted for publication.

Rice, M. (1994, February). *Linguistic markers of language impairment.* Paper presented at the American Association for the Advancement of Science, San Francisco, CA.

Savich, P. A. (1984). Anticipatory imagery ability in normal and language-disabled children. *Journal of Speech and Hearing Research, 27,* 494–501.

Scarborough, H. S., & Dobrich, W. (1990). Development of children with early language delay. *Journal of Speech and Hearing Research, 33,* 70–83.

Shaywitz, S. E., Shaywitz, B. A., Fletcher, J. M., & Escobar, M. D. (1990). Prevalence of reading disability in boys and girls: Results of the Connecticut longitudinal study. *Journal of the American Medical Association, 264,* 998–1002.

Snyder, L. S. (1984). Developmental language disorders: Elementary school age. In A. Holland (Ed.), *Language disorders in children* (pp. 129–158). Boston, MA: College Hill.

Sparrow, S. S., Balla, D. A., & Cicchetti, D. V. (1984). *Vineland adaptive behavior scales.* Circle Pines, MN: American Guidance Service.

Swisher, L., Plante, E., & Lowell, S. (1994). Nonlinguistic deficits of language-impaired children complicate the interpretation of their nonverbal IQ scores. *Language, Speech, and Hearing Services in Schools, 25,* 235–240.

Tallal, P., Ross, R., & Curtiss, S. (1989). Familial aggregation in specific language impairment. *Journal of Speech and Hearing Disorders, 54,* 167–173.

Tomblin, J. B. (1989). Familial concentration of developmental language impairment. *Journal of Speech and Hearing Disorders, 54,* 287–295.

Tomblin, J. B., Freese, P. R., & Records, N. L. (1992). Diagnosing specific language impairment in adults for the purpose of pedigree analysis. *Journal of Speech and Hearing Research, 35,* 832–843.

Wada, J. A., Clarke, R., & Hamm, A. (1975). Cerebral hemispheric asymmetry in humans. *Archives of Neurology, 32,* 239–246.

Werner, E., & Kresheck, J. D. (1983). *Structured photographic expressive language test-II.* Sandwich, IL: Janelle Publications.

Witelson, S. F., & Pallie, W. (1973). Left hemisphere specialization for language in the newborn. *Brain, 96,* 641–646.

COMMENTARY ON
CHAPTER 12

Laurence B. Leonard
Purdue University

There is a timeliness in Plante's chapter that makes its contribution especially significant. Since her important findings of atypically large right perisylvian areas in children with specific language impairment (SLI), it has been tempting to conclude that the classification SLI as a distinct disorder has been validated. However, in this chapter, Plante makes it clear that the findings of a neuroanatomical basis for the problems of children with SLI do not lead to such a neat conclusion. The neuroanatomical and behavioral variability seen, even when measurement problems can be solved, suggests instead that children with SLI are not a homogeneous and distinct group. In my remarks, I try to point out some of the implications of Plante's chapter.

NEUROANATOMICAL VARIABILITY

Plante observed that most (though not all) of the children with SLI and a few controls showed the atypical pattern of the right perisylvian area being larger than expected and approximating the size of the left perisylvian area. It is worthwhile to note that in the classic Geschwind and Levitsky (1968) study, nearly one fourth of the brains studied (presumably from normally-functioning individuals) showed the same pattern. It would be very helpful to learn whether the typical left > right pattern and the atypical right = left pattern form two discrete groups, or whether there is a continuum with ratios reflecting left > right collectively representing the modal pattern.

The significance of such a finding rests in the distinction between biological conditions that disfavor language functioning and conditions that constitute a biological disorder. It seems clear that the work of Plante and her colleagues has revealed a neuroanatomical state that renders the learning of language and related skills quite difficult.[1] What is less clear is the qualitative status of this condition. For example, Leonard (1991) argued that many children diagnosed as SLI have language and related deficiencies attributable to genetic and environmental factors in the same way that many other abilities (e.g., musical, bodily-kinesthetic, introspective abilities; see Gardner, 1983) can show inherited or environmentally induced limitations; impairment in the sense of a damaged, diseased, or disrupted system need not be assumed. Given that different levels of behavioral functioning must be associated with neuroanatomical and neurophysiological differences, it is possible that Plante and her colleagues have identified a less typical, but otherwise natural, neuroanatomical profile that places language and related abilities at a disadvantage. If, on the other hand, this profile falls outside the distribution of left-right perisylvian area ratios (as I suspect Plante believes), there would be strong evidence for characterizing SLI as reflecting an abnormal neuroanatomical condition.

BEHAVIORAL VARIABILITY

Plante's chapter serves as an important reminder that children with SLI can show deficits in nonlinguistic abilities and can vary in their linguistic profiles. The finding that nonlinguistic abilities are affected is consistent with earlier work (see reviews in Bishop, 1992; Johnston, 1992) and makes a great deal of sense given the diversity of functions associated with the perisylvian area. Furthermore, children with SLI have deficits in a range of language areas, and some of these areas, such as the lexicon, seem more closely tied to nonlinguistic cognitive areas. One of the implications of this view is that any account of SLI that is limited to explanations of a linguistic nature will be unparsimonious at best because the work of explaining the nonlinguistic deficits will still remain.

Plante's discussion of variability attributable to the language test instrument was very illuminating and suggests the need for great care in the instruments selected for use with these children. Actually, the variability might be even greater than suggested in the Plante and Vance (1994) study reported by Plante; the original diagnosis of SLI by the referring speech-language pathologist was

---

[1]Locke (1994) has recently suggested the possibility that neurodevelopmental delays might promote language learning via less efficient mechanisms in the other hemisphere, which might, in turn, actually alter the child's neuroanatomy. In this chapter, Plante reports that left-right perisylvian area ratios seem quite stable across time. This would argue against the view that the neuroanatomical profile of children with SLI is simply the product of their language-learning difficulty.

based on standardized test results and clinical judgment, which themselves are associated with a certain amount of variability. The latter source of variability is unavoidable; however, this variability could be heightened by the speech-language pathologist's choice of a poorly constructed standardized test during the diagnostic process.

But, as noted by Plante, it is clear that some of the variability seen in the linguistic profiles of children with SLI is not attributable to the test instrument, but to the children themselves. And this suggests that many of the group results in studies reporting areas of special weakness in children with SLI are probably masking data from children who show the pattern to a lesser degree or not at all. For example, although many English-speaking children with SLI seem to be more limited in their use of grammatical morphemes than their mean length of utterance would predict, there are probably individual children in some of these group studies who do not show this pattern. A major task of future investigators will be to account for linguistic variability without losing precision in the prediction of which profiles should and should not occur in these children.

REFERENCES

Bishop, D. (1992). The underlying nature of specific language impairment. *Journal of Child Psychology and Psychiatry, 33*, 3–66.

Gardner, H. (1983). *Frames of mind: The theory of multiple intelligences.* New York: Basic Books.

Geschwind, N., & Levitsky, W. (1968). Human brain: Left-right asymmetries in temporal speech region. *Science, 161*, 186–187.

Johnston, J. (1992). Cognitive abilities of language-impaired children. In P. Fletcher & D. Hall (Eds.), *Specific speech and language disorders in children* (pp. 105–116). London: Whurr.

Leonard, L. (1991). Specific language impairment as a clinical category. *Language, Speech, and Hearing Services in Schools, 22*, 66–68.

Locke, J. (1994). Gradual emergence of developmental language disorders. *Journal of Speech and Hearing Research, 37*, 608–616.

Plante, E., & Vance, R. (1994). Selection of preschool language tests: A data-based approach. *Language, Speech, and Hearing Services in Schools, 25*, 15–24.

13

WHAT GENETICS CAN AND CANNOT LEARN FROM PET STUDIES OF PHONOLOGY

David Poeppel

University of California, San Francisco

In the investigation of the biological foundations of language, two sources of evidence promise to be especially valuable. The *functional neuroimaging* methods can elucidate where and how language is represented and processed in the brain. *Behavioral genetics* can work out the inheritance properties of those aspects of language that are probably mediated by dedicated neural circuitry and can also identify the components of the human genome that are essential to the integrity of the language faculty. Work in these two areas of research has proceeded relatively independently. A major aim of this chapter is to bring into focus an important point of contact between the two domains. That point of contact is the relevance of psycholinguistics and linguistic theory for experimental design (in neuroimaging) and diagnostic adequacy (in behavioral genetics).

This chapter is organized as follows: First, I review some claims about language localization on the basis of functional neuroimaging using positron emission tomography (PET). Comparison of PET studies reveals that the results do not converge, and an analysis of why that might be the case is developed. Second, I explore how this analysis of the neuroimaging nonconvergence finding is relevant to questions of diagnostic adequacy within behavioral genetics. The concrete problems that investigators face turn out to be quite similar across the two domains. In both areas of research, the importance and usefulness of the analytic machinery of psycholinguistics and linguistic theory may be insufficiently appreciated. Consequently, the task analyses and diagnostic categories used in the studies have sometimes been underspecified.

FUNCTIONAL NEUROIMAGING USING PET

Structural and Functional Approaches

The study of where and how language and speech are represented and processed in the brain investigates the interaction of structure and function. The research investigating the neural structure underlying language primarily uses such techniques as computerized tomography and magnetic resonance imaging. Sometimes, it is possible to correlate general structural properties of the brain with behaviorally defined functions. For example, research done by Galaburda (1991) and Plante (1991a, 1991b, this volume) showed that there are cortical structures that appear to be selectively affected in individuals with language pathologies such as dyslexia or specific language impairment (SLI). Plante, for instance, has documented a systematic relationship between specific language impairment and abnormal cortical asymmetry in certain perisylvian regions of the temporal lobes. Similarly, Galaburda has described abnormal asymmetries in individuals diagnosed with dyslexia.[1]

The research on structure has been careful not to attribute any linguistic process to a particular region of the brain. In contrast, the activation approach, although complementary with structural-anatomical research, takes an explicitly functional localizationist tack. Until recently, the only widely available method of functional neuroanatomy has been the deficit-lesion correlation approach. In the past few years, however, a number of new functional neuroimaging methods, which generate images of the brain while normal subjects or patients are performing specially designed experimental tasks, have been introduced.[2] These include

---

[1]These results on abnormal asymmetry of areas in perisylvian cortex are important facts. However, we do not yet know how general structural properties of the brain (like hemispheric asymmetry) interact with physiological properties of the brain, nor do we understand the relationship between physiological properties and computational faculties of the mind, such as language (or anything involving representation of knowledge, for that matter). The facts are not yet interpretable in a larger framework of how the brain represents language, and the results are, therefore, difficult to integrate with other knowledge we have about language and the brain. One would expect such macroscopic effects as significant cortical asymmetry to have grave consequences across many aspects of the language system; instead, the structural abnormality correlates with very specific effects. This is surprising, but presumably we will eventually understand how the brain represents and processes language and other cognitive faculties. Then, given a motivated, theoretically grounded account of the phenomena, these facts may turn out to be crucial. At this point, it is problematic to invoke a relatively general anatomical property like atypical asymmetry of a large area of the brain as an explanation of a specific deficit, such as dyslexia or SLI.

[2]There exist other techniques to study the human brain in vivo, but they are much more invasive. Grid electrodes and deep penetrating electrodes, for example, are being used to map cortex in patients who need to be evaluated before or during neurosurgical procedures. These methods promise to yield particularly high-resolution functional-anatomic results, but they will continue to be limited because of the very invasive nature of the process. For descriptions and discussions of this type of research see Lesser et al. (1987) and Ojemann (1991).

positron emission tomography (PET), functional magnetic resonance imaging (fMRI), single photon emission computerized tomography (SPECT), and magneto-encephalography/magnetic source imaging (MEG/MSI). The logic underlying the different imaging methods is the same: If cognitive processes are mediated by specific parts of the brain, and if increased activity of a brain area correlates with some physiological process (such as increased regional cerebral blood flow at that area), one can derive an image of the brain based on some physiological index while it is engaged in a specific task. A number of studies on various cognitive processes have been performed using the different neuroimaging methodologies. This chapter compares some PET studies on phonological processing in an attempt to help researchers to better appreciate the promise and limitations of the method for the study of language processing. (For a detailed critical review of the PET studies of language see Poeppel, 1994, in press).

A PET Primer

PET studies take advantage of the fact that increased neural activity in an area is reflected by increased regional cerebral blood flow (rCBF) to that particular region of the brain (for methodological details and for explanation of the technical aspects of PET see Roland, 1993). rCBF is computationally derived in PET by detecting gamma rays. A radiolabeled tracer (typically oxygen, because of its short half-life of 123 seconds) is administered to a subject during the continuous performance of an experimental task. As positron-emitting radio-nuclide O^{15} reaches the tissue, positrons are emitted at an increased level where there is increased blood flow and, therefore, increased tissue activity. When a positron is emitted from the radionuclide and collides with an electron, the annihilation reaction yields two quanta of a particular energy (511 keV each) that travel at a path 180° opposite each other. A circular array of photosensitive gamma-ray detectors placed around the subject's head detects the annihilation reaction, and the origin of a given positron emission is computed, based on the temporal coincidence (and other factors) of the two quanta. From the spatial distribution of increased positron emission, one can then infer the locus of increased neuronal activity. In particular, it is possible to reconstruct the three-dimensional distribution of positron activity and, hence, regional activation, in the brain.

Because PET depends on hemodynamics and radioactive decay, the temporal resolution of positron emission tomography is low, on the order of tens of seconds. The fastest image acquisition time is greater than ten seconds (Demonet, Wise, & Frackowiak, 1993). Because the cognitive processes at stake in language processing are at least an order of magnitude faster (say, in the range of tens of milliseconds), this technical feature has led to a special experimental design. Stimuli are presented in blocks of the same stimulus type to attempt to ensure homogeneity of cognitive activation and neuronal activation over time. It has proved helpful to use subtractive methodology for the localization of elementary computations underlying a given task (Fox, Mintun, Reiman, & Raichle, 1988; Posner, Petersen, Fox, & Raichle, 1988).

In the subtractive method, images are acquired during the performance of nested tasks, a simple task A and a more complex task B. The optimal tasks are minimally different and are, ideally, in a subset relation, such that task B encompasses all the components (and, by assumption, the relevant cortical areas) used in task A. Suppose task A has n cognitive components and the image taken during performance of task A shows m neural components. By assumption, task B has $(n + x)$ cognitive components and the image acquired during performance of B shows $(m + y)$ cerebral areas of activation. Subtraction of image A from image B yields a number of cerebral areas (y); these are argued to be the neural substrate for exactly those computations that differ between tasks A and B (namely factor x, whatever that may be). The problematic assumptions built into this methodology are discussed later.[3]

Prospects

Success of the PET method in the study of cognitive processes demands that experiments designed to isolate the same cognitive computation also implicate the same neural substrate. The practical implication is that the results should minimally overlap in some meaningful way. If an experiment implicates some cerebral area or set of areas in the mediation of some cognitive process, a different experiment investigating the same cognitive process should implicate at least some of the areas of the first study. The desideratum is, obviously, that the same cerebral region or set of regions will always be implicated when the same computational subroutine is selectively engaged. Comparative analysis of five recent PET studies of phonetic and phonological processing shows that the results do not overlap in the expected way. In fact, there is no region of the brain that is always implicated in the studies. This meta-analytic result is surprising insofar as the task demands were quite comparable across the studies (cf. Poeppel, in press).

FIVE PET STUDIES OF PHONOLOGICAL PROCESSING

There have been several experimental reports examining phonetic or phonological processing with PET. The experiments reviewed in this chapter are compared for two reasons: First, each report makes a claim about the neuronal basis of phonological processing and, second, each experiment makes relatively similar demands of the subjects. Zatorre, Evans, Meyer, and Gjedde (1992) asked subjects to compare (judge as same or different) the final consonants in auditorily

---

[3]The rationale for the subtractive method is discussed, for instance, by Chertkow and Bub (1994), Posner, Petersen, Fox, and Raichle (1988) and Raichle et al. (1994) for PET studies and by Sternberg (1969) as a more general issue in experimental psychology.

presented pairs of consonant-vowel-consonant (CVC) strings. Sergent, Zuck, Levesque, and MacDonald (1992) required a rhyme judgment on single visual letters rhyming with [i]. Paulesu, Frith, and Frackowiak (1993) combined a letter-rhyming task with a short-term letter memory task to isolate phonological processing. The study by Petersen, Fox, Posner, Mintun, and Raichle (1989) required a rhyme judgement on visual word-pairs. Finally, Demonet et al. (1992) used a phonetic monitoring task in which the subject had to detect a [d]-[b] sequence in an auditorily presented nonword string. All the studies required subjects to make an articulatory-perceptual judgement about a phonetic string. Because making that type of judgement is, by hypothesis, independent of the modality of stimulus presentation, one expects relatively similar results.

A cross-study comparison of areas showing rCBF increase reveals that the results are nonoverlapping. No single area is consistently implicated across the five studies on phonetic/phonological processing. In fact, no area is implicated in more than three experiments. I call this the *no-overlap result*. This meta-analytic result suggests that PET studies on phonological processing to date are contradictory. Figure 13.1 summarizes the results schematically; notice that the brain

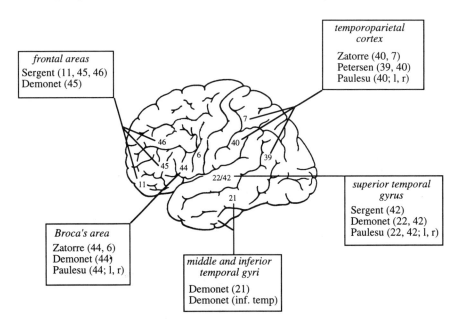

FIG. 13.1. Left hemisphere cortical areas implicated in phonological processing. The areas shown are argued to mediate phonetic/phonological processing in the five studies reviewed. Areas are indicated by Brodmann number and verbal labels of the region of cortex, as in the chapter. The names of authors associated with a particular area indicate which cortical regions are considered primary by each group for the performance of the experimental task in particular, and various aspects of phonological processing in general.

imaging results implicate areas over a broad range of the left hemisphere, including canonical language areas like Broca's area and Wernicke's area. In what follows, each experiment is summarized and the no-overlap finding is analyzed.

Zatorre, Evans, Meyer, and Gjedde (1992)

The experiment tested which cortical areas respond to nonspeech stimulation and which to auditorily presented consonant-vowel-consonant (CVC) strings. The speech stimuli were CVC pairs, half of which were words and half of which were pseudowords (possible words). In the relevant experimental condition, subjects heard CVC pairs and had to judge whether the final consonant for a given pair was the same or different. The control condition used for subtraction was a passive listening condition in which subjects were presented with CVC pairs to which they had to attend, but not generate a response. The stimulus set for both conditions was identical. The authors call the cognitive process isolated by the subtraction 'phonetic processing'.

Table 13.1 summarizes the brain activation results for the five experiments. Table 13.1A lists all areas that showed significant activation in Zatorre et al.'s phonetic judgment condition, after subtraction of the condition in which CVC pairs were heard passively. Special importance is attributed to the activation foci in the frontal lobe (44, 6) and the parietal lobe (7, 40).

The authors' interpretation of the findings focuses on the observation that a significant local increase in blood flow was detected around Broca's area in the left hemisphere in the phonetic condition, with an additional focus in the left parietal region. They argue that lesions around Broca's area often result not only in nonfluent aphasia and articulatory disorders but also in phonetic perceptual deficits (see Blumstein, Baker, & Goodglass, 1977). They assume the framework of the motor theory of speech perception (Liberman & Mattingly, 1985) and suggest that Broca's area is crucially involved in the reconstruction of the intended articulatory gestures that are necessary for speech perception.

Can the authors persuasively argue on the basis of the experimental tasks and the subtraction performed that phonological processing has been isolated? First, let us make explicit what goes into the execution of the task; call this the *task decomposition*. The experimental task has a number of components. Consider that a subject has to accomplish at least the following tasks in the experimental condition: remembering the instructions of the task, listening to and remembering a pair of CVC strings, decomposing the strings (minimally into an onset and a final consonant each), comparing two final consonants, and selecting a response. These components, in turn, can be broken down into finer-grained subcomputations. Clearly, phonetic/phonological processing is involved because the task involves both speech input and judgements on auditorily presented word and pseudoword letter strings. However, it is not obvious that the task isolates phonological processing, or selectively engages phonological proc-

TABLE 13.1

Areas Showing rCBF Increase in Five Experiments on Phonological Processing[a]

A. Zatorre et al. (1992)	B. Petersen et al. (1989)	C. Sergent et al. (1992a)	C. Sergent et al. (1992b)	D. Demonet et al.	E. Paulesu et al. (1993)
phonetic condition minus passive CVCs	judgement minus passive word pairs	letter-sound minus letter-spatial	letter-sound minus objects	phoneme-task minus tones task	rhyming judgment for letters
Area (name or Brodmann number)	Area (name or Brodmann number)	Area (name or Brodmann number)	Area (name or Brodmann number)	Area (name or Brodmann number)	Area (name or Brodmann number)
• 44/6	• 39/40[b]	• 46	• 45	• sup. temporal gyr[c] (BA 42,22)	• SMA (left and right)
• 31 (midline)		• caudate nucleus	• 42	• middle temporal gyr (BA 21)	• (44) (l,r)
• 17 (right)		• 11	• 11	• inf. temporal gyr and fusiform gyr	• 40 (l,r)
• 7/40		• 45	• 6/8	• middle temporal gyr (right)	• 22/42 (l,r)
• 24 (midline)		• 21	• 46	• Broca's area (44)	• 18
• 20			• 24 (right)	• ant. to Broca's (45)	• insula (l,r)
					• cerebellum (l,r)

Note. All areas listed are in the left hemisphere, unless otherwise noted.

[a]Note that results obtained with positron emission tomography are reported in a standardized coordinate system (Tailarach & Tournoux, 1988); the names of regions and the corresponding designation of Brodmann's areas merely serve as mnemonics.

[b]The condition was run as a control, and there is no other data available on which areas were activated during the task.

[c]Excluding primary auditory cortex.

347

essing in a way that is relevant beyond the specific demands of this particular experiment. Listeners are not in the habit of breaking down the speech stream into pairs of temporally adjacent CVC strings, remembering them, and maintaining them in a format that allows the comparison of two particular segments in the pair. There is also no special reason in spoken language to attend exclusively to the final consonant of any given pair. In other words, although the task does require that the subject recruits phonological representations, it also has very specific attentional demands and other extralinguistic factors that probably play a significant role in the execution of the experimental task.

Next consider the dimensions along which the test condition differed from the control: In the passive speech condition, subjects were instructed to press a key to every CVC pair, alternating between yes and no responses. To what extent subjects actually processed each CVC pair is not determined. The control condition was, therefore, underconstrained relative to the test and, consequently, how extensively subjects did in fact process the stimuli in the passive condition is not known. Subjects may automatically make a word/nonword judgement, access lexical-semantic information, activate other lexical entries, and so on. The test condition makes a specific attentional demand (discrimination along a particular phonological dimension); subjects not only have to remember instructions and the CVC pairs, but also must maintain the stimuli in a format that allows comparison of the final consonants (something like the articulatory loop; cf. Baddeley, 1992). A straightforward alternative interpretation of the function of Broca's area is possible, given that the most salient difference between test and control conditions is the fact that subjects had to remember each pair in the test condition and there was no memory component in the control condition. Rather than mediating articulatory representations, one could argue that the Broca's area activation reflects some aspect of short-term verbal memory[4] (e.g., rehearsal). The local blood-flow increase in the test relative to the control condition could be attributed to the task's requirement of the selective engagement of verbal working memory.

Petersen, Fox, Posner, Mintum, and Raichle (1989)

The experiment on which Petersen et al. based their case about the sites of phonological processing was one of a number of conditions in a large experiment. In the test condition, subjects were presented visually with two words, one above and one below a fixation point. The required response was a yes/no rhyme

---

[4]This is, in fact, the interpretation that Paulesu, Frith, and Frackowiak (1993) have for the activation of Broca's area. Moreover, Awh et al. (1994) have substantiated that result in an experiment designed to engage components of working memory. Both groups conclude that verbal short-term memory (perhaps the rehearsal aspect) is mediated by Broca's area and that phonological encoding (as Baddeley's model calls it) selectively activates parietal areas around 7 and 40.

judgement, and the presented pairs consisted of rhyming or nonrhyming, visually similar or dissimilar words. In the control task, subjects were presented with the same stimulus set but had to maintain fixation and generate no response. Although a number of areas showed an increase in regional cerebral blood flow, the authors focus on the left temporoparietal cortex (specifically Brodmann's areas 7 and 40).

Consider first the decomposition analysis for this experiment: To what extent does the task selectively recruit phonological processing? What computations go into the experimental task and may, therefore, account for the activations? Subjects must first analyze the visual scene into written words, perhaps consulting an iconic memory representation because of the brief exposure of the stimuli and the noncanonical arrangement of the written words. Subsequently, at least the following computations must occur: grapheme to phoneme recoding, all the processing routines that occur automatically in word recognition, reactivation of the task demands, consultation of the phonological representation relevant to the execution of the task, comparison of the two words, and selection of the response. Most of these computations are unique to the test condition and, therefore, presumably do not subtract out.

Second, consider how closely test and control conditions are matched. The subjects did not have to respond in the control condition. Rather, subjects fixated while the same words were presented in the array. This constitutes a good control for the visual aspect of the task because everything in the tasks was identical from that point of view. However, one cannot ascertain that subjects did not engage in any type of phonological processing during the control condition. Because no response was required, subjects' behavior was unconstrained in the control task, so one cannot rule out phonological processing of the stimuli in that condition. In fact, the interpretation depends on one's model of language processing. On some accounts (cf. Fodor, 1983), the null hypothesis is that a language stimulus (e.g., a visually presented word) is computed automatically by the linguistic system, and perhaps along a variety of dimensions, including the item's grammatical category, phonological form, similarity to items stored nearby in the mental lexicon, and so on. On such a view, one might argue that—because subjects were not prevented from computing a phonological representation—they computed at least some aspect of a phonological representation automatically. Therefore, the difference between test and control does not isolate the phonological aspect of processing sufficiently. Another concern might be that other linguistic representations (e.g., grammatical category, items nearby in the mental lexicon, etc.) are computed, even if they are not germane to the task.

Sergent, Zuck, Levesque, and MacDonald (1992)

Sergent et al.'s experiment is a response to the studies by Zatorre et al. (1992) and Petersen et al. (1989). Sergent et al. suggested that to isolate a particular linguistic code one should use stimuli that are not susceptible to interference from other linguistic codes. Therefore, they used letters as visual stimuli, arguing that one can selectively engage the visual and phonological codes of a letter without interference from semantic, lexical, orthographic, or syntactic codes.

One test condition and two control conditions were used for the subtraction paradigm. In one control, an object task, the stimulus set consisted of 24 line-drawing objects, half of which represented living things and half of which represented nonliving things. The task was a forced-choice categorization along the "living" dimension. A letter-spatial task used 12 capital-letter, visually asymmetric consonants, half of which rhyme with /i/ (B, C, D, G, P, Z) and half of which do not (F, J, K, L, N, R) when the letter is named. Letters appeared either in their canonical orientation or in an upright mirror-reversed orientation. Subjects had to make a forced-choice orientation judgement. The main experimental condition, the letter-sound task, was a rhyme judgment (does the presented letter rhyme with [i]?). The stimulus-set was identical to the letter-spatial condition. Table 13.1C (a and b) lists the areas activated for the two separate contrasts, rhyme (letter-sound) minus letter-spatial and rhyme minus object.

The results regarding phonological processing differ from those found by Zatorre et al. (1992) and Petersen et al. (1989). Neither Broca's area nor temporoparietal cortex (specifically the supramarginal and angular gyri in the left hemisphere) were significant foci. The authors' interpretation of phonological processing centers on the observation that in both subtractions from the letter-sound condition three foci in the prefrontal cortex showed significant activation: areas 11, 45, and 46 in the left hemisphere (see Table 1Ca, b). They argued that this result "suggests the recruitment of these areas for the programming of articulatory patterns for the generation of the actual sound of the letters" (Sergent et al., p. 76).

Task decomposition and the issue of matching test and control conditions within the paired-image subtraction paradigm can be illustrated very clearly using Sergent et al.'s study as an example. To appreciate the experimental design difficulties that face the investigators, consider Fig. 13.2. Figure 13.2 reconstructs schematically the subroutines that are contained in the execution of Sergent et al.'s experimental task. The stages shown are the minimal number of computations required to execute the experimental task. The two experimental tasks, letter/sound (rhyme judgment) and letter/spatial (orientation judgment), are closely matched with respect to the number of computational components putatively required. Each task consists roughly of 8 stages; and with two exceptions (stages 5 and 7), the computations are similar, at least on the surface. In fact, stages 1 through 3 are probably identical. Later stages are macroscopically similar in that they bring into play similar cognitive processes (a memory component in stage 6 of both tasks, a decision component in stage 8 of both tasks, and response selection in stage 9 of both tasks).

The careful matching along number of computational components notwithstanding, there remain further substantive differences between the two tasks.

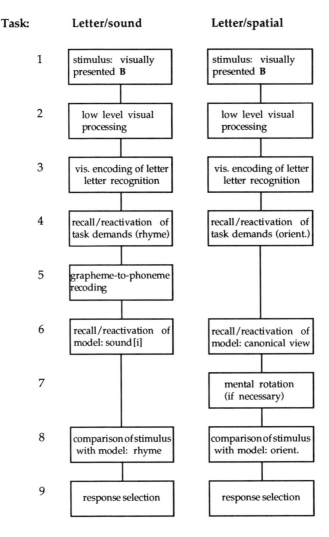

Task:	**Letter/sound**	**Letter/spatial**
1	stimulus: visually presented **B**	stimulus: visually presented **B**
2	low level visual processing	low level visual processing
3	vis. encoding of letter letter recognition	vis. encoding of letter letter recognition
4	recall/reactivation of task demands (rhyme)	recall/reactivation of task demands (orient.)
5	grapheme-to-phoneme recoding	
6	recall/reactivation of model: sound [i]	recall/reactivation of model: canonical view
7		mental rotation (if necessary)
8	comparison of stimulus with model: rhyme	comparison of stimulus with model: orient.
9	response selection	response selection

FIG. 13.2. Task decomposition for Sergent et al. (1992). Task decomposition for two conditions of the experiment reported by Sergent et al. (1992). In the letter/sound condition subjects had to judge whether a visually presented consonant rhymed with [i]. In the letter/spatial condition, subjects had to judge whether a visually presented consonant was in the canonical or mirror-reversed orientation. The letter/spatial condition acted as a control, and the PET scan image generated during this task was subtracted from the letter/sound condition.

Both experimental tasks require that subjects remember what to do; this aspect of the two experimental tasks is likely to be similar in some respects (memory component) and different in others (different content of memory, distinct modalities required for task execution). Stage 5, grapheme-to-phoneme recoding, exists only in the letter/sound condition. One might argue that stage 7 is a corresponding operation for the letter/spatial task and, therefore, the tasks are matched on this dimension. They differ, however, in that mental rotation is an operation on the same mental representation whereas recoding involves a change into a different representational format. Stage 6 requires that subjects recall the model stimulus for comparison. Although there is a well-defined memory component, in one case the subject must reactivate the representation of a sound; in the other, the subject must reactivate a visual representation. Stage 8, the comparison and decision stage, presumably differs between the two tasks given that the comparisons range over visual representations in one case and sound representations in the other. This analysis shows that even in a study as carefully designed as Sergent et al.'s, one must be quite cautious about the interpretation. Although one can match experimental tasks very closely, the decomposition reveals that there are still substantive computational distinctions among the tasks, and all of the distinctions must inform the interpretation of the PET results, especially after subtraction.

Demonet et al. (1992)

Demonet et al. (1992) also investigated the "neural structures which mediate phonological, lexical and semantic processing of heard words" (p. 1753). They "address[ed] directly whether there are separate areas dedicated to phonological and semantic processing of heard linguistic stimuli" (p. 1754). There were three experimental conditions in the study: (a) in a nonspeech auditory control in which the stimuli were triplets of pure tones, the task was to respond to rising pitch in the third tone (e.g., 500Hz-500Hz-1000Hz: yes response); (b) in a phoneme monitoring task in which the stimuli were multisyllabic nonwords, subjects had to detect a specific phoneme sequence ([b] preceded by a [d]) occurring in a target; (c) a word task in which adjective-noun pairs were presented required subjects to categorize the noun phrases into arbitrarily determined groups. Targets were positive adjectives combined with nouns denoting animals smaller than a chicken or a cat, e.g., *kind mouse* was a target, but *horrible horse* was a distractor.

The language tasks differed in that they were designed to engage phonological or semantic processing at the level of single words. The relevant result derived from the subtraction of the tones task (acoustic control) from the phoneme monitoring task. The areas of rCBF increases for this comparison are listed in Table 13.1D. A regional increase in activation was seen in the superior, middle, and inferior temporal gyri in the left hemisphere and Broca's area. Primary importance was attributed to the activation of the superior temporal cortex: The authors argued that the "neural structures associated with the phonological processes include the associative auditory cortex in the left superior temporal gyrus and the anterior part of Wernicke's area" (p. 1761). They also suggested a role for Broca's area in phonological processing.

There is an aspect of the task that engages the extralinguistic computational machinery quite strongly. The phoneme-monitoring task appears to be difficult, an assumption that is reflected in the relatively poor performance of subjects (see Demonet et al., 1992). In contrast, a typical feature of natural language processing under normal conditions is its speed, ease, and automaticity. This is true at the level of phonological, morphological, syntactic, and semantic analysis (for a number of references see Marslen-Wilson, 1989). If the control conditions are substantially easier from the point of view of the computational load of extralinguistic processing factors, the subtraction that has been set up to isolate phonological processing may not isolate the process of interest in a crisp enough way. Demonet et al.'s control (tones task), beyond differing from the test condition along a number of other factors, is so much easier that it is presumably also different with respect to general resource allocation of attention and memory.

Paulesu et al. (1993)

Paulesu, Frith, and Frackowiak (1993) discussed the neural basis of the verbal component of working memory. The research is motivated by Baddeley's memory model (Baddeley, 1992). The experiment tried to find structures corresponding to the processing components postulated within that framework, specifically the articulatory loop. By hypothesis, the articulatory loop consists of two components, subvocal rehearsal and phonological encoding. The work is of interest for this chapter because a claim is made about the neural basis of phonological processing.

Experiment 1 contrasted a verbal short-term memory task with a visual control. The stimuli for the test condition were a sequence of six randomized, visually presented consonants that are phonologically dissimilar (rate: 1 per second; the consonants used were not listed). Subjects were instructed to maintain the string in memory by rehearsing it subvocally. The control stimuli consisted of a set of unfamiliar Korean letters that subjects were instructed to remember (visual short-term memory; presentation rate also 1 per second). This verbal task was designed to require phonological encoding and subvocal rehearsal. Experiment 2 was the same rhyme judgment task as used by Sergent et al.: Subjects had to make a rhyme judgment (target rhymes with [i]) on visually presented letters. In this condition the control again consisted of visually presented unfamiliar Korean letters that had to be judged for visual similarity. The rhyming task was argued on independent grounds to require subvocal rehearsal but not phonological encoding. That allowed a separate comparison of the two language tasks; specifically, the comparison was designed to isolate phonological encoding. This is, by hypothesis, the only difference between the memory and rhyming conditions of the experiment.

The authors examined the phonological processing by comparing the two experimental tasks with the controls. Table 13.1E summarizes the areas showing rCBF increase in the language tasks as compared to the visual tasks.[5]

Based on these activation results, the authors made several claims: (a) the functional anatomy of the articulatory loop includes (bilaterally) Broca's area (Brodmann's area 44), superior temporal gyrus (BA 22/42), supramarginal gyrus (BA 40), and the insulae—in other words most of the typical perisylvian language areas and their right hemisphere homologues; (b) subvocal rehearsal is mediated by Broca's area in the left hemisphere; (c) phonological encoding can be attributed to area 40 (supramarginal gyrus) of the left hemisphere; (d) the areas in superior temporal gyrus (BA 22/42) are "probably involved in phonological processing independent from memory" (p. xx). Importantly, recent PET experiments by Smith and Jonides and their colleagues (see Awh et al., 1994) converged in important ways with the result of Paulesu et al. Awh et al. (1994) designed tasks to selectively engage different components of working memory, and their results are consistent with Paulesu et al.'s claims with respect to verbal working memory.

The precise nature of the contrast is somewhat unclear for both the memory and the rhyming test conditions. An important experimental desideratum is a well-defined and minimal difference between conditions. In this experiment, the difference in task demands is substantive. Although both test and control have a strong memory component, the tasks are otherwise not comparable; one is a language task whereas the other is a visual memory task. Beyond short-term memory, there is no overlap. Moreover, the relevant component of working memory is purportedly specifically linguistic in this model, so one would expect some other cerebral area to be mediating visual short-term memory. The question is, therefore, is it plausible to compare tasks typical of language with tasks designed to be purely visual? Although the subtracted control conditions are well-matched to the test conditions with respect to extralinguistic performance factors, they are the visual analogue of a control that would perhaps better be anchored in language processing itself.

THE NO-OVERLAP META-RESULT

Direct Comparison Across Five Studies

Although the five papers argued that their results provide evidence for the neural substrate of phonological processing, they were designed and executed with different questions in mind. Paulesu et al. (1993), for instance, investigated

---

[5]Notice that with the exception of area 18 (an extrastriate visual area often also called V2) all activations in this study were bilateral.

verbal short-term memory, but Petersen et al. (1989) used their condition as an additional task embedded in a larger study of language processing. Demonet et al. (1992), in contrast, explicitly investigated phonological processing. The designs are different in that distinct sensory modalities come into play, different stimuli are used, and subtly different tasks are required of the subjects. There are, however, important similarities as well:

- Each group argues that they have isolated a computation called "phonological processing."
- Four of the five studies use tasks one could broadly call "rhyming tasks" to isolate phonetic/phonological processing.
- Stimuli are typically pairs of items (e.g., CVC pairs, word pairs, individual letters compared to a target letter).
- There are a number of similarities across studies concerning the conceptual decomposition of language. It is assumed that it is possible to isolate components of language processing such as phonetics and phonology in the brain using subtractive methodology. Furthermore, none of the reports addressed (at least directly) the issue that phonological processing is not a unitary process but consists of a number of distinct computational components.

Given these similarities, one can reasonably compare the results directly. Ideally, an area or set of areas will always be implicated in phonological processing, because some underlying elementary computation required for phonetic/phonological processing is presumably consistently activated during these types of tasks. Are there one or more areas that show significant rCBF increases in all five studies? The answer, surprisingly, is that there is no single area or set of areas implicated in all PET studies of phonetic/phonological processing. This observation is called the *no-overlap result*.

Table 13.1 lists the areas that show activation in the relevant experimental conditions. Across the 5 studies, 22 different areas show rCBF increase in response to the phonetic/phonological condition (counting bilateral activation of homologues just once). Of these 22, only 8 areas show significant rCBF increase in more than one study; however, no areas show significant activation across all 5 studies and no areas are active in 4 out of the 5 studies. Of the 8 areas that do show activation in more than one experiment, only 3 are implicated in 3 out of 5 studies; these are discussed further in a later section.

The "maximal overlap" occurred with the following three areas that showed increased activation in three studies: (a) Broca's area (specifically, Brodmann's Area 44): activation seen by Zatorre et al. (1992), Demonet et al. (1992), and Paulesu et al. (1993); (b) secondary auditory cortex (BA 42): Sergent et al. (1992), Demonet et al. (1992), Paulesu et al. (1993); and (c) the supramarginal gyrus (BA 40): Zatorre et al. (1992), Petersen et al. (1989), Paulesu et al (1993).

Interestingly, each study is able to adduce supporting deficit-lesion correlation evidence. Because there are five different PET results, this means that the authors were able to find deficit-lesion evidence supporting each distinct conclusion. This leaves the reader with the neuropsychologically established assumption that (left hemisphere) perisylvian cortical areas are important to language processing. The PET results, therefore, do not increase the precision of the map that has been established independently with the methods of cognitive neuropsychology.

To each of the three 'candidate areas' a different function is attributed. Broca's area (BA 44), for instance, was implicated in phonological processing in three PET studies, but received a rather different functional interpretation each time: For Zatorre et al. (1992), Broca's area was central to the task of speech perception. Paulesu et al. (1993) found Broca's area to be relevant to an entirely different component of processing, that of the subvocal rehearsal system. Demonet et al. (1992) made no particular claim about Broca's area, but consider areas in auditory cortex (22/42) to be the critical neural substrate. They, thus, merely point out that there is a "possible role of Broca's area in phonological processing" (p. 1762). Similar discord obtains when comparing secondary auditory cortex (specifically, areas 22/42) and the supramarginal gyrus (specifically, area 40). Functions attributed to the supramarginal gyrus, for instance, included phonological encoding (Petersen et al., 1989), phonological storage (Paulesu et al., 1993), and possible nonspecific participation in phonetic/phonological processing (Zatorre et al., 1992). The different interpretations constitute ordinary disagreement about the interpretation of the results. This fact, however, should not detract from the core observation that there is no overlap to begin with.

From the perspective of the experimental tasks, two situations occur in the comparison among studies: Different test procedures generate similar results (overlap) or very similar procedures have different results (no overlap). Consider the case where there is overlap among the three studies that implicate area 40 in the supramarginal gyrus. The task used by Petersen et al. (1989) required rhyme judgments on visually presented word pairs that appeared at a rate of 48 per minute, the task used by Paulesu et al. (1993) required rhyme judgments on visually presented letters at a rate of 60 per minute, and the task used by Zatorre et al. (1992) required phoneme judgments on auditorily presented nonword CVC syllables at a rate of 14–18 per minute. The three tasks are rather different in terms of task demands, sensory modality, and rate, yet they all implicated area 40. This may turn out to be the most robust result; three distinct experimental tasks designed to tap one language computation consistently activated (among others) one specific area. One could interpret these three studies as successfully replicating each other with respect to the involvement of area 40 for some computation underlying the processing of speech sounds. Precisely which computation is being engaged requires more analysis of the task demands, but a detailed decomposition of the tasks yields some predictions that make reference to the computations involved in the execution of the experimental tasks. For instance, one might argue that the three tasks (rhyming and phonetic judgement) all crucially require the parsing of a stimulus into an onset and a rhyme to be able to do the task (comparison of syllable rhymes or ends of phonetic strings). An alternative interpretation may make reference to the recruitment of articulatory information, that is, perhaps some model predicts that rhyme judgments require articulatory information. There are a number of plausible arguments to be made, but the analysis will have to use concepts more specified than 'phonological processing'.

Consider, in contrast, the more problematic comparison in which there is almost no overlap despite similar tasks. Paulesu et al. and Sergent et al. both used visually presented letters that had to be judged on whether they rhymed with [i]; the crucial difference between the experiments was in the rate of presentation (20/minute versus 60/minute). This difference (and possible other, even more subtle differences), then, should account for the fact that there was only one area (for only one of the subtractions reported in Sergent, as reported in Table 13.1Cb) that showed up as significant in both studies (area 42), although the studies were otherwise almost identical. The region of overlap (area 42), incidentally, is a plausible candidate area for some aspect of speech processing. Nevertheless, it is surprising that two such similar experimental tasks fail to converge more dramatically in PET activations. This result could mean that minor differences in rate of presentation can lead to serious differences as reflected by PET results. A possible account for the disparity between the results is that at lower presentation rates, subjects' behavior is less constrained, leaving computational resources for the processing of representations orthogonal to those that are supposed to be selectively engaged.[6]

Another recent PET language study by Mazoyer et al. (1993) added another cortical region to our list of candidate areas for the mediation of phonological computations. The studies I have considered all explore the language system using single-word processing tasks. Mazoyer et al. investigated the fractionation of the language system with PET using sentence-level stimuli. These authors also investigated phonological processing. Based on their findings, they argued for "a role for the left middle temporal gyrus in language-specific phonological analysis" (p. 472). This result suggests that sentence-level stimuli may yield yet more different results from word-level stimuli in experiments attempting to selectively engage the phonological system.

---

[6]There is some overlap between the studies of Sergent et al. (1992) and Zatorre et al. (1992), namely area 24 (left anterior cingulate) and area 6 (left supplementary motor area, SMA). Because neither anterior cingulate nor SMA are typically associated with phonological processing or other language processes (because these areas appear to be recruited for many tasks across domains and may be involved in attentional modulation, planning of responses, etc.), it seems reasonable not to attribute special significance to their activations with regard to phonological processing.

How can the no-overlap finding be interpreted? There are at least three possibilities:

1. All studies correctly identify at least one area that is crucial to some aspect of phonological processing. Phonological processing is, consequently, performed by a number of cortical regions, each of which mediates a particular elementary computation underlying phonological processing, namely that aspect necessary for the specific experimental task in question. However, because there is no information on what phonological or phonetic concepts are operative and relevant for each experiment, we cannot sharpen our understanding of how the brain processes and computes the sounds of language.

2. No study is right, that is, the area or areas actually underlying phonological processing have not been identified in any of the experiments. Although this is a logical possibility, it is somewhat unlikely because essentially all (cortical perisylvian) language areas identified at one time or another by other methods have been implicated by PET studies.

3. One of the experiments could be the preferable one. However, there is no reason to prefer any one experiment over the others. Because the experiments rely on underspecified and intuitive notions of phonology and phonological processing, they do not effectively isolate phonological processing. If any of the results and interpretations turn out to be correct, it is for coincidental reasons, and not because the experiment has engaged some predicted component of phonological processing based on independent theoretical and psycholinguistic evidence.

Diagnostic Speculations

Two factors contribute in important ways to the no-overlap problem. First, none of the studies gives an adequately detailed decomposition of the experimental tasks. As a consequence, it is not sufficiently clear what computations go into a given task and what subroutines are subtracted out by the control tasks. Both the explicit contents of tasks and the explicit relationship between tasks and their respective controls remain poorly articulated. Each experimental condition, however, consists of a number of subroutines that must be performed in the execution of the experimental task. These unstated computational subroutines could account for important aspects of the results. They include nontrivial factors, such as when a task is automatic versus controlled and attentionally modulated. A careful and detailed decomposition of the experimental task is crucial with PET because the temporal resolution is so low. Presumably, it is essential to know whether subjects can perform the task easily, freeing up resources for other, task-external computations or whether subjects' behavior is tightly constrained.

Second, the experiments do not look to concepts beyond intuitive notions of phonetic/phonological processing. There is no contact with any theoretical framework or model, and there is only sparse reference to any psycholinguistic work that investigates speech or phonological processes. From the perspective of psycholinguistics, the experiments proceed in a relatively unguided fashion, without explicitly breaking down tasks into well-defined linguistic processing components. This is surprising in view of the large literature on lexical and phonological processing. (For a number of articles reviewing aspects of sublexical and lexical processing from various perspectives see Marslen-Wilson, 1989; examples of detailed models of processing during production can be found in Levelt, 1989.)

The most straightforward interpretation of the no-overlap finding is probably the one previously stated as the first of three logical possibilities: All five studies engage some region relevant to phonological processing. A possible explanation of the no-overlap result is simply that different aspects of phonological processing have been isolated. Although it is not made clear in the papers discussed in this review, processing the sounds of one's language is, of course, not unitary. Phonological processing is a broad concept that includes a number of computations. Consider that as a listener/speaker one must minimally perform such computations as segmental (featural) analysis, syllabic analysis, and stress assignment (metrical analysis) in the processing of single words. These computations may require access to distinct representations that are processed in different cortical or subcortical areas. A consequence of (a) a more detailed task analysis and (b) contact with a model or theory of phonetic/phonological processing would be that the experiments could connect with aspects of phonological processing that linguists know something about independently. What looks like a failure to replicate from the perspective taken in this review is simply a consequence of the selective recruitment of distinct phonological processes that have remained unarticulated due to the insufficiently detailed analysis of language processing.

PRINCIPLED PROBLEMS OF PET

Three factors seem to account for the discrepancy among the PET results: (a) The decomposition of the experimental tasks is not specific enough: Precisely what cognitive processes are engaged typically remains unstated. Therefore, it is hard to know what subjects are doing while performing the experiments. (b) There are no specific psycholinguistic or theoretical reasons offered for what the experiments are about and what theoretical constructs go into an experiment. (c) When researchers study cognitive processes, the temporal resolution of PET mandates a special design, paired-image subtraction. This particular design has a number of problems associated with it. This section addresses issues that are independent of the particular experiments, including the issue of subtraction.

There are some systemic problems with PET methodology that might contribute to the *no-overlap result*. Even in the face of the hypothetical ideal psycholinguistic experiment that perfectly isolates a processing component and perfectly controls for differences between test and control conditions, there are principled limitations that contribute in systematic ways to the difficulty of doing PET studies of cognition.[7]

Some differences in experimental task parameters may be sufficient to account for some of the variance. Raichle et al. (1994) discussed an important finding that can have dramatic import for experimental design: Practicing and repeating an experimental task (e.g., a verbal task) can affect the blood flow results over just a few trials. One presumably must control for that effect in a given test-control pairing by allowing for practice effects (or whatever response attenuation is relevant) to occur in both test and control conditions. Raichle et al., thus, cautioned that small changes in stimulus parameters can lead to dramatic changes as seen in the CBF image. Raichle (1991) mentioned the possible effect of stimulus presentation rates. This is, in fact, a parameter along which the studies I compared differed. The stimulus presentation rate ranged from 14 items per minute (Zatorre et al., 1992) to 60 items per minute (Paulesu et al., 1993). If stimulus timing differences of this sort contribute significantly to the results, some portion of the no-overlap result can be accounted for by appealing to different temporal properties of the experiments, even though they may isolate the same linguistic computation.

Because of the low temporal resolution of PET, the paired-image subtraction paradigm (Fox et al., 1988), has turned out to be quite useful. The temporal resolution of positron emission tomography is on the order of tens of seconds. This forces the experimenter to give the stimuli of a given condition in blocks of the same type. The block-presentation of stimuli and the sensitivity of the instrument in view of the signal-to-noise ratio make it helpful to use paired-image subtraction when investigating cognitive tasks. There are some problems associated with the method (see Sergent et al., 1992; Demonet et al., 1993; and Raichle et al., 1994 for a similar discussion of the problems arising with the subtractive paradigm).

First, even if the more complex cognitive computation is almost identical to the allegedly simpler, nested task, the difference may be qualitative. There is no a priori reason to believe that cognitive processes can be put in a subset relation and subtracted from one another, yielding a component process. One assumption underlying the subtractive analysis is, therefore, that the tasks

---

[7]There are, naturally, principled technical limitations to the method that have been recognized for some time. For discussion of the technical issues see the CIBA Foundation volume on PET: Ciba Foundation Symposium 163, 1991; Roland, 1993; or Greitz, Ingvar, and Widen, 1983.

involved are processed in serially ordered, discrete stages. But this may not be true.

Second, feedback relations must be excluded. One could imagine a system of discrete stages that do not allow for feedback relations. The added components of a more complex task would then not feed back on alleged prior subroutines. Subtraction must have the built-in assumption of forward-only processing to be interpretable. There exists, however, ample evidence (anatomical and experimental) that feedback is widespread.

Third, the experimental tasks implicitly presuppose that only the representations relevant to the task are activated. Experimental evidence indicates, however, that subjects do, in fact, activate codes that are not explicitly tapped in a given experimental task (see Tanenhaus, Flanigan, & Seidenberg, 1980). It is quite difficult to control for subjects attending only to the relevant stimulus dimension.[8] Given that conditions must be presented in blocks, subjects may get better at the task and be able to allocate resources to other, orthogonal computations, or they may employ strategies or heuristics that are different from the way they performed the task at the outset (cf. Raichle et al., 1994). It is virtually impossible to guarantee that subjects performed nothing but the cognitive computation in question. Therefore, it is rather difficult to interpret what the result of a subtraction represents in terms of cognitive processes.

Consider as an example of the last problem what could occur when nonword stimuli are used to fractionate phonetic/phonological processing from other linguistic computations. Nonword stimuli are used in the studies reviewed in various forms: as possible words (Demonet et al.), as syllables (Zatorre et al.), and as single letters (Paulesu et al.; Sergent et al.). There is an advantage to using these kinds of stimuli in that they do not explicitly require other linguistic representations. It is not obvious beyond argument, though, why the processing of these stimuli isolates a computational component of the language system. An experimental task using nonword stimuli engages not just phonological processes but also generates at least (a) a search through the lexicon to find the item, (b) a morphological analysis of the item, and (c) a comparison to the closest neighbors in the lexicon (i.e., items that are very similar) as the item is not listed itself. Regardless of the particular processes that actually occur, it may plausibly be the case that a nonword task bootstraps the subject into the lexicon and that it is just not possible to isolate a language processing component.

Suppose that the computational system that generates linguistic representations is a dedicated, automatic cognitive system, and that when you en-

---

[8]Note that there are positive results: Some experiments show that within-modality selective attention can lead to differences in PET results. Corbetta, Miezin, Dobmeyer, Shulman, & Petersen (1990) have demonstrated, for instance, that selective attention to particular stimulus parameters like color, shape, or velocity of visual displays engages specific and different cortical regions. Whether this kind of selective engagement holds for language processing remains to be shown.

counter a linguistic stimulus, the language apparatus is automatically engaged (Fodor, 1983). That is, all computational modules relevant to language processing are engaged. In such a system one could not isolate any single component performing independently. There must be ways, of course, to engage or access particular subroutines or modules more than others. For example, very broken (ungrammatical, elliptical) speech is often interpretable. Similarly, there are Jabberwocky-like sentences and conceptually impossible sentences that allow syntactic analysis but are uninterpretable. However, even insofar as these kinds of stimuli can access the representations of a specific module of the system more than others (at least phenomenologically), that fact neither implies nor requires that the other processing subroutines are silent or disengaged. Rather, it may be the case that even those computational processes not explicitly invoked by an experimental task are activated if the stimulus is linguistic. What modules or processing subroutines are engaged may be modulated by the type of linguistic information available, by attentional requirements, by memory limitations, or by other internal system states. The point is that one could imagine a plausible conceptual framework that questions, on principled grounds, whether one can isolate particular components of the linguistic system. Because there are no a priori reasons to rule out either viewpoint, the alternatives must be motivated and, given the state of the art, considerable argument is required.

A final point concerns one of the assumptions implicit in this research. The studies characteristically reflect a commitment to localizing specific functions to specific areas of the brain. This is an obvious consequence of the method, but one wonders to what extent unitary localizationism will improve the understanding of the functional organization of the brain. In this sense, PET studies are conceptually closely related to the deficit-lesion approach. PET studies have a number of advantages, the most important one being able to look at intact brains in vivo. But the grain level of analysis and the logic underlying the style of experimentation yields explanations that are very similar to the neuropsychological explorations of the same questions. There is, of course, nothing wrong with trying to contact the results of cognitive neuropsychology. But, arguably, one goal is to extend and clarify the results from cognitive neuropsychology, not just replicate its findings. It is a linguist's charge to find ways to give PET a chance to demonstrate why it is a real departure from and advance over the deficit-lesion approach. If PET studies require deficit-lesion evidence post hoc to validate their results, the method does not really constitute progress.

STUDYING THE GENETICS OF LANGUAGE

The analysis of the no-overlap PET result converges on two explanations, the decomposition problem (i.e., insufficiently detailed information-processing psychology) and the relative insensitivity of the research to the results of linguistics and psycholinguistics. In what way are that analysis and its implicit recommendations for functional neuroimaging relevant to behavioral genetics?

The Problem of Phenotype Refinement and Diagnostic Adequacy

In the terminology of behavioral genetics, the point of contact is the issue of diagnosis and heterogeneity. The problem of the adequacy of the diagnosis is the most immediate and pragmatically relevant connection to the no-overlap result. Specifically, the equivalent of the task-decomposition problem in PET would be an insufficiently detailed (psycho)linguistic analysis and assessment of probands in the case of the genetic study of language. Insofar as the models of language used in behavior genetics research are as underspecified as the models underlying the neuroimaging work, it would be unsurprising if results in the genetic domain did not converge in a satisfactory way. Consequently, the same analytic machinery that can and should inform PET research can and should inform the detailed diagnosis and assessment of language in behavior genetics research.

There are at least two ways in which the analytic machinery of linguistics and psycholinguistics is relevant. First, the careful description of the language phenotypes in question is readily accomplished because linguistic theory and psycholinguistics offer the tools for a rich characterization of the phenomena of interest. There exist different and detailed theories of virtually every domain of language ranging from phonetics and phonology to morphology, properties of the lexicon including lexical access and storage, syntax, semantics, and pragmatics, and extending to discourse phenomena. The available categories allow one to be maximally explicit in description and assessment. These detailed, independently established categories, can guide both phenotype refinement and diagnosis. Second, subtypes can be identified and studied more consistently and with more detail if the evidence for subtypes derives both from the analytic machinery of (psycho)linguistic analysis and the results of genetic analysis. The inclusion of theory and evidence from both domains, thus, considerably broadens the empirical basis available to both clinical and basic research.

That the problems surrounding diagnosis and its adequacy are being seriously considered is attested in recent programmatic statements in the behavioral genetics literature. For example, Plomin and Rende (1991) suggested in a review article that "[p]erhaps the single most important issue in current behavioral-genetic research ... is heterogeneity—whether our diagnoses split nature at its joints" (p. 169). Plomin and Rende pointed out that there has been less genetic research as of yet in domains like perception, learning, and language (among others); they attributed this to the fact that there is little research on individual differences in those domains. This is not a necessary consequence of studying those topics; fine-grained analytical work can make it straightforward to detect even subtle differences and, importantly, subtle interindividual similarities as well. Indeed, one of the most interesting features of contemporary psycholinguistics and linguistic theory is that very precise predictions are made about what components of the system are innate and what components are likely to be crucially dependent on specific information available in the input to the learner. Presumably, this kind of information is quite helpful in the context of a behavioral genetics study insofar as one can target features of the language system that are (by hypothesis) biologically mediated or not. (For an example of specific claims about innate biological language features see de Villiers, this volume.) What de Villiers calls the *closed program* for acquisition constitutes an explicit proposal about what aspects of language are genetically determined and innate. In fact, there are detailed hypotheses about which features show more or less interindividual and crosslinguistic variation, which features are relatively buffered from other components of linguistic and extralinguistic systems, which features are subject to formal triggers for learning, and so on.

In a recent editorial on the genetics of stuttering, Cox (1993) articulated views congruent with Plomin and Rende's. She suggested that "linkage studies on complex disorders have been more frustrating than successful"(p. 177) and that the "most critical components of linkage studies in complex disorders [is] diagnosis" (p. 177). Cox argued that "much of the current lack of success of linkage studies in complex disorders stems from our failure to adequately address the related issues of misdiagnosis and heterogeneity" (p. 178). The diagnostic task is made particularly difficult for behavior genetics by the complicating factors (methodological difficulties and caveats) that are discussed by Brusztowicz (this volume).

Consider, finally, the following example of research where an explicit synergy between behavioral genetics and linguistics could be helpful. Smith and Pennington and their colleagues (see Smith, Pennington, Kimberling, & Ing, 1990) used linkage data to define subtypes in a study of familial dyslexia. Suppose that the subtypes derived from the genetic work converge (or do not converge) with the predictions of a motivated psycholinguistic model of reading disability. More than one source of evidence would then support the distinctions, the distinctions drawn would be independently motivated, and one could argue with some confidence that one is describing components (and their breakdown) that are actually part of the language system. The analytic tools both of genetics and of linguistics are sufficiently explicit to make such a convergence of results a realistic possibility.

On the Nature of the Phenotype

Beyond the broad interdisciplinary connection between genetics and neuroscience sketched earlier—the results of linguistics must play an integral role in both disciplines when the nature of language is at stake—the specific results of neuroscience are, of course, directly relevant to the concerns of behavior genetics.

Neuroanatomy and physiology and genetics are intimately causally related: Genes build the nervous system and regulate various aspects of its function. However, although behavioral output reflects cortical architecture and function relatively directly (at many levels of description ranging from the molecular to the macroscopic-anatomical), the influence of genes is rather more distal. To put it crudely: Genes do not make sentences, brains do.

What is then the optimal level of description of the phenotype? In what units is the phenotype best expressed—units of brain or units of linguistics? There is no a priori reason to believe that the most accurate analysis of the phenotype is at the level of linguistic output rather than at the level of neural circuitry. It turns out that presently a lot more is known about language from the perspective of linguistic theory than is known about the biological end of things (claims to the contrary notwithstanding). There is no principled reason, however, why neuroscience cannot develop equally detailed models for cognitive faculties. Future accounts will certainly involve reference to the neural circuitry that underlies the linguistic computations studied. Behavioral genetics will, thus, need to be increasingly sensitive to neural architecture because, perhaps, the description of the phenotype is ultimately best captured by the correspondences that must exist between the brain and the linguistic knowledge it represents. The ideal case would naturally be a one-to-one-to-one correspondence: The genomic results correlate perfectly with the neuroscientific analysis that, in turn, correlates perfectly with the linguistically described phenotype. For this type of correspondence, the genetic level of analysis is in place, as is the arsenal of linguistics. Neuroscience (in the guise of neuroimaging and other subdisciplines studying language) will have to catch up. It will become crucial, though, to bridge the gap from gene to language, and bridging that gap will necessarily involve reference to the neural circuitry underlying the representation of language. A particularly exciting possibility for future research is, thus, the interaction of behavioral genetic work and morphological and functional analysis of the relevant neural circuitry.[9]

A Hypothetical Scenario

Suppose a gene is implicated in the mediation of a specific linguistic computation. What would one expect the gene to do? Presumably, the gene affects at

---

[9]Of course, not all problems are magically solved in such a hypothetical collaboration. An example of a serious problem that would then have to be addressed concerns the problem that the mapping from brain states to linguistic output may be many-to-one, that is, more than one configuration (brain state) can yield the same linguistic output. In other words, one would have to resolve how the issue of phenocopies plays out in this way of thinking. If a gene is found to correlate with some linguistic property, what is the mapping from gene to anatomy/physiology of neural circuitry to linguistic output? The type of problems that come up show that interdisciplinary thinking will be required to address these issues.

least those neural circuits that compute (some aspect of) the linguistic process in question. Suppose, then, that the research effort succeeds optimally. That is, exactly the right theoretically, psycholinguistically, and cognitively motivated conceptual tools and analyses that yield perfectly predictive and reliable diagnoses have been used. To work through an example, consider for the sake of illustration the analysis of specific language impairment (SLI) discussed by Rice and Wexler (this volume). SLI is best characterized as follows: Probands use infinitives in declarative (and other) utterances a large proportion of the time whereas controls supply inflected (finite) forms all the time. Now, suppose I work out the genetics, and I get evidence completely consistent from the point of view of familial analysis, twin and adoption studies, segregation/pedigree analysis, linkage analysis, and molecular analysis. Moreover, suppose the factors that make something complex (see Brusztowicz, this volume) do not apply or have been optimally controlled for, that is, there is no reduced penetrance, variable expressivity, allelic or nonallelic heterogeneity, or epistasis; the inheritance pattern is not polygenic, nor is there reason to suspect phenocopies—in short, gentle reader, delude yourself for a moment: I have identified a gene the expression of which correlates perfectly with a disproportionately frequent use of infinitives by SLI probands. What kind of interpretation would such a result suggest for a biological model of language?

If a number of neural circuits compute various aspects of morphology and syntax (and other linguistic representations), then at least one circuit computes something that is characterized as *verbal inflection*, for example, the attachment of an affix to a verbal stem.[10] Call the responsible gene *infl*. The gene *infl* regulates the development and function of the circuit that computes the representation of inflection. In other words, there exists a gene/RNA/protein pathway such that when the gene is expressed, the specific circuit that computes inflection is functionally intact. Intact *infl* correlates with adult-like morphosyntactic inflection by, say, 3 years of age.

What if gene *infl* were corrupted? Presumably the *infl*-regulated circuit would not function properly and the linguistic representation computed by that circuit would be nonexistent or damaged. The probands would be deficient at syntactic inflection and never fully recover (although they may be able to develop alternative strategies for attaching the affixes that represent inflection, even explicit strategies like conscious retrieval of a memorized form). This characterisation of SLI is one in which the representations of the language system are actually deviant. Notice that it is just inflection that is deviant. Other aspects of the language system are intact.

This straightforward model is already too simplisitic to account for Rice and Wexler's conjecture. They argued that SLI is an instance of extended develop-

---

[10]Inflection has a number of consequences for sentence interpretation. For a discussion of inflection and its syntactic manifestation and consequences see Wexler, this volume.

ment, as opposed to deviant linguistic representations. Specifically, SLI is the significant temporal extension of a feature of normal development, the stage of optional infinitives (see Rice, this volume). This suggests that the circuit is eventually fully intact although with significant delay. How might one account for this? *Infl* cannot be corrupted because it is ultimately expressed and the circuit is functional. Perhaps in SLI, *infl* exists as a mutant that is expressed at a later stage of ontogenesis. However, when it is expressed, inflection is fully functional. Consequently, there must be *another* gene, say gene *infl-dev* (inflectional development), that delays the expression of *infl*. In other words, two genes regulate the circuit: *infl* maintains the integrity of the circuit, and *infl-dev* regulates the expression of *infl*. SLI is a mutation not in *infl* but in *infl-dev* such that *infl* gets expressed much later but is functionally indistinguishable from its normal expression.

There is yet a further complication: Why does it look like the circuit is sometimes intact? Both in normal syntactic development and in SLI the correct inflection is supplied some of the time. The question why the circuit looks like it operates some of the time thus remains unsolved. Why is there optionality in the system? The existence of optionality turns out to be an extremely challenging problem for theoretical linguistics and psycholinguistics. A possible explanation is that, initially, inflection is computed differently, and the relevant circuit is indeed not in place. The way finiteness is initially computed is through some other mechanism that is already in place. Then, in the course of ontogenesis, the steady-state adult circuit takes over the computation. This type of account begs the question in that one now owes an explanation of how inflection is sometimes (correctly) supplied through a completely different mechanism at earlier stages of acquisition.

Is there any evidence that a gene can disrupt a very specific aspect of a complex computation? Is the hypothesis of a gene like *infl* misguided? Let me give a brief example illustrating that such a model is not completely crazy. The example concerns the worm Caenorhabditis elegans (C. elegans), a creature whose nervous system and genome have been studied in exhaustive detail (see Thomas, 1994). C. elegans individuals have a precisely worked out nervous system: Each worm has 302 neurons that have been characterized structurally, morphologically, and functionally. Also, over 250 genes have been identified that disrupt some aspect of the behavior of C. elegans. Important for our speculations is the fact that C. elegans has a very carefully worked out behavior called the defecation motor program (DMP). The worm's DMP is one well-defined behavior, but consists of three motoric computations that are easily identified and independently regulated genetically. Compromising a gene that is crucial to the integrity of one of the three circuits will disrupt only that circuit. In other words, the DMP is a complex computation that is wired such that three circuits generate the entire behavior, but each circuit can be corrupted individually. Thus, a DMP can be irregular as a consequence of a single mutation. This research on C.

elegans shows, on a general level, the success of the combination of genetics with detailed neuroscientific and behavioral analysis and more specifically provides evidence that a single mutation can influence the integrity of a particular, neurally mediated subroutine of a complex computation.

CONCLUSION: WHAT ONE CAN AND CANNOT LEARN FROM PET STUDIES OF LANGUAGE

One cannot learn from the PET studies of phonological processing reviewed in this chapter which cerebral areas mediate which aspects of language processing, in particular phonological processing, because the results conflict. This is disappointing given that the major goal of neuroimaging studies of cognition is the identification of neural areas mediating some cognitive process.

One can, however, learn:

- That although the results do not overlap, phonological processing is probably mediated by multiple areas (as is semantic processing; see Poeppel, 1994).
- That the canonical functional neuroanatomy of language (see Geschwind, 1970) will have to be carefully reevaluated in view of the recent neuroimaging results. Although Broca's and Wernicke's areas continue to be consistently implicated in language processing, the PET results extend the amount of brain one will have to look at when studying language.
- That psycholinguistics and linguistic theory are not just sources of more or less orthogonal evidence, but rather can and must provide crucial constraints for experimental design and language assessment.

When studying language, one should know something about language. Although this sounds like a triviality to some it is, in fact, not adhered to by a number of investigators who, nevertheless, articulate claims about the human language system. In the failure to acknowledge explicitly that the study of language, by itself, requires considerable technical expertise and that language is not a unitary, unanalyzed concept but a complex computational system mediated by many different parts of the brain, one is likely to be seriously misled by one's assumptions about the system. When construed as a phenomenon of the natural world that is investigated with the methods of the sciences, an entirely different picture may emerge. It is the scientific view of language that will unite behavioral genetics and neuroimaging, two critical disciplines in the elucidation of the biological foundations of language. The essence of the enterprise, then, is not to be guided and then misled by one's intuitions about language, but to respect scientific inquiry, even if the results are counterintuitive.

REFERENCES

Awh, E., Schumacher, E., Smith, E., Jonides, J., Koeppe, R., & Minoshima, S. (1994, March). *Investigation of verbal working memory using PET.* Cognitive Neuroscience Society Meeting, San Francisco.

Baddeley, A. (1992). Working memory. *Science, 255,* 556–559.

Blumstein, S. E., Baker, E., & Goodglass, H. (1977). Phonological factors in auditory comprehension in aphasia. *Neuropsychologia, 15,* 19–30.

Chertkow, H., & Bub, D. (1994). Functional activation and cognition: the 15O PET subtraction method. In A. Kertesz (Ed.), *Localization and neuroimaging in neuropsychology* (pp. 152–184). New York: Academic Press.

Ciba Foundation (1991). Ciba Foundation Symposium 163. In *Exploring brain functional anatomy with positron tomography.* London: Wiley.

Corbetta, M., Miezin, F., Dobmeyer, S., Shulman, G., & Petersen, S. (1990). Attentional modulation of neural processing of shape, color, and velocity in humans. *Science, 248,* 1556–1559.

Cox, N. J. (1993). Stuttering: A complex disorder for our times? *American Journal of Medical Genetics, 48,* 177–178.

Demonet, J.-F., Chollet, F., Ramsay, S., Cardebat, D., Nespoulous, J.-L., Wise, R., Rascol, A., & Frackowiak, R. (1992). The anatomy of phonological and semantic processing in normal subjects. *Brain, 115,* 1753–1768.

Demonet, J.-F., Wise, R., & Frackowiak, R. (1993). Language functions explored in normal subjects by positron emission tomography: A critical review. *Human Brain Mapping, 1,* 39–47.

Fodor, J. (1983). *The modularity of mind.* Cambridge, MA: MIT Press.

Fox, P. T., Mintun, M. A., Reiman, E. M., & Raichle, M. E. (1988). Enhanced detection of focal brain responses using inter-subject averaging and change-distribution analysis of subtracted PET images. *Journal of Cerebral Blood Flow and Metabolism, 8,* 642–653.

Galaburda, A. M. (1991). Neuropathologic correlates of learning disabilities. *Seminars in Neurology, 11*(1), 20–27.

Geschwind, N. (1970). The organization of language and the brain. *Science, 170,* 940–944.

Greitz, T., Ingvar, D. H., & Widen, L. (Eds.). (1983). *Positron emission tomography.* New York: Raven.

Lesser, R., Lüders, H., Klem, G., Dinner, D., Morris, H., Hahn, J., & Wyllie, E. (1987). Extraoperative cortical functional localization in patients with epilepsy. *Journal of Clinical Neurophysiology, 4*(1), 27–53.

Levelt, W. (1989). *Speaking: from intention to articulation.* Cambridge, MA: MIT Press.

Liberman, A., & Mattingly, I. (1985). The motor theory of speech perception revised. *Cognition, 21,* 1–36.

Marslen-Wilson, W. (Ed.). (1989). *Lexical representation and process.* Cambridge, MA: MIT Press.

Mazoyer, B. M., Dehaene, S., Tzourio, N., Frak, V., Murayama, N., Cohen, L., Levrier, O., Salamon, G., Syrota, A., & Mehler, J. (1993). The cortical representation of speech. *Journal of Cognitive Neuroscience, 5*(4), 467–479.

Ojemann, G. (1991). The cortical organization of language. *Journal of Neuroscience, 11,* 2281–2287.

Paulesu, E., Frith, C. D., & Frackowiak, R. S. J. (1993). The neural correlates of the verbal component of working memory. *Nature, 362,* 342–345.

Petersen, S., Fox, P., Posner, M., Mintun, M., & Raichle, M. (1989). Positron emission tomographic studies of the processing of single words. *Journal of Cognitive Neuroscience, 1*(2), 153–170.

Plante, E. (1991a). MRI findings in the parents and siblings of specifically language-impaired boys. *Brain and Language, 41,* 67–80.

Plante, E., Swisher, L., Vance, R., & Rapcsak, S. (1991b). MRI findings in boys with specific language impairment. *Brain and Language, 41,* 52–66.

Plomin, R., & Rende, R. (1991). Human behavioral genetics. *Annual Review of Psychology, 42,* 161–190.

Poeppel, D. (1994). A critical review of PET studies of language. MIT Center for Cognitive Science, Occasional Paper #49, Cambridge, MA.

Poeppel, D. (in press). A critical review of PET studies of phonological processing. *Brain & Language.*

Posner, M., Petersen, S., Fox, P., & Raichle, M. (1988). Localization of cognitive operations in the human brain. *Science, 240,* 1627–1631.

Raichle, M. E. (1991). Memory mechanisms in the processing of words and word-like symbols. In D. J. Chadwick & J. Whelan (Eds.), *Ciba Foundation Symposium: Exploring brain functional anatomy with positron tomography, 163* (pp. 198–217). London: Wiley.

Raichle, M. E., Fiez, J. A., Videen, T. O., MacLeod, A. K., Pardo, J. V., Fox, P. T., & Petersen, S. E. (1994). Practice-related changes in human brain functional anatomy during non-motor learning. *Cerebral Cortex, 4,* 8–26.

Roland, P. (1993). *Brain activation.* New York: Wiley.

Sergent, J., Zuck, E., Levesque, M., & MacDonald, B. (1992). Positron emission tomography study of letter and object processing: empirical findings and methodological considerations. *Cerebral Cortex, 2,* 68–80.

Smith, S., Pennington, B., Kimberling, W., & Ing, P. (1990). Familial dyslexia: use of genetic linkage data to define subtypes. *Journal of the American Academy of Child Psychiatry, 29,* 204–213.

Sternberg, S. (1969). The discovery of processing stages: extensions of Donder's method. In W. G. Koster (Ed.), *Attention and Performance II* (pp. 276–315). Amsterdam: North-Holland.

Tanenhaus, M. K., Flanigan, H. P., & Seidenberg, M. S. (1980). Orthographic and phonological activation in auditory and visual word recognition. *Memory and Cognition, 8,* 513–520.

Thomas, J. T. (1994). The mind of a worm. *Science, 264,* 1698–1699.

Zatorre, R., Evans, A., Meyer, E., & Gjedde, A. (1992). Lateralization of phonetic and pitch discrimination in speech processing. *Science, 256,* 846–849.

COMMENTARY ON
CHAPTER 13

Martha Crago
McGill University

When I was a young girl, my father built me a playhouse in my backyard. It had a door but no windows. This windowless house presented me with a problem. If I heard approaching footsteps, I had no way of figuring out whether it was my neighborhood friends or the neigborhood bullies who were about to appear. I could only infer who the visitors might be from the sound of their footsteps. I explained the problem to my father who was an engineer. He came up with an easy, economical, and tactical solution. He cut out a pine knot in the wall, leaving me a small peephole onto the outside world. I no longer had to infer who my interlopers were. On the other hand, I did not have a perfect view either: sometimes glimpsing only a hand, or a right eye, or worse yet, only the snowball about to be thrown. I look back on my peephole as a lesson in imperfect solutions. I knew more than before, but I still had to do some guesswork.

In his chapter, David Poeppel describes another kind of peephole and its imperfections. Positron emission tomography (PET) has provided functional neuroimages of the brain at work. In doing so, it has begun a process that will undoubtedly provide increasingly complex and informative "peeps" into the brain and, more particularly, into the brain and language. Brain imaging moves us away from earlier inferential studies of brain functioning into what is presumably a more direct form of evidence. Just how direct and just what nature of evidence is part of what Poeppel addresses in his chapter.

The chapter reviews some of the controversies and problems within the PET activation literature, concerning language in general and phonological processing, in particular. After an incisive review of the underlying problems of five superficially comparable studies, Poeppel argues that the basic flaw throughout all five of the studies is a lack of detailed control of the psycholinguistic variables being tested. He presents evidence for the importance of specifying with adequate clarity the psycholinguistic components of the task. Such specification should lead to a greater likelihood of converging experimental evidence. Poeppel goes on to warn that similar psycholinguistic rigor needs to become a part of future investigations that are attempting to decipher the underlying functional neuroanatomy of language in individuals with familial language impairment. Indeed, one of the important lessons to be learned from the divergent PET results described by Poeppel, is the need for adequate task analysis and adequate control of variables in any given linguistic task.

In addition to the variability of psycholinguistic tasks described by Poeppel, there is the striking problem of individual variability in the subject populations being studied with PET. Caplan (1992) has pointed out the effects of just this type of variability on the neural localization of psycholinguistic processing components. At the present time, PET studies require the use of intersubject averaging to achieve adequate signal-to-noise ratios for statistical analyses. This means that group data that show discrepancies, for instance, may potentially contain individual results that show similarities (Whittaker, 1995). Results at the individual level cannot yet be detected by PET research methodology. Functional magnetic resonance imaging (fMRI) can, however, be used to provide individual data. Perhaps with the advent of fMRI, some of PET's reliance on group data can be obviated. Indeed, researchers may be forced to rethink PET results that have been derived from what appear to be homogeneous groups, groups that, in fact, may turn out to be quite heterogeneous at the individual level.

Functional magnetic resonance imaging studies also provide another notable improvement over PET studies. PET technology involves the injection of radioactive substances into the brain. This is not true of fMRIs, which means that they involve less potential harm to the person who is involved. This is of particular note in the case of children being studied for research purposes. However, use of the fMRI has one major drawback for language research at the present time. The production of an fMRI is an extremely noisy process. This places severe limitations on the nature of any auditorily delivered stimuli. Restricting language stimuli to the written modality will put constraints on the age of individuals who can be tested and will create interference from the subset of the specifically language-impaired population that has reading problems (see Tomblin, this volume).

There is yet another consideration that applies to both PET and fMRI studies. Poeppel has reviewed studies that are based on demonstrating increased cerebral blood flow during cognitive processing. This suggests another potential problem in brain imaging experimentation, namely, the underlying assumption that cognition will necessarily be accompanied by increases in cerebral blood flow. Because approximately 70% of the cortical neurons are inhibitory, it is possible that certain aspects of cognitive processing (including those involved in language processing) may be accompanied by a decrease in cerebral blood flow. It is, in fact, conceivable that active cognition might entail equal degrees of increased and decreased cerebral blood flow in adjacent areas of the brain. This might produce no appreciable overall change in cerebral blood flow despite very real cognitive processing going on within an area (Whittaker, 1995). Furthermore, in PET studies it has proven difficult to dissociate activation related to linguistic processing from activation due to accompanying attentional effects, working memory, practice effects, and other nonlinguistic aspects of cognition. For instance, recent evidence suggests that simply providing instructions alone produces measurable changes in cerebral blood flow (Chertkow et al., 1993). There is no particular reason to assume that these same types of interference would be any different for functional magnetic resonance imaging. In summary, then, it seems that some scepticism about the nature of the evidence being provided by brain images is in order at this stage in their development.

One may ask what PET and fMRI studies have to do with the genetics of language impairment. As Plante (this volume; Plante et al., 1991) pointed out, there is preliminary anatomical information provided by MRI techniques that suggests familial language impairment may be accompanied by particular neurological properties. Looking into the neurological substrate of this population should be all the more interesting at a functional level if the "peeps" that PET and fMRI can provide us are interpretable. Moreover, some of the indecipherable elements that arise in association with PET and fMRI may actually encroach upon the establishment of a genetic phenotype for individuals with familial language impairment. Will it be possible to dissect out specific hereditary lesions in linguistic processing that do not and are not affected by attentional and cognitive processes? Will it be possible to separate out the effects of compensatory mechanisms that will further obscure the phenotype? Will the problem of individual variability encountered in PET be a problem for behavioral genetics? The history of recent study into genetics of neurological diseases would suggest that individual variability will indeed be a problem. A consistent theme throughout much of recent genetic research is the wide variability in phenotype derived from a specific genotype. In other words, there may be unexpected difficulties in elucidating the neurolinguistic phenotype of subjects with genetic disorders of language that will only serve to amplify the already existing problems in associating linguistic abnormalities with underlying problems of neural circuitry. In that regard, Poeppel's comments about the necessity of behavioral geneticists understanding the psycholinguistic components with an adequate degree of sophistication are well taken. His warnings only add to my previously mentioned concerns about the imperfections in brain imaging. Both psycholinguistic understanding and new imaging techniques will certainly provide us with more knowledge about the brain and language, but as with the peephole in my playhouse, there will still be some guesswork to be done.

ACKNOWLEDGMENTS

I thank Dr. Howard Chertkow for his more than generous contribution to this commentary. Support for our collaboration is provided by an intercouncil grant from the Medical Research Council of Canada and the Social Sciences and Humanities Research Council of Canada. Finally, I probably never thanked my father enough for the peephole, the playhouse, and his inspiration to do science.

REFERENCES

Caplan, D. (1992). *Language: Structure, processing, and disorders.* Cambridge, MA: Bradford Books/MIT Press.

Chertkow, H., Bub, D., Evans, A., Whitehead, V., Hosein, C., & Bruemmer, A. (1993, October). Separate effects of instructions and stimuli on cerebral blood flow: A positron emission tomographic study. Paper presented at the American Academy of Neurology, Toronto.

Plante, E., Swisher, L., Vance, R., & Rapcsak, S. (1991). MRI findings in parents and siblings of specifically language-impaired boys. *Brain and Language, 41*(1), 52–66.

Whittaker, H. (1995, June). *Alternative interpretations of PET measurements.* Paper presented at the Functional Brain Mapping Conference, Paris.

INTERACTIONIST ACCOUNT OF LANGUAGE ACQUISITION

14

TOWARD A RATIONAL EMPIRICISM: WHY INTERACTIONISM IS NOT BEHAVIORISM ANY MORE THAN BIOLOGY IS GENETICS

Catherine E. Snow
Harvard Graduate School of Education

As has been extensively argued elsewhere, understanding any complex behavioral outcome requires considering both biological and environmental influences on that outcome. For language as for other similarly complex domains of psychological functioning, both a well-designed brain and a well-designed environment are prerequisite to normal development. Although presumably no one would dispute this statement, nonetheless, relatively extreme innatist positions are frequently defended, under the assumption that the design features of an adequate environment are so underspecified as to be uninteresting. Although I do not share that view, I do not attempt in this brief chapter to present all aspects of a counterargument. I simply, in the first part of the chapter, review findings about a few design features of the environment that contribute crucially to successful language acquisition, after presenting a general argument for the practical as well as theoretical importance of incorporating information about the environment into theories of language development. I see these arguments, then, as a small contribution to a realistic interactionist position on the role of biology versus environment, not as an attempt to exclude either set of factors from further consideration.

In the second part of the chapter, I address more directly the topic of this book, arguing that not all developments that can be demonstrated to be relatively impervious to environmental variation are, thus, automatically attributable to innate or to genetic factors and, furthermore, that evidence of biological influences on development must be carefully distinguished from genetic explanations. There is much to biology besides genes, and a premature retreat to genetic explanations may be blinding us to nongenetic but biological, as well as experiential, influences on development.

The Value of an Environmentalist Research Strategy

Though genetic explanations for phenomena as varied as intelligence (Herrnstein & Murray, 1994), male sexual orientation (Hamer, Hu, Magnuson, Hu, & Pattatucci, 1993), alcoholism (Blum & Noble, 1990), depression (Kelsoe et al., 1989), and reading disability (Pennington, this volume) have become popular, it is important to recognize how premature it would be to abandon the search for social and environmental predictors of behavioral and cognitive outcomes. Consider, for example, the case of intellectual functioning—one of the domains where heritability estimates have been rather high.[1] Genetic or chromosomal etiology accounts for 5% to 40% of severe retardation in different studies, and a much smaller percentage of mild retardation (see McLaren & Bryson, 1987, for a review). Most of the other 80% to 85% fall into the category officially labeled *social-cultural retardation*, a category that certainly does not exclude some biological involvement (e.g., from poor nutrition, head injury, perinatal illness), but that does suggest that the resultant retardation was preventable (see Landesman & Ramey, 1989 and Ramey & Ramey, 1992, for reports of intervention programs that reduced social-cultural retardation rates). Note that in Sweden the figures for genetic/chromosomal retardation versus social-cultural retardation are reversed, confirming the preventability of much of the second category. It would obviously be a mistake, under normal policy considerations, for research funds and prevention funds to be devoted primarily to seeking causes or cures for retardation in the genome because, even if massively successful, such a policy would leave the society with only a marginal reduction in the incidence of mental retardation. Recognizing and describing the nature of environmental contributions to intellectual deficits under these circumstances makes for better science and more effective intervention. In fact, much more than half the funding through NIH specifically targeted for mental retardation and developmental disabilities goes to biological (primarily molecular biological and genetic) projects, with only approximately 30% to behavioral projects (Braddock, 1986; Mental Retardation and Developmental Disabilities Branch [MRDD], 1993)—some evidence of the danger to sensible policy of the preference for biological solutions to social problems.

---

[1]The notion of heritability as operationalized in human genetics research is, of course, extremely limited. Because heritability estimates are based on adoption or twin studies, always within cultures and typically with limited variability on social class as well, the available environmental variance is severely limited. Because by virtue of the methodology, the total variance observed is simply apportioned between environmental and genetic factors, limitation to populations that are highly homogeneous with regard to environment inflates the heritability variance estimates. If one could take a cross-population perspective on heritability, environmental variance would presumably increase enormously.

A second example of the practical danger of focusing on biological to the exclusion of social factors in explaining development comes from the Home School Study of Language and Literacy Development (Snow, 1991; Snow & Dickinson, 1991), a longitudinal study of 75 low-income children from age 3 through fourth grade. We are attempting to describe how oral language skills relate to literacy achievement for these children and to determine what characteristics of their homes and their classrooms promote oral language and literacy skills. Of the 75 children in the sample, one has a frank language problem, characterized by severe word-finding difficulties, and one is hearing-impaired and showing delayed language development. These problems were, in both cases, identifiable in our first visit to the children when they were aged 3½. Both children have received special services for speech, language, and reading, but continue to struggle at age 9. Another 25 children in the sample, all of whom at age 3½ fell within normal limits on MLU calculated from spontaneous speech samples, and who were not identified by parents or preschool teachers as having any language problems, are receiving special services through their schools for language, reading and learning problems[2]; 38 children in this sample are reading below grade level as second graders, and everything known about the progress of slightly below average readers through elementary school suggests that well over half will be at least a year behind grade level as fourth graders. Eighteen of the children have been retained in a grade at least once, and some more than once. For one of these children, the difficulties can be attributed clearly to language disabilities that are likely of congenital origin. For the other dozens who are in trouble, social, environmental, and interactionist explanations are needed. One might argue that what these low-income children in low-quality schools really need is early identification of their risk status leading to early intervention, but that still implies an environmentalist solution; early identification helps only if it leads to the provision of services designed to mimic the factors that keep middle-class children in better schools from being overly susceptible to eligibility for special services.

What does this suggest? First, it echoes the findings from the demographics of mental retardation in its suggestion that social policy ignores environmental sources of language problems at the cost of missing many chances for remediation, intervention, and prevention. Second, if the problems with oral language performance and with literacy of the 38 children who are having academic problems are included as reasonable targets of study for developmental psycholinguists, then the need to expand the definition of language enormously

---

[2]It is, of course, possible that some of these children would have been identified as language impaired with more rigorous screening at $3^1/_2$, or that they do have identifiable language disabilities subtle enough that they only emerged later. It is also clear, however, that poor and minority children are greatly overrepresented among those receiving special services for speech, language, and reading problems, suggesting again the preventability of the problems that emerge whether or not their origin is constitutional.

beyond lexicon and grammar is conceded. If "knowing a language" is defined as control over the full array of skills needed to communicate effectively, then about half the children in this sample are language disordered.

A first issue to confront in analyzing the effect of biology or environment on language acquisition is the nature of language itself: What are we trying to explain the acquisition of? The concept of language has often been identified with its subdomain of syntax by those most inclined to biological explanations whereas researchers interested in environmental effects have been more likely to study subdomains like the lexicon or the production of extended discourse.[3] The definition of language that I would argue is relevant to developmentalists is one that includes at least all the various rule systems for oral communication that are specific to language communities. This would include, for example, rules for turn-taking in conversation, rules for structuring extended discourse, rules for determining what must be expressed and where ellipsis is possible, rules for address forms, and so on. For many of these areas, no particular models of acquisition have been proposed, so it makes little sense to include them in a discussion of environmental versus biological effects. I focus discussion in this chapter on the domains of lexicon, grammar, and extended discourse, those where serious innatist, as well as environmentalist proposals, have been made. A more complete discussion of the development of communicative competence would, of course, have to include rules for conversational exchange, for topic-extension, for appropriate expression of speech acts, for interpersonal politeness, and so on. For the domains of lexicon, grammar, and extended discourse, I now turn to evidence that the nature of the language learner's environment is crucially important in determining the course of acquisition.

ENVIRONMENTAL EFFECTS

Evidence for Frequency Effects

Frequency effects are among the most reviled of environmental factors, presumably because of some association to behaviorist explanations. Of course, the demonstration that greater frequency of exposure to some environmental factor speeds or eases some aspect of acquisition does not absolve us of the obligation to describe a mechanism by which the greater frequency has its effect; in the domain of language, such proposed mechanisms have ranged from connectionist (MacWhinney, 1989) to highly cognitivist (Harris, 1992). Furthermore, frequency-based analyses must grapple with the question "frequency of what?"; if children are acquiring abstract structures then the frequency of surface structures might well be considered irrelevant.

---

[3]Admittedly some, for example, Markman (Markman & Hutchinson, 1984) or Pinker (1989), would argue that lexical learning, like grammar, is the product of biologically preprogrammed knowledge of or conformity to certain principles, but we have, so far, at least been spared the hypothesis of a word-learning gene.

The most incontrovertible evidence for frequency effects comes in the domain of lexical acquisition. Huttenlocher, Haight, Bryk, Seltzer, and Lyons (1991) reported that the best single predictor of growth in children's vocabulary during the early stages of language learning was the number of words they heard per unit of time from their mothers. This finding suggests that children's natural word-learning strategies are not so powerful that they can overcome the effect of differences in exposure to vocabulary that occur within a middle-class sample. The well-documented differences in vocabulary size between children of middle-class versus working-class parents (Walker, Greenwood, Hart, & Carta, 1994) confirmed this finding, in light of the greater density of talk in middle-class families (Hoff-Ginsberg, 1992). It has, unfortunately, not been documented whether middle-class families expose their children to more words, to the same set of words more often, or both; a work by Hayes and Ahrens (1988) suggested that the greater knowledge of rare or sophisticated lexical items characteristic of individuals who have higher educational levels (because exposure to such words comes almost exclusively through literacy) does not translate itself into striking differences in use of rare words in conversational interaction. Basically, 95% of spoken word tokens are drawn from the 10,000 most common words, even among highly literate speakers. Of course, the average 6-year-old does not have a vocabulary as large as 10,000 words—so middle-class parents who use those same, relatively common, words with greater frequency may be assuring their children large, even if not very sophisticated, vocabularies. On the other hand, the tendency of parents to use rare, relatively sophisticated lexical items (even at lowish frequencies) does relate to children's word knowledge, suggesting that range of words heard may play a role as important as number of words heard (Beals & Tabors, 1995). It is clear that the first words children acquire are highly frequent words in their environments (Hart, 1991), but there is also good evidence that older normally developing children can learn new words from relatively few or even, under ideal circumstances, a single exposure (Carey, 1988). Children with language impairments have been described as having a somewhat higher threshold for successful acquisition; three exposures to novel words did not produce any learning, but after ten exposures per word children with specific language impairment had learned as many new words as MLU controls (Rice, Oetting, Marquis, Bode, & Pae, 1994). Unfortunately, nothing is known about the degree of variation among normally developing children in frequency threshold for novel lexical learning; it is not unlikely that individual differences in sensitivity to frequency effects interact with environmental differences in frequency of exposure. Until careful documentation of natural variation or experimental manipulation can sort out the confound of types and tokens in input, no one will be able to determine the exact mechanism by which middle-class children acquire larger vocabularies, but the role of the environment in determining that outcome is incontrovertible.

Claims about frequency effects are, of course, much harder to prove in domains like morphology or syntax, where most children receive so much information about any particular form that frequency effects can easily be washed out, and where frequency effects may interact massively with readiness effects. The effectiveness of intensive exposure to particular forms (e.g., the passive) under ideal circumstances (when the target structure is presented in a recast, a response that uses at least some lexical items from the previous child utterance) has been amply demonstrated by Nelson and his colleagues in a series of experimental studies (Nelson, 1977, 1981, 1987). Of course, such interventions only work with children at the right stage of development; there is no basis for knowing whether the right stage should be defined based on the sophistication of their language systems, or on some biologically determined readiness. Similarly convincing demonstrations using natural variation in input have been presented by Farrar (1990) for various morphological markers in English. Needless to say, the presence of frequency effects does not mitigate claims of robust acquisition mechanisms any more than the existence of robust acquisition mechanisms presupposes a genetic explanation. Furthermore, frequency effects become much less visible when they take the form of threshold effects, in which impact of increase in frequency can be expected only in a limited frequency range, for example, the difference between one and ten exposures might be measurable whereas any increment in frequency thereafter would not.

An argument that is often offered against frequency effects is the emergence of correct forms or predictable incorrect forms in the face of zero input frequency. Thus, for example, de Villiers (this volume) cites children's correct interpretation of extracted *wh*-words in questions despite the very rare occurrence of this kind of construction in the input. Although that example is perhaps not entirely apposite (because the frequency effects cited earlier help explain children's output, not their comprehension; no one has yet reported 3-year-olds distinguishing in production between forms like, "Mommy, when did daddy think I threw up?" versus "Mommy, when did daddy say where I threw up?") the basic problem remains; but, it remains only if assumed that frequency of exposure is the only mechanism of acquisition, which no credible empiricist theory does. Frequency effects are presumed to coexist with other forms of environmental support to acquisition (see the following). In addition, reorganizations of stored information driven by processes internal to the child certainly occur. If frequency or other environmental supports promote the acquisition, for example, of two somewhat inconsistent forms or structures (e.g., two different sets of case markers), then internal processes of analysis can lead to restructuring (discovery, e.g., of gender as a principle differentiating the two sets) that makes the system more coherent and predictable. Familiar consequences of child-driven changes in the language system include the greater efficiency of later over earlier word learning—once the lexical system is well structured, inserting new items can occur after relatively little input whereas learning the first words requires a high frequency of exposure (Hart, 1991; Rice, Oetting, Marquis, Bode, & Pae, 1994). Emergent analyses result from children's reorganization and reinterpretation of stored knowledge—a process that does not require new input, but that is clearly ultimately input-dependent because input was the source of the stored knowledge. The occurrence of emergent analyses has been documented for many domains, including language, by constructivist theorists of development (e.g., Bowerman, 1985; Karmiloff-Smith, 1979).

Evidence for Interactive Effects

Straight frequency of exposure, although not irrelevant, is clearly less powerful in predicting child language outcomes than is some weighting of frequency with interactive context. Perhaps the most robust finding in the long and dusty search for environmental effects on language acquisition is the power of semantic and/or formal contingency of adult-utterance on child-utterance or behavior. To summarize dozens of empirical results very briefly, children are most likely to learn from adult utterances that are topically contingent on their own activities or utterances and from adult utterances that differ in form from their own utterances but repeat major content words (e.g., Barnes, Gutfreund, Satterly, & Wells, 1983; Cross, 1978; Ellis & Wells, 1980; Nelson, 1987; see Snow, 1977, 1985, 1989, 1995, for reviews). For example, Tomasello found that the best predictor of children's noun acquisition was their opportunity to hear nouns during episodes of joint attention with their mothers (Tomasello & Farrar, 1986), a finding echoed by Harris (1992), who reported that 3 slow talkers in her sample of 30 had mothers who managed to produce a noun that matched the child's current focus of attention only about 50% of the time as opposed to 74% of the time for faster language learners. Note that these figures are not consistent with the notion of a relatively low threshold for frequency effects in this particular domain (see Goldfield, 1987, for confirmatory findings). Verb acquisition, in contrast, was predicted best by maternal use of verbs related to impending rather than ongoing activity (Tomasello & Cale Kruger, 1992; Tomasello & Todd, 1983). After the initial stage of language development during which the major task is lexical, acquisition of morphology and syntax is demonstrably facilitated by the availability of responses that incorporate lexical material from the preceding child utterance, but differ from the child utterance in completeness, correctness or, perhaps, form. Incorporating the major lexical items of the child utterance presumably ensures that the adult response is on the child's topic, that much of it is comprehensible, and that changes in form can be noticed and processed. Such responses were dubbed "recasts" by Nelson (1981), who demonstrated, in a series of experimental studies, their power to influence children's language systems. Recasts tend to co-occur with other adult responses that are also helpful, for example, repetitions and paraphrases of the child's utterance, praise, and encouragement; Camaioni and Longobardi (1994) lumped these various adult behaviors into the category "tutoring" and found high and significant correlations between their use by mothers to 16-month-olds and the children's observed and reported language sophistication at 20 months. Camaioni and Longobardi also found negative outcomes for children whose mothers produced many *asynchronous communications*—overlaps, changes of topic, and failures to respond to child utterances.

Some children, during crucial stages of development at least, have ample access to adult utterances that relate in analyzable and helpful ways to their own preceding utterances. Sokolov (1993), for example, found that maternal use of modals, pronouns, and nouns was highly contingent on child use in preceding utterances and that degree of contingency, furthermore, decreased as child age increased. Much has been made of the finding that recast or expanded replies are not universal, that in some societies adults consider it inappropriate or undesirable to center their own conversation so heavily on children's interests (Heath, 1983; Ochs & Schieffelin, 1984); the nonuniversality of semantic contingency demonstrates that semantic contingency is not necessary for language acquisition, but not that it is not helpful—there is more than enough evidence that it is very helpful. The unavailability of semantic contingency to some language learners would imply the existence of innate structures if, and only if, semantic contingency were the only mechanism of environmental support; considerable evidence suggests that it is one of many possible environmental design features, any subset of which can support language acquisition (see Snow & Gilbreath, 1983).

The power of access to semantically and/or formally contingent utterances in explaining language outcomes relates to a history of research on feedback or negative evidence. Quite apart from questions about what counts as negative evidence, whether all children have access to it, and whether it is crucial to language acquisition, it is clear that children can benefit enormously from the juxtaposition of correct adult forms to incomplete or incorrect child forms, and do learn from such exposure—not necessarily in every case, as the overcited *nobody-don't-like-me* example shows, but in the aggregate (see Moerk, 1994; Sokolov & Snow, 1994). The availability of input somewhat tuned to the child's language level (another much contested feature of input, but one for which the evidence is now quite strong; i.e., Pan, Feldman, & Snow, 1994; Sokolov, 1993) enhances the value of the negative evidence available to the child; the fact that input is tuned to the child's language level means that negative evidence is likely to be about structures the child is close to getting right, rather than those far beyond the current range of skill.

Modeling Desired Behavior

In addition to the interactive effects associated with contingency, it can be documented that interactions involving modeling of forms for later use by the child in similar circumstances are effective in promoting language development (e.g., Snow, Perlmann, & Nathan, 1987), and that children's connected discourse performances conform to the demands expressed by parents through the types of questions they ask (McCabe & Peterson, 1991; Peterson & McCabe, 1992). Thus, for example, children of parents who ask many questions about the setting of, and participants in, reported personal events tell stories that have rich orientations, but children whose parents question about event structure or about personal reactions tell stories that are carefully plotted and highly evaluated, respectively.

Language-Design Effects

A major domain in which environmental effects on language development have received attention recently is the effects of languages themselves on the system acquired. Examples of such effects include those reported by Choi and Bowerman (1991) for the semantics of space; they found that the dimensions of space that are marked consistently within a language are acquired early, and there seems to be little cognitive prestructuring of the system such that some oppositions (*on* versus *in*) are universally acquired early and others (e.g., *tight* versus *loose* attachment) later. Pervasive language effects have also been described by Berman and Slobin (1994) in their crosslinguistic analyses of descriptions of movement through space and time. The children they studied were all telling the same story, but revealed the influence of their language systems in their attention to matters like manner versus path of motion, in their use of relative clauses versus verb satellites to identify place from which or to which locomotion occurred, and in their signaling of simultaneity and durativity of events. The Hebrew-speaking children showed no signs of seeking ways to mark aspect (not grammaticalized in Hebrew) on verbs, any more than English speakers betrayed a sense they thought something was missing in their language because of the absence of gender and case markings on nouns.

SUMMARY

I have touched briefly on a few ways in which environmental factors have been shown to relate to language development. Extensive reviews of related research fields are available in Gallaway and Richards (1994). A truly interactionist point of view, of course, would incorporate communicative intent as a central notion mediating environmental effects on language acquisition (see Bruner, 1983, for a pragmatic approach to language acquisition), a liberty in which I have not indulged in this chapter, but the omission of which might give the impression that the research on environmentalist effects discussed here has all been carried out in reaction to innatist proposals whereas, in fact, much of it has been conceived within quite a different research tradition.

RELATION OF ENVIRONMENTAL TO BIOLOGICAL EXPLANATIONS

Why Is This Such a Contested Issue?

The conflict between social-interactionist and innatist theories of language development might be analyzed as a classic case of miscommunication between two groups who only appear to be speaking the same language, perhaps confirming the adage that people choose to study those phenomena at which they themselves are least competent. Social interactionists have certainly been known to opine that interaction with linguists suggests that linguists are so good at analyzing the structure of language precisely because they are not distracted by attention to its use as a communicative system. Conversely, in light of the nature of much of what has been published under the rubric of social interactionism, linguists might be forgiven for holding the view that social interactionists have focused their energies on studying the nature of mother-child interaction because they are unable or unwilling to grapple with the complexities of grammatical analysis.

Social psychologists claim that a symptom of persistent miscommunication between any two groups is the emergence of negative stereotyping by each group of the other ("snotty, elitist, grubby, data-blind, stuck in an ivory-tower" versus "atheoretical, naive, conformist, detail-oriented, mired in practice"); in the case of innatists versus social interactionists, the negative stereotypes are exacerbated by a slightly paranoid sense within each group that its own views are failing to get a fair hearing and are losing ground. Such conflicting views of reality are characteristic of ethnic and racial, as well as academic, group processes.

I suggest that one basic source of the miscommunication that has led to the current state of affairs is the lack of an agreed upon definition for the term *language*. One might think that the notion of language, the most basic concept in the field, would have been defined by now, but precisely because the notion is so central, its definition has been implicit, even clandestine, and quite different for groups working in different places and theoretical traditions. As alluded to previously, the definition of language that is common for interactionists encompasses several sets of rule systems (for the lexicon, grammar, morphology, conversation, speech acts, extended discourse, and phonology), each theoretically and perhaps empirically independent of the rest. Within this definition of language, notions like "the language gene" are clearly quite misguided. One might seek a gene that is crucial to the normal display of some specific process, but language as a system is the conjunction of so many different processes, and rests on so much knowledge of such disparate sorts, that many thousands of genes and many other biological and social elements as well will no doubt turn out to be part of a reasonable explanation. In other words, the interactionist view rejects extreme modularity and views language as a complex and multifaceted system.

Of course, within the standard paradigm of human genetics the notion of a language gene might seem to be plausible if heritability for language disabilities can be found. But, the fact that a complex system can be disrupted by a mutation in a single gene does not promote that gene to being an explanation for the system. Take the example of the effect of a null mutation (a so-called "knock-out mutation") of the c-fos proto-oncogene on male sexual behavior; male mice homozygous for the c-fos null mutation showed increased latency to mounting an estrous female and reduced mounting rate, a disruption of normal male sexual behavior that is probably explained by the action of fos in conveying olfactory and somatosensory information to the medial preoptic area of the brain (Baum et al., 1994); the interruption of the fos reactivity blocks the normal cascade of gene transcription events mediating the rate at which a male responds to an estrous female. The c-fos proto-oncogene is one of what will probably turn out to be many dozens of genes whose functioning is prerequisite to the normal display of male rodent sexual behavior, yet it would be quite misleading to refer to it as a "male sexual behavior gene." Similarly, heritability for certain kinds of language disorders might be found and the suspect chromosomal material that the affected family members have in common may be identified without having even come close to explaining what is universal, robust, and biologically buffered about language.

Most of the chapters in this volume presuppose the value of seeking the biological origins of language disorders and development. Although I, in fact, deplore the division, even for purposes of discussion, between biological and social explanations, I have been assigned the task of defending the value of the social-interactionist perspective, and do not have the luxury of presenting a more fully articulated, truly interactionist view. I am not here opposing the inclusion of biology in a full-fledged theory of language acquisition, but am raising a warning that naive biologizing is often characteristic of innatist formulations.[4] It is simply wrong to assume that demonstrating some feature of language is universal justifies identifying that feature with genetic determinants.

In this section of this chapter, I discuss some of the most commonly used arguments against environmentalist models and in favor of innatist models of language development, arguing against them and against a simplistic move from them to a genetic (as opposed to alternative biological) explanation.

Arguments From Incredulity

One set of arguments against social-interactionist explanations of language development can be characterized as *arguments from incredulity*. The incredulity argument might be characterized like this:

---

[4]This warning is an echo of comments by Mary-Louise Kean, published in a 1984 review of Chomsky, Huybregts, and van Riemsdijk (1982):

> I found myself wincing mightily over [Chomsky's] discussion of individual differences and developmental neuropsychology. Nothing that he said was exactly wrong; but ... one could never appreciate from Chomsky's discussion that a, if not the, central issue in the ontogeny of linguistic capacity is the fact that a surprisingly uniform capacity is supported by a physical substrate which shows a remarkable range of individual differences. This should be a serious issue for anyone who holds the view that human linguistic capacity is a biological endowment of the species in any non-trivial sense. (p. 603)

I am also grateful to Magdalena Smoczynska, who brought to my attention a 1988 paper by Robin Campbell in which Kean is cited and an impassioned argument against the "tabula omnifera" position is formulated.

Language is abstract whereas primary linguistic data are concrete. Language is extremely complex, but children learn it extremely quickly. There just is not enough time for children to learn so much without specialized prior knowledge. Children interpret complex, abstract structures correctly, though they cannot possibly have had relevant interactive experiences. Children never make certain kinds of errors. Children recover from other errors without help from adults. The system is too complex to be acquired from the information available in input.

I do not mean to ridicule these arguments—they are serious and must be confronted by social interactionists. I would propose, though, that if linguists are to retreat to arguments from incredulity, it is easier to be incredulous about a language gene than about the role of social interaction in language development. I outline here my own arguments from incredulity, which fall into two categories: incredulity about the biological mechanisms that have so far been proposed to explain universal language capacities and incredulity about the uniqueness of language.

Some Pretty Incredible Mechanisms. In addition to balancing innatist explanations of language development with some realism about the importance of social and environmental variation, we should, if we are serious about biological explanations, try to do realistic biology. Several thousand genes play a role in the development of the central nervous system; an overwhelming proportion are presumably operating on the development and function of neural mechanisms controlling evolutionarily central phenomena like homeostasis, movement, sensation, perception of pain, neuroendocrine control, including sexual differentiation, motivation, memory, categorization, and other cognitive functions. After all these genes have done their work, experience in the world is crucial in organizing the somewhat exuberant and overelaborated structure they have created. Experience can, in fact, even organize the brain in ways that might seem to violate biological determinism, recruiting auditory cortex for processing sign language in the deaf, for example (Neville, 1991). More importantly, experience is evidently prerequisite to producing any cortical organization at all. This has been demonstrated in great detail for the development of the brain structures that subserve vision as a result of binocular visual experience (e.g., Shatz, 1992; Wiesel & Hubel, 1963, 1965). In light of the extensive evidence concerning the importance of perceptual and neuronal activity in contributing to brain development, it seems likely it was the absence of language experience that left Genie without left-hemisphere temporal organization whereas her right hemispheric functions were organized more or less normally, presumably because of her access to normal visual experience (Curtiss, 1977). Thus, a direct link between early genetic action determining brain structure and the later brain organization underlying mature language capacity will not be easily visible in this welter of complex developmental forces.

Take a comparison case where crucial genetic events are known to control normal development. Male sexual behavior is a complex system, though certainly not as complex, nor as abstract, as language. Normal male sexual behavior, like language, is dependent on the availability of certain physical structures as well as the development of certain mental capacities. Like language, the system is more robust, less vulnerable to physiological interruption, after having been developed and used than it is during development. Examining the role of genes in determining normal male sexual behavior might provide a template for a realistic discussion of how genes could function in language development. In fact, in the case of male sexual behavior, we are considerably farther along, in that there is one rather small chromosome, the Y chromosome, where it makes sense to begin the search; as early as 1990 the gene responsible for testis development (SRY) had, for example, already been located on the Y chromosome (Koopman, Gubbay, Vivian, Goodfellow, & Lovell-Badge, 1991), and specific disruptions of normal sexual development (e.g., testicular feminization syndrome) have been related to specific gene mutations resulting in the deviant production of androgen receptor protein in brain and somatic tissues (see Brinkmann et al., 1992, for a review). Thus, identification of genetic factors relevant to male sexual behavior is considerably farther along than is identification of any genetic material relevant to language; yet the notion that "the maleness gene" exists has recently been questioned. Researchers have tentatively identified a second Y-chromosome gene (SRA1) whose protein product contributes to the development of testes. Recent findings (Zanaria et al., 1994) suggested that yet another gene (DSS on the X chromosome) can override the action of the SRY gene, such that genetic males with normal Y chromosomes can develop ovaries and feminized external genitalia. In other words, even the somatic structures characteristic of males and females (let alone their sexually differentiated nervous systems, which influence gender-specific behaviors) are the product of action of many genes. No one in the field of reproductive biology has seriously proposed that there is a male sexual behavior gene—on the contrary, it is assumed that the system is multidetermined at the genetic level as well as at the level of neural structure, hormonal influence, and sexual-social experience.

The point here is not to deny the role of genetic determinants in either male sexual behavior or in language, but to ask for a pause to contemplate how genes have to operate in the hope that such contemplation will lead to realistic suggestions about how they might be implicated in language development or language disorders. Individual genes in the nervous system can do a very limited set of things: They control the production of single proteins which, in some cases, influence the transcription of additional neuronal genes or which determine neural structure or chemistry. No known gene product can be related directly to linguistic phenomena like features, tense, the uniqueness principle, or finiteness. It has now been recognized by geneticists (in part because of their forced retreat from proposals about genes for alcoholism and depression; see

Kidd, 1993) that simply identifying chromosomal material common to sufferers of some disorder is a poor basis for hypothesizing a gene specific to the disorder; a mechanism linking the gene to the disorder is required to make the gene identification plausible. Similarly, identifying a behavioral pattern that is specific to a disorder is an insufficient basis for proposing a gene, even if familial patterns of occurrence suggest heritability, without a mechanism linking the behavioral deficit to a possible gene product. Geneticists using DNA linkage techniques to identify genes and behavioral scientists using familial patterns to identify the details of behavior disorders are like engineering teams starting a tunnel from two sides of a wide river; only careful calibration will ensure that their tunnels meet in the middle, rather than missing each entirely.

Though no one would be so naive as to suggest that universal grammatical features are direct gene products, an alternative possibility that has been contemplated is that hypothesized language universals, although not gene products themselves, somehow derive from specific brain structures whose development results from the expression of particular genes. Indeed, if there were not so much evidence that the localization and organization of language in the brain was somewhat variable and, furthermore, itself the product of early linguistic experience, this might be a plausible explanation. It has been reported that certain aspects of normal language processing, for example, the ability to hear rapid temporal transitions, are disrupted in individuals with language disorders (Neville, Coffey, Holcomb, & Tallal, 1993); there is reason to believe that such disruptions, in turn, result from some structural abnormality in the left planum temporale. It is known that neural events mediate auditory processing, and it is feasible that disruption of a certain type of auditory processing could result from some flaw in the neural architecture of the temporal lobe. Plausibility declines, however, when the phenomenon being explained is not a processing mechanism, but a principle like "two forms have two interpretations" or a rule like "root clauses have finite verbs." Yet, it is just this sort of linguistic principle that is typically cited as examples of linguistic universals which, in turn, are presupposed to be universal because they are innate.

Language Is Not So Special. A key argument in support of the innate and, thus, presumably genetic language capacity is the uniqueness of language among cognitive accomplishments. The argument goes like this:

> Language, a system characterized by enormous complexity and an abstract structure not directly accessible from the data available at the surface, is acquired by all children, even many with very low cognitive capacity, without explicit teaching or well-designed instructional support; thus, it must be innate, a point confirmed by its presence in all human societies and the presence of certain similarities in its structure wherever it appears. Language is implicitly or explicitly contrasted, then, to other human, cognitive accomplishments, such as algebra (not universal in human groups, acquired only with instruction), chess (enormous variability in

level of accomplishment after equal exposure), and music (universal in societies, but large individual differences in capacity to perform).

I argue, though, that universal success in language acquisition as compared to the acquisition of these other capacities can be interpreted as plausibly as an argument for social buffering of the acquisition process as for innateness, and that the crucial feature of language that has been offered as proof of its nonlearnability, its abstract nature, is characteristic of many other systems that no one would argue are innate (though, of course, like everything else, they do rely on innate capacities). I briefly discuss two of these systems: the alphabetic principle and politeness.

The alphabetic principle was invented or discovered once in history, and has been disseminated since by contact, in contrast to logographic and syllabic writing systems, which have been invented independently many times over (Sampson, 1985). Alphabetic writing is extremely efficient for the very same reason that it is rather difficult—the abstract nature of the relation between symbols and their referents. As has been extensively discussed and demonstrated (e.g., Liberman, Shankweiler, Fischer, & Carter, 1974), written letters do not typically represent surface structure units; in fact, they represent theoretical, unpronounceable units that constitute an abstraction from perceptual experience of the following sort: The letter *B* represents the sound that is common to the pronunciation of, for example, *bee, bye, boo, bay, tib, tab, tub, tube, table, trouble,* and *tibula.* One might even argue that early reading instruction, in which children are told things like "*B* makes a *buh* sound," in fact, should disrupt their discovery of the alphabetic principle, by providing frankly incorrect information. Nonetheless, children growing up around alphabetic literacy systems typically do acquire the basics of phoneme-grapheme mapping; in fact, some 5% to 10% of children acquire the system spontaneously, before they receive any formal instruction in it, simply from noting relations between spoken and written words (e.g., Davidson & Snow, 1995; Durkin, 1966). Admittedly, the alphabetic principle is somewhat more limited in scope than grammar, but it shares many of the features that have led to the proposition that grammar must be innate: It is abstract, it can be acquired without instruction, and it can be acquired on the basis of very messy data that require considerable reanalysis (at least, for children learning to read in English and French). On the other hand, the alphabetic principle can hardly be said to be universal, given that it emerges only under very specific environmental conditions and is not in any sense an inevitable product of human development.

A second example of a universal phenomenon that cannot plausibly be connected to a gene is the system of rules for politeness. As conceptualized by Brown and Levinson (1987), for example, the rules governing public social interaction among members of a community reflect universal dimensions of analysis of the nature of the social interactive act (the degree to which it threatens negative and positive face of both speaker and hearer, as a function of their social relationship with one another and the nature of the act itself). Because these underlying analytic dimensions are universal, the range of strategies available for engaging in threats to face without disastrous consequences is also widely shared—though each language community makes its own selection of preferred strategies, as it does its own definition of a variety of social relationships on the various relevant dimensions. Would one wish to argue from universality across language communities, from generally successful acquisition and from the absence of explicit teaching about the abstract system that there is a politeness gene?

CONCLUSION

Typically, questions about heredity versus environment in domains such as IQ, have been formulated as attempts to see what percent of variance in the outcome can be explained by each of the sources. It is worth mentioning explicitly, because it may otherwise escape notice, that, for the case of language, the argument has been crafted somewhat differently: biology has typically been proposed as accounting for what is nonvariant or universal in the outcome and environment for that which is variable. Because, in general, most children are quite successful at learning language, this line of reasoning leaves relatively little room for environmental effects. Furthermore, in some views of language, it is precisely those aspects of the language system that are subject to some variability of accomplishment that are considered uninteresting or trivial, thus further reducing the importance of any environmental effects found. Even in the face of the discouraging prospect these positions offer, one goal of this chapter has been to argue that environment plays an important role in determining the course of language acquisition, under any view of the impact on language of innate factors. It could also be argued that universal language outcomes can be taken as evidence for environmental effects as strongly as for biological effects, and that variability in outcomes must similarly be apportioned to environmental and to biological factors based on evidence rather than presupposition. After all, all cultures have developed procedures to ensure child survival and socialization thus creating linguistic environments for children that are universal on certain dimensions, for example, the chance for children to hear interpretable talk about comprehensible topics and opportunities for interaction with affectively important older members of the group. Although work in cultural anthropology has focused on differences among cultures in childrearing, just as work in structural linguistics focused on differences among languages, cultures, like languages, can be analyzed for universal as well as for particularistic features.

In addition, I argued that, although there is clearly room for honest and friendly dispute about the degree to which biological versus environmental factors influence language development, there can be little excuse for catapulting from an acceptance of biological influences to a proposal of language genes. In fact, the identification of biology with heredity is quite unjustified: One's biological capacity is itself a product of environmental factors working in conjunction with heredity, as vast quantities of work showing postnatal, experiential influences on brain structure (e.g., Greenough, Black, & Wallace, 1987; Held, 1985; Neville, 1991) have demonstrated. Although in this chapter I have discussed environmental influences on behavioral rather than on biological outcomes, I do not, thereby, mean to concede the world of biology either to geneticists or those nonbiologists inclined to invoke genetics ritually when discussing biology.

ACKNOWLEDGMENTS

Ideas presented in this chapter were stimulated by conversation with Barbara Pan, Heidi Feldman, Brian MacWhinney, and other collaborators on Foundations for Language Assessment in Spontaneous Speech (HD 23388), as well as with Patton Tabors and David Dickinson, collaborators on the Home School Study of Language and Literacy Development. I express my appreciation to them and to Michael Baum, who has been very helpful in explaining the complexities of male sexual behavior.

REFERENCES

Barnes, S., Gutfreund, M., Satterly, D., & Wells, G. (1983). Characteristics of adult speech which predict children's language development. *Journal of Child Language, 10,* 65–84.

Baum, M. J., Brown, J. J. G., Kica, E., Rubin, B. S., Johnson, R. S., & Papaioannou, V. (1994). Effect of a null mutation of the c-fos proto-oncogene on sexual behavior of male mice. *Biology of Reproduction, 50,* 1040–1048.

Beals, D. E., & Tabors, P. O. (1995). Arboretum, bureaucratic, and carbohydrates: Preschoolers' exposure to rare vocabulary at home. *First Language, 15,* 57–76.

Berman, R., & Slobin, D. (1994). *Relating events in narrative.* Hillsdale, NJ: Lawrence Erlbaum Associates.

Blum, K., & Noble, E. (1990). Allelic association of human dopamine D2 receptor gene in alcoholism. *Journal of the American Medical Association, 263,* 2055–2060.

Bowerman, M. (1985). Beyond communicative adequacy: From piecemeal knowledge to an integrated system in the child's acquisition of language. In K. E. Nelson (Ed.), *Children's language* (Vol. 5). Hillsdale, NJ: Lawrence Erlbaum Associates.

Braddock, D. (1986). *Federal policy toward mental retardation and developmental disabilities.* Baltimore: Brookes.

Brinkmann, A., Jenster, G., Kuiper, G., Ris-Stalpers, C., van Laar, J., Faber, P., & Trapman, J. (1992). Structure and function of the human androgen receptor. In E. Nieschlag & U. Habenicht (Eds.), *Spermatogenesis-fertilization-contraception: Molecular, cellular, and endocrine events in male reproduction.* Berlin: Springer Verlag.

Brown, P., & Levinson, S. (1987). *Politeness: Some universals in language usage.* Cambridge: Cambridge University Press.

Bruner, J. (1983). *Child's talk.* New York: W. W. Norton.

Camaioni, L., & Longobardi, E. (1994, June). *A longitudinal examination of the relationships between input and child language acquisition.* Paper presented at First Lisbon Meeting on Child Language with special reference to Romance, Lisbon.

Campbell, R. N. (1988). *On innateness: Nec rasa est, nec omnia tenet.* Toronto Working Papers in Linguistics, 7.

Carey, S. (1988). Lexical development: The Rockefeller years. In W. Hirst (Ed.), *The making of cognitive science: Esssays in honor of George A. Miller* (pp. 197–209). New York: Cambridge University Press.

Choi, S., & Bowerman, M. (1991). Learning to express motion events in English and Korean: The influence of language-specific lexicalization patterns. *Cognition, 41,* 83–121.

Chomsky, N., Huybregts, R., & van Riesmdijk, H. (1982). *The generative enterprise.* Dordrecht: Foris.

Cross, T. G. (1978). Mothers' speech and its association with rate of linguistic development in young children. In N. Waterson & C. Snow (Eds.), *The development of communication.* New York: Wiley.

Curtiss, S. (1977). *Genie: A psycholinguistic study of a modern-day "wild child."* New York: Academic Press.

Davidson, R., & Snow, C. E. (1994). The linguistic environment of early readers. *Journal of Research in Childhood Education, 10,* 5–21.

Durkin, D. (1966). *Children who read early.* New York: Teachers College Press.

Ellis, R., & Wells, G. (1980). Enabling factors in adult-child discourse. *First Language, 1,* 46–82.

Farrar, J. (1990). Discourse and the acquisition of grammatical morphemes. *Journal of Child Language, 17,* 607–624.

Gallaway, C., & Richards, B. J. (1994). *Input and interaction in language acquisition.* Cambridge: Cambridge University Press.

Goldfield, B. (1987). The contribution of child and caregiver to referential and expressive language. *Applied Psycholinguistics, 8,* 267–280.

Greenough, W. T., Black, J. E., & Wallace, C. S. (1987). Experience and brain development. *Child Development, 58,* 539–559.

Hamer, D., Hu, S., Magnuson, V., Hu, N., & Pattatucci, A. (1993). A linkage between DNA markers on the X chromosome and male sexual orientation. *Science, 261,* 321–323.

Harris, M. (1992). *Language experience and early language development: From input to uptake.* Hillsdale, NJ: Lawrence Erlbaum Associates.

Hart, B. (1991). Input frequency and children's first words. *First Language, 11,* 289–300.

Hayes, D., & Ahrens, M. (1988). Vocabulary simplification for children: A special case of motherese? *Journal of Child Language, 15,* 395–410.

Heath, S. (1983). *Ways with words.* New York: Cambridge University Press.

Held, R. (1985). Binocular vision—behavioral and neuronal development. In J. Mehler & Robin Fox (Eds.), *Neonate cognition: Beyond the blooming buzzing confusion.* Hillsdale, NJ: Lawrence Erlbaum Associates.

Herrnstein, R., & Murray, D. (1994). *The bell curve: Intelligence and class structure in American life.* New York: Free Press.

Hoff-Ginsberg, E. (1992). How should frequency input be measured? *First Language, 12,* 233–244.

Huttenlocher, J., Haight, W., Bryk, A., Seltzer, M., & Lyons, T. (1991). Early vocabulary growth: Relation to language input and gender. *Developmental Psychology, 27,* 236–248.

Karmiloff-Smith, A. (1979). *A functional approach to child language.* Cambridge: Cambridge University Press.

Kean, M.-L. (1984). The generative enterprise. *Language, 60,* 600–604.

Kelsoe, J. R., Ginns, E. I., Egeland, J. A., Gerhard, D. S., Goldstein, A. M., Bale, S. J., Pauls, D. L., Long, R. T., Kidd, K. K., Conte, G., Housman, D. E., & Paul, S. M. (1989). Re-evaluation of the linkage relationship between chromosome 11p loci and the gene for bipolar affective disorder in the Old Order Amish. *Nature, 342,* 238–243.

Kidd, K. (1993). Associations of disease with genetic markers: Déjà vu all over again. *American Journal of Medical Genetics, 48,* 71.

Koopman, P., Gubbay, J., Vivian, N., Goodfellow, P., & Lovell-Badge, R. (1991). Male development of chromosomally female mice transgenic for Sry. *Nature, 351,* 117–121.

Landesman, S., & Ramey, C. (1989). Developmental psychology and mental retardation: Integrating scientific principles with treatment practices. *American Psychologist, 44,* 409–414.

Liberman, I. Y., Shankweiler, K., Fischer, R., & Carter, B. (1974). Explicit syllable and phoneme segmentation in the young child. *Journal of Experimental Child Psychology, 18,* 201–212.

MacWhinney, B. (1989). Competition and teachability. In M. Rice & R. Schiefelbusch (Eds.), *The teachability of language.* Baltimore: Paul Brookes.

Markman, E., & Hutchinson, J. (1984). Children's sensitivity to constraints on word meaning: Taxonomic versus thematic relations. *Cognitive Psychology, 16,* 1–27.

McCabe, A., & Peterson, C. (1991). Getting the story: A longitudinal study of parental styles in eliciting personal narratives and developing narrative skill. In A. McCabe & C. Peterson (Eds.), *Developing narrative structure* (pp. 217–253). Hillsdale, NJ: Lawrence Erlbaum Associates.

McLaren, J., & Bryson, S. (1987). Review of recent epidemiological studies of mental retardation: Prevalence, associated disorders, and etiology. *American Journal of Mental Retardation, 92,* 243–254.

Mental Retardation and Developmental Disabilities Branch. (1993). *Report to council.* National Institute of Child Health and Human Development, Washington, DC.

Moerk, E. (1994). Corrections in first language acquisition: Theoretical controversies and factual evidence. *International Journal of Psycholinguistics, 10,* 33–58.

Nelson, K. E. (1977). Facilitating children's syntax acquisition. *Developmental Psychology, 13,* 101–107.

Nelson, K. E. (1981). Toward a rare event cognitive comparison theory of syntax acquisition: Insights from work with recasts. In P. Dale & D. Ingram (Eds.), *Child language: An international perspective.* Baltimore, MD: University Park Press.

Nelson, K. E. (1987). Some observations from the perspective of the rare event cognitive comparison theory of language acquisition. In K. E. Nelson & A. van Kleeck (Eds.), *Children's language, Vol. 6.* Hillsdale, NJ: Lawrence Erlbaum Associates.

Neville, H. (1991). Neurobiology of cognitive and language processing: Effects of early experience. In K. Gibson & A. Petersen (Eds.), *Brain maturation and cognitive development: Comparative and cross-cultural perspectives.* Amsterdam: De Gruyter.

Neville, H., Coffey, S., Holcomb, P., & Tallal, P. (1993). The neurobiology of sensory and language processing in language-impaired children. *Journal of Cognitive Neuroscience, 5,* 235–254.

Ochs, E., & Schieffelin, B. (1984). Language acquisition and socialization: Three developmental stories and their implications. In R. A. Levine & R. A. Shweder (Eds.), *Culture theory: Essays on mind, self, and emotion* (pp. 276–320). New York: Cambridge University Press.

Pan, B. A., Feldman, H. M., & Snow, C. E. (1994). Parental speech to low-risk and at-risk children. Manuscript submitted for publication.

Peterson, C., & McCabe, A. (1992). Parental styles of narrative elicitation: Effect on children's narrative structure and content. *First Language, 12,* 299–322.

Pinker, S. (1989). Resolving a learnability paradox in the acquisition of the verb lexicon. In M. Rice & R. Schiefelbusch (Eds.), *The teachability of language.* Baltimore: Paul Brookes.

Ramey, C., & Ramey, S. (1992). Effective early intervention. *Mental Retardation, 30,* 337–345.

Rice, M., Oetting, J., Marquis, J., Bode, J., & Pae, S. (1994). Frequency of input effects on word comprehension of children with specific language impairment. *Journal of Speech and Hearing Research, 37,* 106–122.

Sampson, G. (1985). *Writing systems.* London: Hutchinson.

Shatz, C. (1992). The developing brain. *Scientific American, 267*(3), 60–67.

Snow, C. E. (1977). Mothers' speech research: From input to interaction. In C. E. Snow & C. A. Ferguson (Eds.), *Talking to children: Language input and acquisition* (pp. 31–49). Cambridge: Cambridge University Press.

Snow, C. E. (1985). Conversations with children. In P. Fletcher & M. Garman (Eds.), *Language acquisition* (pp. 363–375). London: Cambridge University Press.

Snow, C. E. (1989). Understanding social interaction and language acquisition: Sentences are not enough. In M. Bornstein & J. Bruner (Eds.), *Interaction in human development* (pp. 83–103). Hillsdale, NJ: Lawrence Erlbaum Associates.

Snow, C. E. (1991). The theoretical basis for relationships between language and literacy development. *Journal of Research in Childhood Education, 6,* 5–10.

Snow, C. E. (1995). Issues in the study of input: Fine-tuning, universality, individual and developmental differences, and necessary causes. In B. MacWhinney & P. Fletcher (Eds.), *Handbook of child language.* Oxford: Blackwell.

Snow, C. E., & Dickinson, D. K. (1991). Some skills that aren't basic in a new conception of literacy. In A. Purves & T. Jennings (Eds.), *Literate systems and individual lives: Perspectives on literacy and schooling* (pp. 175–213). Albany: SUNY Press.

Snow, C. E., & Gilbreath, B. J. (1983). Explaining transitions. In R. Golinkoff (Ed.), *The transition from prelinguistic to linguistic communication* (pp. 281–296). Hillsdale, NJ: Lawrence Erlbaum Associates.

Snow, C. E., Perlmann, R., & Nathan, D. (1987). Why routines are different: Toward a multiple-factors model of the relation between input and language acquisition. In K. Nelson and A. van Kleeck (Eds.), *Children's language, Vol. 6* (pp. 65–97). Hillsdale, NJ: Lawrence Erlbaum Associates.

Sokolov, J. (1993). A local contingency analysis of the fine-tuning hypothesis. *Developmental Psychology, 29,* 1008–1023.

Sokolov, J., & Snow, C. E. (1994). The changing role of negative evidence in theories of language acquisition. In C. Gallaway & B. Richards (Eds.), *Input and interaction in language acquisition.* London: Cambridge University Press.

Tomasello, M., & Cale Kruger, A. (1992). Joint attention on actions: Acquiring verbs in ostensive and non-ostensive contexts. *Journal of Child Language, 19,* 311–334.

Tomasello, M., & Farrar, M. J. (1986). Joint attention and early language. *Child Development, 57,* 1454–1463.

Tomasello, M., & Todd, J. (1983). Joint attention and lexical acquisition style. *First Language, 4,* 197–212.

Walker, D., Greenwood, C., Hart, B., & Carta, J. (1994). Prediction of school outcomes based on early language production and socioeconomic factors. *Child Development, 65,* 606–621.

Wiesel, T., & Hubel, D. (1963). Effects of visual deprivation on morphology and physiology of cells in the cat's lateral geniculate body. *Journal of Neurophysiology, 26,* 978–993.

Wiesel, T., & Hubel, D. (1965). Comparison of the effects of unilateral and bilateral eye closure on cortical unit responses in kittens. *Journal of Neurophysiology, 28,* 1029–1040.

Zanaria, E., Muscatelli, F., Bardoni, B., Strom, T. M., Guioli, S. G. W., Lalli, E., Moser, C., Walker, A. P., McCabe, E. R. B., Meitinger, T., Monaco, A. P., Sassone-Corsi, P., & Camerino, G. (1994). An unusual member of the nuclear hormone receptor superfamily responsible for X-linked adrenal hypoplasia congenita. *Nature, 372,* 635–641.

COMMENTARY ON
CHAPTER 14

Jill de Villiers
Smith College

The chapter by Catherine Snow is a very thoughtful analysis of the problems inherent in making an inferential leap from arguments about the lack of environmental influence to the idea of a language gene. It is sobering in the extreme to acknowledge the danger that it will be assumed that arguments favoring genetic contributions to language mean that it is not worth ensuring the adequacy and richness of early environments for language and literacy. In addition, the points made about the distance to be traversed between behavioral and molecular genetics, or between proposals about innateness and discovery of the "gene" for subjacency, are serious matters for the readership of this volume.

The parallel with heritability of IQ would indeed be a serious mistake about the interpretation of most ideas about a genetic contribution to language. In the case of IQ, the questions revolve around the variability in the population: Heritability is an index of what proportion of that variance is due to (unspecified) genetic factors as opposed to environmental factors. Even if heritability turns out to be modestly high, it cannot be concluded that intelligence is largely determined by the genes in an individual human being: Heritability is a population statistic. Furthermore, a high heritability says nothing about immutability by changes in environment: These issues have all been well rehearsed but must be understood again with each generation of arguments. Nevertheless, there are departures from normal intelligence that have been traced to a genetic defect of a very particular sort: phenylketonuria, for instance, or Down syndrome.

In the case of language, it is only if one considers there to be meaningful variation among individuals in language ability that one must be concerned with the source of variance. Researchers primarily concerned with abstract principles of grammar have not, in fact, considered there to be meaningful individual variation in the expression of universal grammar, at least in "normal" populations. Why might this be so?

It is clear that the term *language* is sufficiently encompassing as to allow many types of subject matter and to support several full-blown paradigms, in the sense of Kuhn (1970; see also Atkinson, 1983). The paradigm that underlies my chapter has behind it a richly elaborated theory, a few limited methodologies, and some distinctive metaphysical assumptions, as does the paradigm underlying the chapter by Snow. The two paradigms differ quite markedly in all these aspects, leading to the usual opportunity for prolonged and persistent disagreements about the best use of one's time and resources and graduate training, and the concomitant advantages of focus, theoretical refinements and detailed attention to a limited subject matter of central concern. But one assumes that these paradigms, in fact, do not blind us to the truths of the other, that they are not so incommensurable as to block translation and understanding (Scheffler, 1982). That is, one hopes that a researcher does not have to become born again within the other paradigm's culture, giving up all that they knew before, in order to fully appreciate the other's perspective. In fact, I count myself as evidence that one can cross paradigms and remember the past (though Snow may not agree) (see de Villiers, 1988 for evidence). In what follows, I attempt to make the contrasts in assumptions explicit to enable the readers of this volume to understand why the differences emerge among theorists in this field and to encourage further biculturalism across paradigms.

The paradigm underlying my chapter has its roots in linguistic theory, particularly the generative grammar of Chomsky. As I describe in that chapter, the paradigm has been almost exclusively concerned with grammar, not vocabulary, discourse, speech acts, or literacy development. Among the assumptions about language are that it is a specialized module in the human mind, not shared by any other creatures; that language acquisition is made possible by a prewired set of assumptions about universal grammar; that children's grammars conform to the nature of human grammars in general; that input is necessary only to trigger children's pre-existing grammatical knowledge; and that syntax development is not dependent in any significant way upon the child's cognitive or social development. A most central component of this paradigm has also been the assumption of universality: What is studied is true of all humans, regardless of the language spoken or the culture in which they were raised. This has led researchers to assume generalizability of results from small samples of usually middle-class White children in enriched environments. Attention to cultural variation, to impoverished environments, or to language-delayed children, has been sporadic and only recent within this framework: The dominant trend has been to study the "universal child"—admittedly like the defaults in the Western Canon, such a child was usually privileged, White, and English-speaking, but that was not supposed to matter. It is then inevitable under such a paradigm that the most likely source of explanation for extreme variation in language learning would be seen as genetic.

The paradigm underlying Snow's work, on the other hand, has its roots in psychology, and particularly developmental psychology, in theorists, such as Piaget, Vygotsky, and Bruner. The paradigm contains the belief that the child's social and cognitive development in the first 2 years forms a fundamental basis for language learning; that language learning is a complex social/cognitive skill that the child must acquire; that parents and other caregivers tailor the input and feedback provided to the child and, hence, provide significant environmental support for this learning; that children's early sentences are semantically based rather than abstract; and that induction from particular examples to general rules is the fundamental mechanism of learning. Unlike work under the Chomskyan paradigm, this alternative paradigm has embraced diversity of environments, cultures, and outcome as central to its empirical work. It is not so much uniformity as variation in language development that is the substance of its investigation, because that presents the possibility of explaining the variation along environmental lines. And the variation in normal development can be seen most readily in the very parts of language that the first paradigm neglects—in vocabulary, in discourse, in speech acts, and readiness for literacy.

Whether the variation in outcomes of grammatical acquisition can be attributed as readily to environmental cause is a central matter of concern for researchers and a central issue dividing the two paradigms. Unfortunately, it is unlikely that agreement will be forthcoming until each paradigm concedes an error of its ways: Namely, that researchers on esoteric, but vital, theoretical aspects of grammatical development incorporate a wider range of subjects with respect to environmental backgrounds and outcomes (and we are trying: see references in my chapter), and that researchers from the other paradigm reach beyond counting lexical variation, MLU, and speech styles, and venture into grammar.

I attempt to describe a continuum within grammar in terms of the likely impact of the input, rather than maintaining only the stark contrast between vocabulary (largely environmentally detemined) versus grammar (largely innate knowledge). In doing so, possible avenues of investigation are opened up that concern the environmental sources of variation in age of acquisition of some grammatical features. In other work, I explore the role of cognitive development in the acquisition of complex complement constructions involving mental verbs (de Villiers, 1994). Researchers do need to approach language from all these viewpoints: The price of paradigm incommensurability is too high for the field to bear.

My proposed title for this commentary was "Is Politeness Innate?" I have endeavoured to suggest reasons why Snow's analogy between language and politeness is not an apt one. Even though there may be striking parallels across cultures in the rules governing polite social interaction, even though it may look as if the universals find expression in different ways in different cultures while maintaining root similarities, the parallel to language fails on this one point. Look around you: Are you not struck by the vast individual variations in politeness despite common cultural "rules"? The source of that variation must be explained, and no one would propose looking at the genes.

If, in contrast, we lived in a world in which there were striking uniformity across individuals in the expression of politeness (let's even allow cultural variation), and despite the abstractness of the rules involved, young children were rising from the cradle sensitive to the needs of others, we might wonder what marvelous evolutionary process had paved the way for this innate tact. But if there arose from the very rare cradle a child with no social graces and a stark disregard for the feelings or status of others, might we not wonder if something had gone awry in the genes of such an individual?

Although it was inspired by the question raised in Snow's chapter, my proposed title was stolen from a biting piece of satire published in a British magazine, the *New Scientist*, in 1976 by "Dr. Edward Start," that poked fun at the latest developmental research coming into vogue with infants. The (rather thinly disguised) research team was purportedly studying very young infants by a new research technique called watching. The central claim was that newborn infants upon issuing a burp were observed to make incipient hand movements to cover their mouths, hence, that politeness was innate. I was extremely amused by this piece at the time (actually, I still fell about laughing when I dug it out of my files yesterday) and kept urging it upon students and colleagues to read. But for Freudian reasons clearly related to my own paradigm struggle at that time, every time I tried to say the title of the piece, I would accidentally say that it was called "Is innateness polite?"

REFERENCES

Atkinson, M. (1983). FLATS and SHARPS: The role of the child in language acquisition. In J. Durand (Ed.), *A Festschrift for Peter Wexler*. University of Essex Department of Language and Linguistics Occasional Papers: 27.

de Villiers, J. G. (1988). Faith, doubt, and meaning. In F. Kessel (Ed.), *The development of language and language researchers. Essays in honor of Roger Brown* (pp. 51–63). Hillsdale, NJ: Lawrence Erlbaum Associates.

de Villiers, J. G. (1994, November). Questioning minds and answering machines. Plenary address, Boston University Conference on Language development, Boston.

Kuhn, T. (1970). *The structure of scientific revolutions* (2nd ed.). Chicago: Chicago University Press.

Scheffler, I. (1982). *Science and subjectivity*. Indianapolis: Hackett.

Start, E. (1976). Is politeness innate? *New Scientist, 71,* 586–587.

AUTHOR INDEX

SUBJECT INDEX

413